Exploring Animal Behavior

Readings *from* American Scientist

SECOND EDITION

207
228

EDITED BY

Paul W. Sherman
Cornell University

John Alcock
Arizona State University

Sinauer Associates, Inc. *Publishers*
Sunderland, Massachusetts

The Cover

White-fronted bee eaters in Kenya. As described by Emlen et al. in this volume, these birds live in extended families, and their complex interactions with family members are based on status and genetic relatedness. Photograph courtesy of Marie Read.

Library of Congress Cataloging-in-Publication Data
Exploring animal behavior: readings from American scientist / edited by Paul W. Sherman and John Alcock. —2nd ed.
       p.   cm.
Includes index.
ISBN 0-87893-766-8 (paper)
1. Animal behavior. I. Sherman, Paul W., 1949-   . II. Alcock, John, 1942-   . III. American scientist.
QL751.6.E96   1997
591.5—dc21
                                  97-27624
                                    CIP

Printed in Hong Kong
8 7 6 5 4 3 2 1

CONTENTS

PREFACE

In the five years since the first edition of *Exploring Animal Behavior*, many more fine articles on animal behavior have appeared in *American Scientist*, the journal of the scientific society Sigma Xi. We have selected some of these articles for inclusion in this second edition, integrating the new choices with some of the original papers to create an anthology of 30 articles that covers the entire field of animal behavior.

We have assembled this collection as supplementary reading for students in animal behavior courses. We believe the articles can be profitably employed in classrooms in several ways, but especially as material for discussion and debate on key concepts. The articles also illustrate how behavioral scientists conduct research, providing greater depth of coverage than the generally brief textbook accounts of this process.

This anthology should be particularly useful for classes that use John Alcock's textbook, *Animal Behavior: An Evolutionary Approach*, because the articles are organized in a sequence complementary to that text. However, the collection also can enrich courses based on other textbooks. Indeed, this reader can stand alone as a sampler of the diversity of topics and approaches that constitute the modern study of animal behavior.

The articles have been grouped into five parts: Part I examines various aspects of science as a profession; Part II focuses on investigations into the proximate mechanisms that underlie animal behavior; Part III shifts to studies that trace the evolutionary histories of selected behavioral traits; Part IV examines the adaptive significance of various reproductive behaviors; and Part V deals with the adaptive value of social behavior. We provide a brief introduction to each part. We hope that students and teachers alike will benefit from this collection of enjoyable and instructive articles.

ACKNOWLEDGMENTS

This book is a reality because of the goodwill and hard work of many individuals, most notably the authors of the articles, who graciously provided permission to reprint their work, and the many photographers and illustrators whose images brighten so many pages. Michelle Press, who was editor of *American Scientist* when many of the articles were first published, deserves our gratitude for her keen interest in the field of animal behavior and for her efforts to see that behavior was prominently featured in the journal. Happily, Rosalind Reid, current editor of *American Scientist*, has continued that tradition. The small size of this volume belies the number of details that had to be attended to in its production. Rosalind Reid and Lil Chappell at *American Scientist* worked hard to see that everything was in order before we went to press. At Sinauer Associates, Peter Farley, Kerry Falvey, and Christopher Small made sure that our two goals of a handsome book and a low price could be met without compromise. At home, our families kept things lively and interesting. Thanks to all.

Paul W. Sherman
John Alcock

PART I
Doing Science

We begin our collection with several articles that examine what it means to be a scientist. Scientists can be defined as persons who use something called the "scientific method" when trying to explain how or why a particular phenomenon occurs in nature. The scientific method is a set of procedures that comes into play when someone develops one or more potential explanations, or *hypotheses,* for an interesting natural occurrence. Each hypothesis is then examined to see what *predictions,* or expected results, can be derived from it so that these can be checked against reality.

For example, male langurs sometimes attempt to kill the infants of females in their band (see the articles by Sarah Blaffer Hrdy and by Richard Curtin and Phyllis Dolhinow). One possible evolutionary explanation for this unpleasant behavior is that infanticide enables the killer male to eliminate the offspring of a rival male and gain sexual access to the females who have lost their offspring. We can make some predictions about when males will practice infanticide in nature if this hypothesis is correct—predictions that can be tested by further observation. Thus, infanticide "should" (is predicted to) occur only at times when a new male langur ousts the previous resident and takes charge of a band of females, some of whom will have dependent youngsters sired by the previous male. The discovery that newcomer males really do attack only the infants of other males and not their own offspring (when these come on the scene months later) supports the prediction, and so constitutes evidence for the hypothesis that infanticide is a behavior that males use to raise their reproductive success in competition with other males.

The ability of scientists to logically deduce what they "should" see in nature if a particular hypothesis is correct enables them to test their predictions by collecting "actual" results to be matched against the expected or predicted ones. Hypotheses whose predictions are apparently incorrect are discarded, whereas hypotheses that yield predictions that are supported by the results of experiments, by additional observations, or by comparisons among species, can be accepted. Scientists claim that by winnowing alternative ideas in this manner, keeping the "good" ones and getting rid of those that produce wrong predictions, they make progress, getting closer and closer to the truth about nature.

But, as Lewis Branscomb points out, scientists are like everyone else in wanting economic and social rewards in addition to the satisfaction that comes from doing their jobs well. These rewards go disproportionately to the first persons to publish important new findings, which creates pressures that could lead researchers to deceive themselves into thinking that they had done an adequate job of testing their favorite hypothesis when they had not. Branscomb believes that this kind of self-deception is probably common in science, much more so than outright fraud, and that the "honest mistakes" of scientists harm the scientific community at large because this community has no way "to protect itself from sloppy or deceptive literature." He argues that scientists therefore have a special ethical responsibility to be aware of the risks of self-deception and to check and double-check their results, while also considering all possible alternative explanations, before reporting their discoveries in the scientific literature.

James Woodward and David Goodstein come to very different conclusions about what comprises ethical scientific behavior. They note that prescriptions similar to the ones Branscomb advocates are widely accepted in theory, but not in practice. They go further, arguing that many ethical principles that sound good on paper would actually do harm if followed. They believe, for example, that the pressure on scientists to get their novel results out quickly has the important benefit of making information available promptly and efficiently. Likewise, if one scientist strongly advocates his or her pet idea, no real harm is done, because other scientists are likely to remain skeptical and may soon publish findings that expose the defects associated with a rival's argument.

Readers of Branscomb's and Woodward and Goodstein's articles can compare how they think these different authors would explain the apparent rarity with which researchers simply invent data and publish totally fraudulent papers. Which approach is more persuasive in accounting for the infrequency of outright cheating?

Woodward and Goodstein would probably also claim that even the "honest mistakes" that Branscomb says are common are unlikely to be numerous, thanks to competition among scientists for recognition and high status, which do not go to those whose conclusions are overthrown by others. Do you agree? Or are Woodward and Goodstein, and others like them, engaged in a form of self-deception when they speak of science as a self-correcting enterprise?

In the light of the first two articles, you can read the next two as a test of the positions staked out by Branscomb on the one hand and by Woodward and Goodstein on the other. What kind of approach has been taken by the authors of these articles on infanticide? Are they committed advocates for a particular hypothesis, or do they appear to be totally impartial and objective about the competing explanations for infanticide by male langurs? Would controversy exist in science if Branscomb's views were not only widely accepted, but actually practiced? Would the elimination of controversy in favor of reaching "genteel agreement" be a good thing in terms of getting closer to the truth about infanticide, or anything else of interest? Or does progress in science arise from controversies of this sort? Is there a more sensible way for scientists to evaluate competing explanations for behaviors that interest them? Do you agree with Woodward and Goodstein that "Popperian falsification" is the best tool for scientists to use in this process, and that there is no place for "Baconian inductivism" in science? Or is it possible that good science proceeds through sequential stages, from inductivism to falsification?

As noted above, publishing one's findings is a central element of doing science, a point captured in the academic admonition "Publish or perish!" Writing about their conclusions and the means by which they were reached enables scientists to communicate with a broad audience, which can then evaluate the message and learn from it. Yet the training that most scientists receive rarely includes any formal instruction in how to communicate effectively, which may contribute to the widespread impression that scientific writing is generally turgid, close to incomprehensible, and no fun at all to read. Happily, there are numerous exceptions to this "rule," as the articles contained in this collection demonstrate. Moreover, useful formal instruction on how to write scientifically does exist. The suggestions provided here by George Gopen and Judith Swan strike us as superbly helpful—not just for scientists, but for anyone who wants to write in ways that readers will appreciate. After reading what Gopen and Swan have to say, analyze and dissect a paragraph or two from any other article in this collection—and, after pulling the writing apart, see if you can put it back together in an improved form, taking advantage of what you have learned about effective writing. Then apply Gopen and Swan's suggestions to your own writing assignments in other classes and see whether they work for you.

Integrity in Science

Lewis M. Branscomb

Much of the problem of honor—or lack of honor—in science stems not from malice but from self-deception

In 1945 a physics graduate student at Harvard began a Ph.D. thesis project involving the use of molecular spectroscopy to determine the temperature of the atmosphere 1,000 km above the earth, at that time quite unknown. The Schumann-Runge bands of molecular oxygen had been observed as very weak emissions from the upper atmosphere. It was thought that they could be used as a thermometer, subject to verification in laboratory studies. But the bands had been observed only in absorption at very high pressures. Then in 1948 there appeared in *Nature* a report that stated that the Schumann-Runge bands had been observed in emission, excited at low pressure in a high-frequency discharge. The author also analyzed the molecular constants of the states involved (1).

Delighted to find from the literature that his thesis problem could be successfully attacked, the student set about reproducing the experiment described in *Nature*. After months of fruitless effort, he became suspicious that the results reported were in error and even that the photograph published with the text was not a picture of the Schumann-Runge spectrum at all. Indeed, it appeared that the results might have been fabricated from the proverbial whole cloth. In any case, six months of a predoctoral fellowship were lost, and another way to tackle the thesis problem had to be found.

I was that graduate student, and I have always felt sorry for the author of the article in *Nature*, who must have been under terrible pressure to show something for his efforts. I doubt that he had any intent to injure anyone, certainly not an unknown student thousands of miles away.

I believe that there are very few scientists who deliberately falsify their work, cheat on their colleagues, or steal from their students. On the other hand, I am afraid that a great many scientists deceive themselves from time to time in their treatment of data, gloss over problems involving systematic errors, or understate the contributions of others. These are the "honest mistakes" of science, the scientific equivalent of the "little white lies" of social discourse. But unlike polite society, which

Lewis M. Branscomb is Vice President and Chief Scientist of the IBM Corporation, President of Sigma Xi, and a past-president of the American Physical Society. He joined IBM in 1972 after a 21-year career at the National Bureau of Standards, of which he became director in 1969. He taught physics at several universities, was editor of Reviews of Modern Physics, and conceived and chaired for several years the Joint Institute for Laboratory Astrophysics at the University of Colorado. Dr. Branscomb was appointed by President Carter to the National Science Board in 1979 and was elected chairman the following year, serving until 1984. Address: IBM Corporation, Armonk, NY 10504.

easily interprets those white lies, the scientific community has no way to protect itself from sloppy or deceptive literature except to learn whose work to suspect as unreliable. This is a tough sentence to pass on an otherwise talented scientist.

The pressures on young science faculty are often fierce, not so much from tenure committees or even from peers, but from within. A young untenured scientist has all his emotional eggs in one basket. He picks a research problem and invests a year or more in its pursuit. Getting a successful start is important to the opportunity to do research. A lifetime career hinges on nature's cooperation as well as his own diligence and ingenuity. As we are reminded on television, it is dangerous to trifle with Mother Nature. Scientists run that risk every day. It takes a very self-confident young scientist to laugh at Tom Lehrer's "Lobachevsky" without a twinge of fear.

The Sigma Xi project on Honor in Science must deal with the broader question of the integrity of scientists' behavior, not just with the morality of what is admittedly the more serious evil, deliberate cheating (2–5). Unless science students are thoroughly inculcated with the discipline of correct scientific process, they are in serious danger of being damaged by the temptation to take the easy road to apparent success. And outright cheating can best be contained if the standards in all disciplines are held at high levels.

When is an experiment complete?

The reader may feel that the rules are simple and easy to follow for those who care about the integrity of their work. That is not necessarily so. Take, for example, the problem of knowing when an experiment is finished and the results are ready to publish. In 1953, building on the work of Wade Fite and profiting from his help at a critical time, I succeeded in making the first laboratory measurement of the photodetachment cross section for a negatively charged atomic ion in vacuum (6). The absorption of light by the negative ion of hydrogen ($H-$) was believed by Rupert Wildt to dominate the opacity of the solar photosphere. Simply put, the temperature of the sun, and thus the wavelengths to which human eyes are sensitive, is determined by this cross section. No one knew how accurate the quantum calculation of this three-body problem might be.

In order to test the calculation, Stephen J. Smith and I undertook an experiment requiring an absolute measurement in a very complex crossed-beam apparatus. After several years of preparation, the experiment began to yield data, and we made a reasonably diligent search for sources of systematic errors. The results differed from the quantum calculations by about 15%, a not unreasonable percentage considering the challenge of the three-body problem at the time. Stephen Smith and I were writing up the paper and making some final tests on the radiometric calibration system when the apparatus gave us a hint that something was amiss. We put the paper aside, tore the experiment down, and started over again on the calibrations. Three months later we had done everything necessary to quantify the limits of systematic error. Only then did we convert the results to cross-section units. We discovered to our utter amazement that the corrections we had introduced measured exactly 15%, bringing the experiment and the theory into an agreement so exact as to be clearly fortuitous.

At that point we were faced with a tough decision. What to do now? The experiment was finished. But we had thought it was finished once before. Were we in danger of stopping when we liked the answer? I realized then, as I have often said since, that Nature does not "know" what experiment a scientist is trying to do. "God loves the noise as much as the signal" (7). I decided to spend another three months looking for more sources of systematic error—a time exactly equal to the time we had spent on the last effort, which resulted in bringing experiment into agreement with theory. Fortunately, no additional sources could be found, and Steve and I felt we were ready to publish (8).

Perhaps this degree of conservatism is not necessary in every case, but it is certainly crucial in the case of absolute as opposed to relative measurements. The most severe requirement for such care is in the measurement of the fundamental constants of nature, a major interest of scientists at the National Bureau of Standards. Ever since the 1960s, scientists measuring atomic constants have adopted the policy of never reducing their data to final form (permitting comparison with the work of others) until all error analysis has been completed and the experiment is over.

Why, if the scientists are both honest and disciplined, is this necessary? Because the temptation to get a "good" (i.e., "safe" or "significant") result by stopping when the data pass through the desired coordinates is ever present. Some excellent scientists may have succumbed to the temptation. Back in the 1930s, for example, there was a long series of measurements of the universal constant of nature c, the speed of light in a vacuum. Following the pioneering measurement of Michelson and his co-workers, who used a rotating polygonal mirror to chop a beam of light passing between Mt. Wilson and Mt. Baldy in California, subsequent experimenters found more precise results using better equipment. In 1941, Birge's review of all the work concluded that the best weighted average of all the prior work was $c = 299,776 \pm 4$ km/sec (9).

Then came World War II. New technology and new people came into science. Very low frequency radio navigation (Loran) had been developed for military use, and electrical engineers realized that this system could be used to measure the speed of propagation of those 16 KHz waves in ways totally independent of the prewar optical methods. Within a few years, microwave cavity methods and free-space microwave interferometry gave consistent values with much higher precision.

Froome found $c = 299,793 \pm 0.3$ km/sec (10). There had been a shift of 17 km/sec, yet the stated accuracy of most of the previous measurements was 4 km/sec or better.

In their review of this mystery, Cohen and DuMond concluded that "two things contributed strongly to mislead [Birge in 1941] and would have misled anyone else in the same circumstances. These were the great prestige of Michelson's name as an expert in the field, and the fact that . . . two measurements . . . in 1937 and 1941 agreed quite well with the Michelson-Pease-Pearson result" (11). Writing in 1957, Birge said: "In any highly precise experimental arrangement there are initially many instrumental difficulties that lead to numerical results far from the accepted value of the quantity being measured. . . . Accordingly, the investigator searches for the source or sources of such error, and continues searching until he gets a result close to the accepted value. Then he stops! . . . In this way one can account for the close agreement of several different results and also for the possibility that all of them are in error by an unexpectedly large amount" (12). Cohen and DuMond credit Peter Franken with labeling this tendency "intellectual phase locking."

Commitment to quality

What might be done to reduce these "honest mistakes," to support the quality and thus the integrity of science? It takes the concerted efforts of teachers and research mentors, of promotion and tenure committees, of journal editors and referees. Above all it takes renewed commitment by the working scientist.

Young scientists should understand all the subtle ways in which they can delude themselves in the design of observations and the interpretation of data and statistics. They should understand metrology and should know what tendencies to manipulate information are built into their digital signal processors. They should also get to know the algorithms used in their favorite computers, which may under certain circumstances give strange results. Above all they should be trained in the detection and control of systematic errors.

The responsibility of the gatekeepers of scientific careers, the tenure committees, deans, and laboratory directors, is a heavy one. To reward people solely on the basis of numbers of papers published is destructive of the quality of science. Publication is of course the conventional method of making one's work available for critical appraisal by one's peers, but it is not the only way. And while perhaps even a necessary way, it is most emphatically not sufficient.

Journal editors and referees are, of course, the stewards of scientific quality, and they face a very difficult task. No journal can afford to publish all the evidence required to support an author's experimental conclusions. But how can a referee approve publication, when information necessary to proof is missing? The traditional answer is that authors use a certain shorthand to refer to procedures used which are either common practice or documented elsewhere. The reader has to trust the author to invoke those procedures properly. Thus one's reputation for trustworthiness, call it intellectual integrity if not honesty, is crucial to a scientific career. Are young people entering the world of scientific research as aware of this as they should be?

The quality of science places another burden on the scientist: not only to ensure that his own work meets the highest standards, but to participate in both the peer review of primary literature and the authorship of reviews of areas of work in which he is competent. Maurice Goldhaber, when he was director of the Brookhaven National Laboratory, encouraged his staff to write scholarly reviews. He felt that the review literature was a special responsibility of scientists at national laboratories; his motto was, "A good review is the moral equivalent of teaching."

During the last two decades substantial organized efforts at professional reviews of the literature of physical science have been undertaken. Groups of research experts have undertaken critical evaluations of original literature, usually dealing with properties of matter and materials. The goal of such reviews is to increase the density of useful information in the literature. Information that is wrong is not useful. And information lacking evidence revealing whether it is right or wrong is scarcely more so. Quality control in original research is the responsibility of the individual, part of the duty, if not the honor, expected from each of us.

To make the literature worth reviewing, authors of original papers must give the reader quantitative estimates of the amount by which the values given may be in error, and scientific justification for their conclusions. Scientists must demand of others and of themselves a revival of sound scholarship, instead of the cream-skimming and large numbers of hastily written papers with which we are all too familiar.

Commitment to integrity

The broader view of honor in science that I have discussed here should help everyone understand that this is not someone else's problem and is not just the problem of fraud in science. Most of us will never encounter a piece of truly fraudulent research. But concerns about scientific integrity permeate every piece of research we do, every talk we hear, every paper we read. A revitalization of interest in scientific honesty and integrity could have an enormous benefit both to science and to the society we serve.

First of all, integrity is essential for the realization of the joy that exploring the world of science can and should bring to each of us. Beyond that, the integrity of science affects the way the public looks at the pronouncements of scientists and the seriousness with which it takes our warnings, whether they relate to acid rain, the loss of genetic materials from endangered species, or the possibilities for science to help solve the global problems facing mankind. The users of our results, the decision-makers who need our advice, will always press us to be more sure of ourselves than our data permit, for it would make their jobs easier. The pressures to take shortcuts in science come from outside, as well as inside, the community.

We must help the public understand the rules of scientific evidence, just as we insist on rules of judicial evidence in our courts. A precondition for success in this endeavor is to refine and apply those rules with great rigor in our own work and literature. The future of mankind hangs in no small measure on the integrity, and thus the credibility, of science.

References

1. L. Lal. 1948. *Nature* 161:477.
2. C. I. Jackson and J. W. Prados. 1983. *Am. Sci.* 71:462.
3. *Honor in Science.* 1984. Sigma Xi.
4. R. N. Hall. 1968. Gen. Elec. rep. no. 68-C-035.
5. R. P. Feynman. 1974. *Engineering and Science* 37(7):10.
6. L. M. Branscomb and W. L. Fite. 1954. *Phys. Rev.* 93:651.
7. L. M. Branscomb. 1980. *Phys. Today* 33(4):42.
8. L. M. Branscomb and S. J. Smith. 1955. *Phys. Rev.* 98:1028.
9. R. T. Birge. 1941. *Rep. Progr. Phys.* 8:90.
10. K. D. Froome. 1954. *Proc. Royal Soc. London* A223:195.
11. E. R. Cohen and J. DuMond. 1965. *Rev. Modern Phys.* 37:537.
12. R. T. Birge. 1957. *Nuovo Cimento*, supp. 6:39.

Conduct, Misconduct and the Structure of Science

Many plausible-sounding rules for defining ethical conduct might be destructive to the aims of scientific inquiry

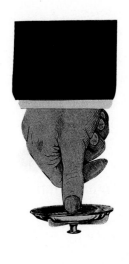

James Woodward and David Goodstein

In recent years the difficult question "what constitutes scientific misconduct?" has troubled prominent ethicists and scientists and tied many a blue-ribbon panel in knots. In teaching an ethics class for graduate and undergraduate students over the past few years, we have identified what seems to be a necessary starting point for this debate: the clearest possible understanding of *how science actually works*. Without such an understanding, we believe, one can easily imagine formulating plausible-sounding ethical principles that would be unworkable or even damaging to the scientific enterprise.

Our approach may sound so obvious as to be simplistic, but actually it uncovers a fundamental problem, which we shall try to explore in this article. The nature of the problem can be glimpsed by considering the ethical implications of the earliest theory of the scientific method. Sir Francis Bacon, a contemporary of Galileo, thought the scientist must be a disinterested observer of nature, whose mind was cleansed of prejudices and preconceptions. As we shall see, the reality of science is radically different from this ideal. If we expect to find scientists who are disinterested observers of nature we are bound to be disappointed, not because scientists have failed to measure up to the appropriate standard of behavior, but because we have tried to apply the wrong standard of behavior. It can be worse: Rules or standards of conduct that seem intuitively appealing can turn out to have results that are both unexpected and destructive to the aims of scientific inquiry.

In drafting this article, we set out to examine the question of scientific ethics in light of what we know about science as a system and about the motivations of the scientists who take part in it. The reader will find that this exercise unearths contradictions that may be especially unpleasant for those who believe clear ethical principles derive directly from the principles of scientific practice. In fact, one can construct a wonderful list of plausible-sounding ethical principles, each of which might be damaging or unworkable according to our analysis of how science works.

Ideals and Realities

We can begin where Sir Francis left off. Here is a hypothetical set of principles, beginning with the Baconian ideal, for the conduct of science:

1. Scientists should always be disinterested, impartial and totally objective when gathering data.
2. A scientist should never be motivated to do science for personal gain, advancement or other rewards.
3. Every observation or experiment must be designed to falsify an hypothesis.
4. When an experiment or an observation gives a result contrary to the prediction of a certain theory, all ethical scientists must abandon that theory.
5. Scientists must never believe dogmatically in an idea nor use rhetorical exaggeration in promoting it.
6. Scientists must "lean over backwards" (in the words of the late physicist Richard Feynman) to point out evidence that is contrary to their own hypotheses or that might weaken acceptance of their experimental results.
7. Conduct that seriously departs from that commonly accepted in the scientific community is unethical.
8. Scientists must report what they have done so fully that any other scientist can reproduce the experiment or calculation. Science must be an open book, not an acquired skill.

James Woodward and David Goodstein are professors at Caltech, where, in addition to the Research Ethics course they teach together, Woodward teaches philosophy and Goodstein teaches physics. Woodward has served Caltech as Executive Officer for the Humanities; Goodstein serves as vice provost and is the Frank J. Gilloon Distinguished Teaching and Service Professor. Address for both: California Institute of Technology, Pasadena, CA 91125.

9. Scientists should never permit their judgments to be affected by authority. For example, the reputation of a scientist making a given claim is irrelevant to the validity of the claim.

10. Each author of a multiple-author paper is fully responsible for every part of the paper.

11. The choice and order of authors on a multiple-author publication must strictly reflect the contributions of the authors to the work in question.

12. Financial support for doing science and access to scientific facilities should be shared democratically, not concentrated in the hands of a favored few.

13. There can never be too many scientists in the world.

14. No misleading or deceptive statement should ever appear in a scientific paper.

15. Decisions about the distribution of resources and publication of results must be guided by the judgment of scientific peers who are protected by anonymity.

Should the behavior of scientists be governed by rules of this sort? We shall argue that it should not. We first consider the general problems of motivation and the logical structure of science, then the question of how the community of scientists actually does its work, showing along the way why each of these principles is defective. At the end, we offer a positive suggestion of how scientific misconduct might be recognized.

Behavior that may seem at first glance morally unattractive can, in a properly functioning system, produce results that are generally beneficial.

Motives and Consequences

Many of the provocative statements we have just made raise general questions of motivation related to the issue explicitly raised in principle 2, and it is worth dealing with these up front. We might begin with a parallel: the challenge of devising institutions, rules and standards to govern commerce. In economic life well-intentioned attempts to reduce the role of greed or speculation can turn out to have disastrous consequences. In fact, behavior that may seem at first glance morally unattractive, such as the aggressive pursuit of economic self-interest, can, in a properly functioning system, produce results that are generally beneficial.

In the same way it might appear morally attractive to demand that scientists take no interest in obtaining credit for their achievements. Most scientists are motivated by the desire to discover important truths about nature and to help others to do so. But they also prefer that they (rather than their competitors) be the ones to make discoveries, and they want the recognition and the advantages that normally reward success in science. It is tempting to think that tolerating a desire for recognition is a concession to human frailty; ideally, scientists should be interested only in truth or other purely epistemic goals. But this way of looking at matters misses a number of crucial points.

For one thing, as the philosopher Philip Kitcher has noted, the fact that the first person

to make a scientific discovery usually gets nearly all the credit encourages investigators to pursue a range of different lines of inquiry, including lines that are thought by most in the community to have a small probability of success. From the point of view of making scientific discoveries as quickly and efficiently as possible, this sort of diversification is extremely desirable; majority opinion turns out to be wrong with a fairly high frequency in science.

Another beneficial feature of the reward system is that it encourages scientists to make their discoveries public. As Noretta Koertge has observed, there have been many episodes in the history of early modern science in which scientists made important discoveries and kept them private, recording them only in notebooks or correspondence, or in cryptic announcements designed to be unintelligible to others. The numerous examples include Galileo, Newton, Cavendish and Lavoisier. It is easy to see how such behavior can lead to wasteful repetition of effort. The problem is solved by a system of rewards that appeals to scientists' self-interest. Finally, in a world of limited scientific resources, it makes sense to give more resources to those who are better at making important discoveries.

We need to be extremely careful, in designing institutions and regulations to discourage scientific misconduct, that we not introduce changes that disrupt the beneficial effects that competition and a concern for credit and reputation bring with them. It is frequently claimed that an important motive in a number of recent cases of data fabrication has been the desire to establish priority and to receive credit for a discovery, or that a great deal of fraud can be traced to the highly competitive nature of modern science. If these claims are correct, the question becomes, how can we reduce the incidence of fraud without removing the beneficial effects of competition and reward?

The Logical Structure of Science

The question of how science works tends to be discussed in terms of two particularly influential theories of scientific method, *Baconian inductivism* and *Popperian falsification,* each of which yields a separate set of assumptions.

According to Bacon's view, scientific investigation begins with the careful recording of observations. These should be, insofar as is humanly possible, uninfluenced by any prior prejudice or theoretical preconception. When a large enough body of observations is accumulated, the investigator generalizes from these, via a process of induction, to some hypothesis or theory describing a pattern present in the observations. Thus, for example, an investigator might inductively infer, after observing a large number of black ravens, that all ravens are black. According to this theory good scientific conduct consists in recording all that one observes and not just some selected part of it, and in asserting only hypotheses that are strongly inductively supported by the evidence. The guiding ideal is to avoid any error that may slip in as a result of prejudice or preconception.

How can we reduce the incidence of fraud without removing the beneficial effects of competition and reward?

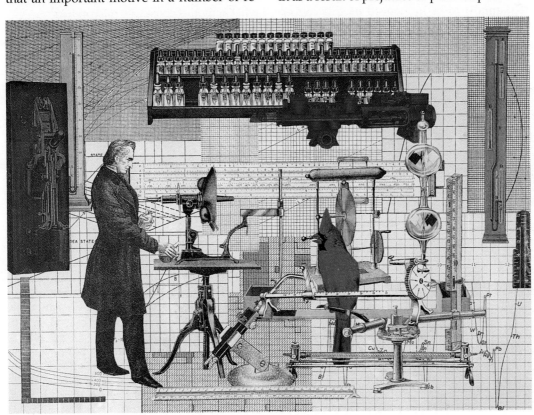

Historians, philosophers and those scientists who care are virtually unanimous in rejecting Baconian inductivism as a general characterization of good scientific method. The advice to record all that one observes is obviously unworkable if taken literally; some principle of selection or relevance is required. But decisions about what is relevant inevitably will be influenced heavily by background assumptions, and these, as many recent historical studies show, are often highly theoretical in character. The vocabulary we use to describe the results of measurements, and even the instruments we use to make the measurements, are highly dependent on theory. This point is sometimes expressed by saying that all observation in science is "theory-laden" and that a "theoretically neutral" language for recording observations is impossible.

The idea that science proceeds only and always via inductive generalization from what is directly observed is also misguided. Theories in many different areas of science have to do with entities whose existence or function cannot be directly observed: forces, fields, subatomic particles, proteins and other large organic molecules, and so on. For this and many other reasons, no one has been able to formulate a defensible theory of inductive inference of the sort demanded by inductivist theories of science.

The difficulties facing inductivism as a general conception of scientific method are so well known that it is surprising to find authoritative characterizations of scientific misconduct that appear to be influenced by this conception. Consider the following remarks by Suzanne Hadley, at one time acting head of what used to be called the Office of Scientific Integrity (now the Office of Research Integrity),

the arm of the U.S. Public Health Service charged with investigating allegations of scientific misconduct. In a paper presented at the University of California, San Diego, in October 1991, Hadley wrote: "...it is essential that observation, data recording, and data interpretation and reporting be veridical with the phenomena of interest, *i.e.,* be as free as humanly possible of 'taint' due to the scientist's hopes, beliefs, ambitions, or desires." Elsewhere she writes: "Anything that impinges on the veridical perception, recording and reporting of scientific phenomena is antithetical to the very nature of science." She also says, "...it is the human mind, which based on trained observations, is able to form higher-order conceptions about phenomena."

Hadley's view may not be as rigidly inductivist as these remarks imply. She adds, "I hasten to say that I am not suggesting that a scientist can or should be relegated to a mechanistic recording device." But this more nuanced view is not allowed to temper excessively her oversight responsibility as a government official: "The really tough cases to deal with are the cases closest to the average scientist: those in which 'fraud' is not clearly evident, but 'out of bounds' conduct is: data selection, failure to report discrepant data, over-interpretation of data...."

The idea that data selection and overinterpretation of data are forms of misconduct seem natural if one begins with Hadley's view of scientific method. A less restrictive view would lead to a different set of conclusions about what activities constitute misconduct.

Although relatively few contemporary scientists espouse inductivism, there are many scientists who have been influenced by the falsificationist ideas of Karl Popper. According to falsificationists, we test a hypothesis by deducing from it a testable prediction. If this prediction turns out to be false, the hypothesis from which it is deduced is said to be falsified and must be rejected. For example, the observation of a single nonblack raven will falsify the hypothesis H: "All ravens are black." But if we set out to test H and observe a black raven or even a large number of such ravens we cannot, according to Popper, conclude that H is true or verified or even that it is more probable than it was before. All that we can conclude is that H has been tested and has not yet been falsified. There is thus an important asymmetry between the possibility of falsification and the possibility of verification; we can show conclusively that an hypothesis is false, but not that it is true.

Because of this asymmetry it is a mistake to think, as the inductivist does, that good science consists of hypotheses that are proved or made probable by observation, whereas bad science does not. Instead, according to Popper, good science requires hypotheses that might be falsified by some conceivable observation. For example, the general theory of relativity predicts that starlight passing sufficiently close to the sun will be deflected by a certain measurable amount E. General relativity is a falsifiable theory because observations of starlight deflection that differ substantially from E are certainly conceivable and had they been made, would have served to falsify general relativity. By contrast, writes Popper, Freudian psychology is unfalsifiable and hence unscientific. If, for example, a son behaves in a loving way toward his mother, this will be attributed to his Oedipus complex. If, on the contrary, he behaves in an hostile and destructive way, this will be attributed to the same Oedipus complex. No possible empirical observation constitutes a refutation of the hypothesis that the son's behavior is motivated by an Oedipus complex.

According to Popper, bad scientific behavior consists in refusing to announce in advance what sorts of evidence would lead one to give up an hypothesis, in ignoring or discarding evidence contrary to one's hypothesis or in introducing ad hoc, content-decreasing modifications in one's theories in order to protect them against refutation. Good scientific method consists in putting forward highly falsifiable hypotheses, specifying in advance what sorts of evidence would falsify these hypotheses, testing the hypotheses at exactly those points at which they seem most likely to break down and then giving them up should such evidence be observed. More generally (and moving somewhat beyond the letter of Popper's theory) we can say that to do science in a Popperian spirit is to hold to one's hypothesis in a tentative, nondogmatic fashion, to explore and draw to the attention of others various ways in which one's hypothesis might break down or one's experimental result may be invalid, to give up one's hypothesis willingly in the face of contrary evidence, to take seriously rather than to ignore or discard evidence that is contrary to it, and in general not to exaggerate or overstate the evidence for it or suppress problems that it faces. Richard Feynman, in a commencement address at Caltech some years ago, recommended a recognizably Popperian attitude in the following remarks:

> [There is an] idea that we all hope you have learned in studying science in school—we never explicitly say what this is, but just hope that you catch on by all the examples of scientific investigation.... It's a kind of scientific integrity, a principle of scientific thought that corresponds to a kind of utter honesty—a kind of leaning over backwards. For example, if you're doing an experiment, you should report everything that you think might make it invalid—not only what you

think is right about it; other causes that could possibly explain your results; and things you thought of that you've eliminated by some other experiment, and how they worked—to make sure the other fellow can tell they have been eliminated.

....In summary, the idea is to try to give all the information to help others to judge the value of your contribution, not just the information that leads to judgment in one particular direction or another.

These views form the basis of principles 3–6 above.

Although falsificationism has many limitations (see below), it introduces several corrections to inductivism that are useful in understanding how science works and how to characterize misconduct. To begin with, falsificationism rejects the idea that good scientific behavior consists in making observations without theoretical preconceptions. For Popper, scientific activity consists in attempting to falsify. Such testing requires that one have in mind a hypothesis that will indicate which observations are relevant or worth making. Rather than something to be avoided, theoretical preconceptions are essential to doing science.

Inductivists attach a great deal of weight to the complete avoidance of error. By contrast, falsificationists claim that the history of science shows us that all hypotheses are falsified sooner or later. In view of this fact, our aim should be to detect our errors quickly and to learn as

For science to advance, scientists must be free to be wrong.

efficiently as possible from them. Error in science thus plays a constructive role. Indeed, according to falsificationists, putting forward a speculative "bold conjecture" that goes well beyond available evidence and then trying vigorously to falsify it will be the strategy that enables us to progress as efficiently as possible. For science to advance, scientists must be free to be wrong.

Despite these advantages, there are also serious deficiencies in falsificationism, when it is taken as a general theory of method. One of the most important of these is sometimes called the Duhem-Quine problem. We claimed above that testing a hypothesis H involved deriving from it some observational consequence O. But in most realistic cases such observational consequences will not be derivable from H alone, but only from H in conjunction with a great many other assumptions A (auxiliary assumptions, as philosophers sometimes call them). For example, to derive an observational claim from a hypothesis about rates of evolution, one may need auxiliary assumptions about the processes by which the fossil record is laid down. Suppose one hypothesizes that a certain organism has undergone slow and continuous evolution and derives from this that one should see numerous intermediate forms in the fossil record. If such forms are absent it may mean H is false, but it may also be the case that H is true but fossils were preserved only in geological deposits that were laid down at widely separated times. It is possible that H is true and that the reason that O is false is that A is false.

One immediate result of this simple logical fact is that the logical asymmetry between falsification and verification disappears. It may be true, as Popper claims, that we cannot conclusively verify a hypothesis, but we cannot conclusively falsify it either. Thus, as a matter of method, it is sometimes a good strategy to hold onto a hypothesis even when it seems to imply an observational consequence that looks to be false. In fact, the history of science is full of examples in which such anti-Popperian behavior has succeeded in finding out important truths about nature when it looks as though more purely Popperian strategies would have been less successful.

Anti-Popperian strategies seem particularly prevalent in experiments. In doing an experiment one's concern is often to find or demonstrate an effect or to create conditions that will allow the effect to appear, rather than to refute the claim that the effect is real. Suppose a novel theory predicts some previously unobserved effect, and an experiment is undertaken to detect it. The experiment requires the construction of new instruments, perhaps operating at the very edge of what is technically possible, and the use of a novel experimental design,

It is sometimes a good strategy to hold onto a hypothesis even when it seems to imply an observational consequence that looks to be false.

which will be infected with various unsuspected and difficult-to-detect sources of error. As historical studies have shown, in this kind of situation there will be a strong tendency on the part of many experimentalists to conclude that these problems have been overcome if and when the experiment produces results that the theory predicted. Such behavior certainly exhibits anti-Popperian dogmatism and theoretical "bias," but it may be the best way to discover a difficult-to-detect signal. Here again, it would be unwise to have codes of scientific conduct or systems of incentives that discourage such behavior.

Social Structure

Inductivism, falsificationism and many other traditional accounts of method are inadequate as theories of science. At bottom this is because they neglect the psychology of individual scientists and the social structure of science. These points are of crucial importance in understanding how science works and in characterizing scientific misconduct.

Let us begin with what Philip Kitcher has called the division of cognitive labor and the role of social interactions in scientific investigation. Both inductivism and falsificationism envision an individual investigator encountering nature and constructing and assessing hypotheses all alone. But science is carried out by a community of investigators. This fact has important implications for how we should think about the responsibilities of individual scientists.

Suppose a scientist who has invested a great deal of time and effort in developing a theory is faced with a decision about whether to continue to hold onto it given some body of evidence. As we have seen, good Popperian method requires that scientists act as skeptical critics of their own theories. But the social character of science suggests another possibility. Suppose that our scientist has a rival who has invested time and resources in developing an alternative theory. If additional resources, credit and other rewards will flow to the winner, perhaps we can reasonably expect that the rival will act as a severe Popperian critic of the theory, and vice versa. As long as others in the community will perform this function, failure to behave like a good Popperian need not be regarded as a violation of some canon of method.

There are also psychological facts to consider. In many areas of science it turns out to be very difficult, and to require a long-term commitment of time and resources, to develop even one hypothesis that respects most available theoretical constraints and is consistent with most available evidence. Scientists, like other human beings, find it difficult to sustain commitments to arduous, long-term projects if they spend too much time contemplating the various ways in which the project might turn out to be unsuccessful.

A certain tendency to exaggerate the merits of one's approach, and to neglect or play down, particularly in the early stages of a project, contrary evidence and other difficulties, may be a necessary condition for the success of many scientific projects. When people work very hard on something over a long period of time, they tend to become committed or attached to it; they strongly want it to be correct and find it increasingly difficult to envision the possibility that it might be false, a phenomenon related to what psychologists call *belief-perseverance*. Moreover, scientists like other people like to be right and to get credit and recognition from others for being right: The satisfaction of demolishing a theory one has laboriously constructed may be small in comparison with the satisfaction of seeing it vindicated. All things considered, it is extremely hard for most people to adopt a consistently Popperian attitude toward their own ideas.

Given these realistic observations about the psychology of scientists, an implicit code of conduct that encourages scientists to be a bit dogmatic and permits a certain measure of rhetorical exaggeration regarding the merits of their work, and that does not require an exhaustive discussion of its deficiencies, may be perfectly sensible. In many areas of science, if a scientist submits a paper that describes all of the various ways in which an idea or result might be defective, and draws detailed attention to the contrary results obtained by others, the paper is likely to be rejected. In fact, part of the intellectual responsibility of a scientist is to provide the best possible case for important ideas, leaving it to others to publicize their defects and limitations. Studies of both historical and contemporary science seem to show that this is just what most scientists do.

If this analysis is correct, there is a real danger that by following proposals (like that advocated by Hadley) to include within the category of "out-of-bounds conduct" behavior such as overinterpretation of data, exaggeration of available evidence that supports one's conclusion or failure to report contrary data, one may be proscribing behavior that plays a functional role in science and that, for reasons rooted deep in human psychology, will be hard to eliminate. Moreover, such proscriptions may be unnecessary, because interactions between scientists and criticisms by rivals may by themselves be sufficient to remove the bad consequences at which the proscriptions are aimed. Standards that might be optimal for single, perfectly rational beings encountering nature all by themselves may be radically deficient when applied to actual scientific communities.

Rewarding Useful Behavior

From a Popperian perspective, discovering evidence that merely supports a hypothesis is easy to do and has little methodological value; therefore one might think it doesn't deserve much credit. It is striking that the actual distribution of reward and credit in science reflects a very different view. Scientists receive Nobel prizes for finding new effects predicted by theories or for proposing important theories that are subsequently verified. It is only when a hypothesis or theory has become very well established that one receives significant credit for refuting it. Unquestionably, rewarding confirmations over refutations provides scientists with incentives to confirm theories rather than refute them and thus discourages giving up too quickly in the face of contrary experimental results. But, as we have been arguing, this is not necessarily bad for science.

Conventional accounts of scientific method (of which there are many examples in the philosophical literature) share the implicit assumption that all scientists in a community should adopt the same strategies. In fact, a number of government agencies now have rules that define as scientific misconduct "practices that seriously deviate from those that are commonly accepted within the scientific community…" (see principle 7). But rapid progress will be more likely if different scientists have quite different attitudes toward appropriate methodology. As noted above, one important consequence of the winner-takes-all (or nearly all) system by which credit and reward are allocated in science is that it encourages a variety of research programs and approaches. Other features of human cognitive psychology—such as the belief-perseverance phenomenon described above—probably have a similar consequence. It follows that attempts to characterize misconduct in terms of departures from practices or methods commonly accepted within the scientific community will be doubly misguided: Not only will such commonly accepted practices fail to exist in many cases, but it will be undesirable to try to enforce the uniformity of practice that such a characterization of misconduct would require. More generally, we can see why the classical methodologists have failed to discover "the" method by which science works. There are deep, systematic reasons why all scientists should not follow some single uniform method.

Our remarks so far have emphasized the undesirability of a set of rules that demand that all scientists believe the same things or behave in the same way, given a common body of evidence. This is not to say, however, that "anything goes." One very important distinction has to do with the difference between claims and behavior that are open to public assess-

A certain tendency to exaggerate the merits of one's approach, and to neglect or play down contrary evidence and other difficulties, may be a necessary condition for the success of many scientific projects.

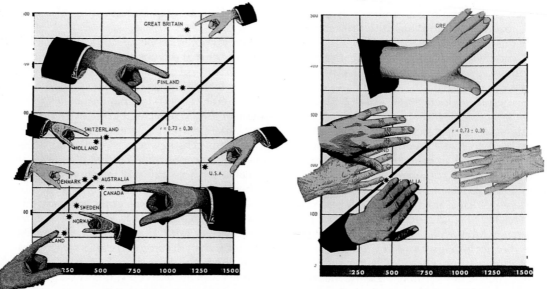

Standards that might be optimal for single, perfectly rational beings encountering nature all by themselves may be radically deficient when applied to actual scientific communities.

ment and those that are not. Exaggerations, omissions and misrepresentations that cannot be checked by other scientists should be regarded much more harshly than those that can, because they subvert the processes of public assessment and intellectual competition on which science rests. Thus, for example, a scientist who fabricates data must be judged far more harshly than one who does a series of experiments and accurately records the results but then extrapolates beyond the recorded data or insists on fitting some favored function to them. The difference is the fact that in the case in which there is no fabrication nothing has been done to obstruct the critical scrutiny of the work by peers; they can look at the data themselves and decide whether there is support for the conclusions. By contrast, other scientists will not be able to examine firsthand the process by which the data have been produced. They must take it on trust that the data resulted from an experiment of the sort described. Fabrication should thus be viewed as much more potentially damaging to the process of inquiry and should be more harshly punished than other forms of misrepresentation.

Science as Craft

Contemporary scientific knowledge is so vast and complex that even a very talented and hardworking scientist will be able to master only an extremely small fragment well enough to expect to make contributions to it. In part for this reason scientists must rely heavily on the authority of other scientists who are experts in domains in which they are not. A striking example of this is provided by the sociologist Trevor Pinch's recent book, *Confronting Nature*, which is a study of a series of experiments that discovered in the solar-neutrino flux far fewer neutrinos than seemed to be predicted by accepted theory. Pinch found that what

he called the "personal warrant" of the experimenters involved in this project played a large role in how other scientists assessed the experimental results. According to Pinch, other scientists often place at least as much weight on an experimentalist's general reputation for careful, painstaking work as on the technical details of the experiment in assessing whether the data constitute reliable evidence.

One reason why such appeals to personal warrant play a large role in science has to do with the specialized character of scientific knowledge. There is, however, another related reason which is of considerable importance in understanding how science works and how one should think about misconduct. This has to do with the fact that science in general—and especially experimentation—has a large "skill" or "craft" component.

Conducting an experiment in a way that produces reliable results is not a matter of following algorithmic rules that specify exactly what is to be done at each step. As Pinch put it, experimenters possess skills that "often enable the experimenter to get the apparatus to work without being able to formulate exactly or completely what has been done." For the same reason, assessing whether another investigator has produced reliable results requires a judgment of whether the experimenter has demonstrated the necessary skills in the past. These facts about the role of craft knowledge may be another reason why the general rules of method sought by the classical methodologists have proved so elusive.

The importance of craft in science is supported by empirical studies. For example, in a well-known study, Harry Collins investigated a number of experimental groups working in Britain to recreate a new kind of laser that had been successfully constructed elsewhere. Collins found that no group was able to reproduce a working laser simply on the basis of detailed written instructions. By far the most reliable method was to have someone from the original laboratory who had actually built a functioning laser go to the other laboratories and participate in the construction. The skills needed to make a working device could be acquired by practice *"without necessarily formulating, enumerating or understanding them."* Remarks on experimental work by working scientists themselves often express similar claims, not withstanding principle 8 above. If claims of this sort are correct, it often will be very difficult for those who lack highly specific skills and knowledge to assess a particular line of experimental work. A better strategy may be to be guided at least in large part by the experimenter's general reputation for reliability.

These facts about specialization, skill and authority have a number of interesting consequences for understanding what is proper scientific conduct. For example, a substantial amount of conduct that may look to an outsider like nonrational deference to authority may have a serious epistemological rationale. When an experimentalist discards certain data on the basis of subtle clues in the behavior of the apparatus, and other scientists accept the experimentalist's judgment in this matter, we should not automatically attribute this to the operation of power relationships, as is implied by principle 9 in our list above.

A second important consequence has to do with the responsibility of scientists for the misconduct or sloppy research practices of collaborators. It is sometimes suggested that authors should not sign their names to joint papers unless they have personally examined the evidence and are prepared to vouch for the correctness of every claim in the paper (principle 10). However, many collaborations bring together scientists from quite different specializations who lack the expertise to evaluate one another's work directly. This is exactly why collaboration is necessary. Requiring that scientists not collaborate unless they are able to check the work of collaborators or setting up a general policy of holding scientists responsible for the misconduct of coauthors would discourage a great deal of valuable collaboration.

Understanding the social structure of science and the operation of the reward system within science also has important ethical implications. We consider three examples: the Matthew effect, the Ortega hypothesis and scientific publication.

Matthew *vs.* Ortega

The sociologist Robert K. Merton has observed that credit tends to go to those who are already famous, at the expense of those who are not. A paper signed by Nobody, No-

body and Somebody often will be casually referred to as "work done in Somebody's lab," and even sometimes cited (incorrectly) in the literature as due to "Somebody *et al.*" Does this practice serve and accurately depict science, or does the tendency to elitism distort and undermine the conduct of science?

It is arguable that what Merton called the Matthew Effect plays a useful role in the organization of science; there are so many papers in so many journals that no scientist has time to read more than a tiny fraction of those in even a restricted area of science. Famous names tend to identify those works that are more likely to be worth noticing. In certain fields, particularly biomedical fields, it has become customary to make the head of the laboratory a coauthor, even if the head did not participate in the research. One reason for this practice is that by including the name of the famous head on the paper, chances are greatly improved that the paper will be accepted by a prestigious journal and noticed by its readers. Some people refer to this practice as "guest authorship" and regard it as unethical (as would be implied by principle 11 above). However, the practice may be functionally useful and may involve little deception, since conventions regarding authorship may be well understood by those who participate in a given area of science.

"In the cathedral of science," a famous scientist once said, "every brick is equally important." The remark (heard by one of the authors at a gathering at the speaker's Pasadena, California home) evokes a vivid metaphor of swarms of scientific workers under the guidance, perhaps, of a few master builders erecting a grand monument to scientific faith. The speaker was Max Delbrück, a Nobel laureate often called the father of molecular biology. The remark captures with some precision the scientists' ambivalent view of their craft. Delbrück never for an instant thought the bricks

A substantial amount of conduct that may look to an outsider like nonrational deference to authority may have a serious epistemological rationale.

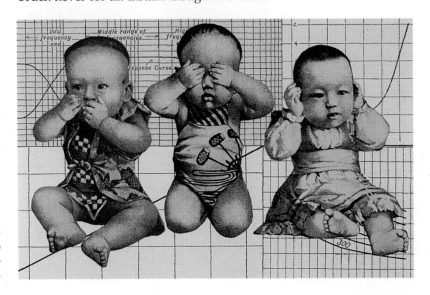

If the elitist view is right, then since science is largely financed by the public purse, it is best for science and best for society to restrict our production to fewer and better scientists.

he laid were no better than anyone else's. If anything, he regarded himself as the keeper of the blueprints, and he had the fame and prestige to prove it. It was exactly his exalted position that made it obligatory that he make a ceremonial bow to the democratic ideal that many scientists espouse and few believe. In fact it is precisely the kind of recognition that Delbrück enjoyed that propels the scientific enterprise forward.

The view expressed by Delbrück has been called the Ortega Hypothesis. It is named after Jose Ortega y Gasset, who wrote in his classic book, *The Revolt of the Masses,* that

> …it is necessary to insist upon this extraordinary but undeniable fact: experimental science has progressed thanks in great part to the work of men astoundingly mediocre. That is to say, modern science, the root and symbol of our actual civilization, finds a place for the intellectually commonplace man and allows him to work therein with success. In this way the majority of scientists help the general advance of science while shut up in the narrow cell of their laboratory like the bee in the cell of its hive, or the turnspit of its wheel.

This view (see principle 12) is probably based on the empirical observation that there are indeed, in each field of science, many ordinary scientists doing more or less routine work. It is also supported by the theoretical view that knowledge of the universe is a kind

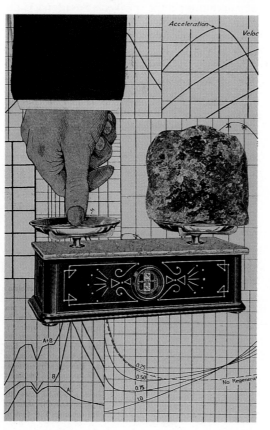

of limitless wilderness to be conquered by relentless hacking away of underbrush by many hands. An idea that is supported by both theory and observation always has a very firm standing in science.

The Ortega Hypothesis was named by Jonathan and Steven Cole when they set out to demolish it, an objective they pursued by tracing citations in physics journals. They concluded that the hypothesis is incorrect, stating:

> It seems, rather, that a relatively small number of physicists produce work that becomes the base for future discoveries in physics. We have found that even papers of relatively minor significance have used to a disproportionate degree the work of the eminent scientists.…

In other words, a small number of elite scientists produce the vast majority of scientific progress. Seen in this light, the reward system in science is a mechanism evolved for the purpose of identifying, promoting and rewarding the star performers.

One's view of the Ortega Hypothesis has important implications concerning how science ought to be organized. If the Ortega Hypothesis is correct, science is best served by producing as many scientists as possible, even if they are not all of the highest quality (principle 13). On the other hand, if the elitist view is right, then since science is largely financed by the public purse, it is best for science and best for society to restrict our production to fewer and better scientists. In any case, the question of whether to produce more or fewer scientists involves ethical issues (what is best for the common good?) as well as policy issues (how to reach the desired goal).

Peers and Publication

In a classic paper called "Is the scientific paper a fraud?" Peter Medawar has argued that typical experimental papers intentionally misrepresent the actual sequence of events involved in the conduct of an experiment, the process of reasoning by which the experimenter reached various conclusions and so on. In general, experimentalists will make it look as if they had a much clearer idea of the ultimate result than was actually the case. Misunderstandings, blind alleys and mistakes of various sorts will fail to appear in the final written account.

Papers written this way are undoubtedly deceptive, at least to the uninitiated, and they certainly stand in contrast to Feynman's exhortation to "lean over backward." They also violate principle 14. Nevertheless, the practice is virtually universal, because it is a much more efficient means of transmitting results than an accurate historical account of the scientist's activities would be. Thus it is a simple fact that, contrary to normal belief, there are types of misrepresentation that are condoned and accepted in scientific publications, whereas other types are harshly condemned.

Nevertheless, scientific papers have an exalted reputation for integrity. That may be because the integrity of the scientific record is protected, above all, by the institution of peer review. Peer review has an almost mystical role in the community of scientists. Published results are considered dependable because they have been reviewed by peers, and unpublished data are considered not dependable because they have not been. Many regard peer review to be (as principle 15 would suggest) the ethical fulcrum of the whole scientific enterprise.

Peer review is used to help determine whether journals should publish articles submitted to them, and whether agencies should grant financial support to research projects. For most small projects and for nearly all journal articles, peer review is accomplished by sending the manuscript or proposal to referees whose identities will not be revealed to the authors.

Peer review conducted in this way is extremely unlikely to detect instances of intentional misconduct. But the process is very good at separating valid science from nonsense. Referees know the current thinking in a field, are aware of its laws, rules and conventions, and will quickly detect any unjustified attempt to depart from them. Of course, for precisely this reason peer review can occasionally delay a truly visionary or revolutionary idea, but that may be a price that we pay for conducting science in an orderly way.

Peer review is less useful for adjudicating an intense competition for scarce resources. The pages of prestigious journals and the funds distributed by government agencies have become very scarce resources in recent times. The fundamental problem in using peer review to decide how these resources are to be allocated is obvious enough: There is an intrinsic conflict of interest. The referees, chosen because they are among the few experts in the author's field, are often competitors with the author for those same resources.

A referee who receives a proposal or manuscript to judge is being asked to do an unpaid professional service for the editor or project officer. The editor or officer thus has a responsibility to protect the referee, both by protecting the referee's anonymity and by making sure that the referee is never held to account for what is written in the report. Without complete confidence in that protection, referees cannot be expected to perform their task. Moreover, editors and project officers are never held to account for their choice of referees, and they can be confident that, should anybody ask, their referees will have the proper credentials to withstand scrutiny.

Referees would have to have high ethical standards to fail to take personal advantage of

their privileged anonymity and to make peer review function properly in spite of these conditions. Undoubtedly, most referees in most circumstances do manage to accomplish that. However, the fact is that many referees have themselves been victims of unfair reviews and this must sometimes influence their ability to judge competing proposals or papers fairly. Thus the institution of peer review seems to be suffering genuine distress.

Once again, this analysis shows that science is a complex enterprise that must be understood in some detail before ethical principles can be formulated to help guide it.

Conclusions

We have put forth arguments in this article that indicate why each of the principles listed above may be defective as a guide to the behavior of scientists. However, our repeated admonition that there are no universal rules of scientific conduct does not mean that it is impossible to recognize distinctive scientific misconduct. We would like to conclude with some thoughts on how scientific misconduct might be distinguished from other kinds of misconduct.

We propose that distinctively scientific forms of misconduct are those that require the expert judgment of a panel of scientists in order to be understood and assessed. Other forms of misconduct may take place in science, but they should not constitute scientific misconduct. For example, fabricating experimental data is scientific misconduct, but stealing scientific instruments is not. Similarly, misappropriation of scientific ideas is scientific misconduct, but plagiarism (copying someone else's words) is not. Stealing and plagiarism are serious misdeeds, but there are other well-established means for dealing with them, even

There are types of misrepresentation that are condoned and accepted in scientific publications, whereas other types are harshly condemned.

Fabrication or covert and unwarranted manipulation of data is an example of the kind of deceptive practice that cannot be tolerated because it undermines the mutual trust essential to the system of science.

when they are associated with science or committed by scientists. No special knowledge is required to recognize them.

On the other hand, only a panel of scientists can deal with matters such as data fabrication that require a detailed understanding of the nature of the experiments, the instruments used, accepted norms for presenting data and so on, to say nothing of the unique importance of experimental data in science. In a dispute over an allegation that a scientific idea has been misappropriated, the issues are likely to be so complex that it is difficult to imagine a lay judge or jury coming to understand the problem from testimony by expert witnesses or any other plausible means. Similarly, expert judgment will usually be required to determine whether an experimenter's procedures in selecting or discarding data constitute misconduct—the conventions governing this vary so much across different areas of science that judgments about what is reasonable will require a great deal of expert knowledge, rather than simply the application of some general rule that might be employed by nonscientists.

In the section on the logical structure of science, we drew a sharp distinction between advocacy, which is permitted or encouraged in science, and deception that is not open to public assessment, which is judged very harshly in science. Fabrication or covert and unwarranted manipulation of data is an example of the kind of deceptive practice that cannot be tolerated because it undermines the mutual trust essential to the system of science that we have described. Similarly, misappropriation of ideas undermines the reward system that helps motivate scientific progress. In both cases, a panel of scientists will be required to determine whether the deed occurred and, if so, whether it was done with intent to deceive or with reckless disregard for the

truth. Should these latter conditions be true, the act may be judged to be not merely scientific misconduct but, in fact, scientific fraud.

Acknowledgment

The authors wish to thank Kathy Cooke, Ph.D., for her valuable assistance in thinking through the problems discussed in this article.

Bibliography

Cole, J., and S. Cole. 1972. The Ortega hypothesis. *Science* 178:368–375.

Collins, H. 1974. The TEA-set: Tacit knowledge and scientific networks. *Science Studies* 4:165–186.

Collins, H. 1975. The seven sexes: A study in the sociology of a phenomenon, or the replication of experiments in physics. *Sociology* 9:205–224.

Duhem, P. 1962. *The Aim and Structure of Physical Theory.* New York: Athenaeum.

Feyerabend, P. 1975. *Against Method: Outline of an Anarchistic Theory of Knowledge.* London: New Left Books.

Feynman, R. 1985. *Surely You're Joking, Mr. Feynman.* New York: W. W. Norton.

Galison, P. 1987. *How Experiments End.* Chicago: University of Chicago Press.

Hadley, S. "Can Science Survive Scientific Integrity?" Paper presented on October 17, 1991, at the University of California, San Diego. Unpublished.

Kitcher, P. 1990. The division of cognitive labor. *Journal of Philosophy* 87:5–22.

Koertge, N. 1991. The function of credit in Hull's evolutionary model of science. *Proceedings of the Philosophy of Science Association* 2:237–244.

Kuhn, T. 1970. *The Structure of Scientific Revolutions.* Chicago: University of Chicago Press.

Lakatos, I. 1974. Falsification and the methodology of scientific research programmes. In *Criticism and the Growth of Knowledge,* ed. I. Lakatos and A. Musgrave. Cambridge, U.K.: Cambridge University Press. Pp. 91–196.

Medawar, P. 1963. Is the scientific paper a fraud? *The Listener* (12 September) 70:377–378.

Merton, R. 1968. The Matthew Effect in science. *Science* 159:56–63.

Ortega y Gasset, J. 1932. *The Revolt of the Masses.* New York: W. W. Norton.

Pinch, T. 1986. *Confronting Nature.* Dordrecht, Holland: P. Reidel.

Popper, K. 1968. *The Logic of Scientific Discovery.* New York: Harper and Row.

Popper, K. 1969. *Conjectures and Refutations.* London: Routledge and Kegan Paul.

Quine, W. V. O. 1961. *From a Logical Point of View.* New York: Harper and Row.

Sarah Blaffer Hrdy

Infanticide as a Primate Reproductive Strategy

Conflict is basic to all creatures that reproduce sexually, because the genotypes, and hence self-interests, of consorts are necessarily nonidentical. Infanticide among langurs illustrates an extreme form of this conflict

The Hanuman langur, *Presbytis entellus,* is the most versatile member of a far-flung subfamily of African and Asian leaf-eating monkeys known as Colobines. Langurs are traditionally classified as arboreal, but these elegant monkeys are built like greyhounds and can cover distances on the ground with speed and agility. Far more omnivorous than "leaf-eater" implies, Hanuman langurs feed on fully mature leaves, leaf flush, seeds, sap, fruit, insect pupae, and whatever delicacies might be fed them or left unguarded by local people. In forests, langurs spend much of their days in trees, but near open areas the adaptable Hanuman descends to the ground to feed and groom and may spend as much as 80 percent of daytime there. Monkeys are considered sacred by Hindus. This tolerance and their flexibility of diet and locomotion combine to make the Hanuman langur the most widespread primate other than man on the vast subcontinent of India. Ranging from as high

Sarah Blaffer Hrdy received her Ph.D. from Harvard in 1975 and was appointed a lecturer in biological anthropology there. Five years of research on langurs are chronicled in her forthcoming book, The Langurs of Abu: Male and Female Strategies of Reproduction *(Harvard Univ. Press). Currently she is doing research on monogamous primates. Dr. Hrdy wishes to acknowledge her debt to the community of langur fieldworkers, most especially to P. (Jay) Dolhinow, S. M. Mohnot, and Y. Sugiyama, and to other primatologists, J. Fleagle, D. Fossey, G. Hausfater, S. Kitchener, J. Oates, T. Struhsaker, R. Tilson, and K. Wolf, who allowed her access to unpublished findings. D. Hrdy, J. Seger, and R. Trivers made valuable comments on the manuscript. Dr. Blaffer Hrdy is also author of* The Black-man of Zinacantan *(Univ. of Texas Press, 1972), an analysis of myths of Maya-speaking people. Address: Department of Anthropology, Harvard University, Cambridge, MA 02138.*

as 400 meters in the Himalayas down to sea level, and living in habitats that grade from moist montane forest to semidesert, this flexible Colobine occurs in pockets and in connected swaths from Nepal, down through India, to the island country of Sri Lanka.

The stable core of langur social organization is overlapping generations of close female relatives who spend their entire lives in the same matrilineally inherited 40 hectare plot of land. Troops have an average of 25 individuals, including as many as three or more adult males, but more often only one fully adult male is present. Whereas females remain in the same range and in the company of the same other females throughout their lives, males typically leave their natal troop or are driven out by other males prior to maturity. Loose males join with other males (in some cases brothers or cousins) in a nomadic existence. These all-male bands, containing anywhere from two to 60 or more juvenile and fully adult males, traverse the ranges of a number of female lineages. They will not return again to troop life unless as adults they are successful in invading a bisexual troop and usurping resident males.

With the exception of male invasions, langur troops are closed social units. Troops are spaced out in separate ranges with some areas of overlap between them. When troops meet at the borders of their ranges, both males and females participate in defending their territory. Males are especially active, relying on a wide repertoire of impressive audiovisual displays, such as whooping, canine grinding, and daring leaps that create a swaying turmoil in the treetops. Despite chases and lunges, the ap-

parent aggressiveness of intertroop encounters is largely bravado and almost never results in injuries. Serious fighting among langurs is largely confined to the business of defending troops against invading males; invasions are the only encounters in which males have actually been seen to inflict injuries on one another.

Because of the close association between man and langurs in a part of the world where monkeys are considered sacred, the earliest published accounts of their behavior date back before the time of Darwin and provide us with extraordinary descriptions of langur males battling among themselves for access to females and of females going to great lengths to defend their own destinies. In the 1836 issue of the *Bengal Sporting Magazine,* for example, we are told that in langur society, males compete for females and "the strongest usurps the sole office of perpetuating his species" (Hughes 1884). Another account (see also Hughes 1884) was written by a Victorian naturalist who witnessed invading males attack and kill a resident male followed by a counterattack against the invaders by resident females, who—if we are to believe the account—castrated and mortally wounded one of the invaders:

In April 1882, when encamped at the village of Singpur . . . my attention was attracted to a restless gathering of Hanumans. . . . Two opposing troops [were] engaged in demonstrations of an unfriendly character. Two males of one troop . . . and one of another—a splendid looking fellow of stalwart proportions—were walking round and displaying their teeth. . . . It was some time—at least a quarter of an hour—before actual hostilities took place, when, having got within striking distance, the two monkeys made a rush at their adversary. I saw their arms

19

Mohnot—that led me to reject my initial crowding hypothesis in favor of the theory that infanticide is adaptive behavior, extremely advantageous for the males who succeed at it.

The langurs of Abu

The forested hillsides of Mt. Abu rise steeply from the parched Rajasthani plains. The town itself is an Indian pilgrimage and tourist center 1,300 m above sea level. My study concentrated on five troops in the vicinity of the town, but I will focus here on just two of these: the small Hillside troop and its neighbor, the Bazaar troop, whose name derives from the fact that these langurs spent a portion of almost every day scavenging in the bazaar (see Fig. 2).

In June 1971 the Hillside troop contained one adult male, seven adult females, six infants, and one juvenile male. In August of that year, Mug was replaced by a new male, Shifty Leftless—named for a bite-sized chunk missing from his left ear. At the time of the takeover, one adult female and all six infants disappeared from the troop. Soon after, mothers who had lost infants came into estrus and solicited the new male. Local inhabitants witnessed the killing of two infants by an adult male. Each killing took place at a site well within the range of the Hillside troop; in fact, one occurred at a location used exclusively by that group. It seemed highly probable that the missing infants had been killed, and that the usurping male Shifty was the culprit. (These events are discussed in greater detail in Blaffer Hrdy 1974 and 1975 diss.)

On my return to Abu in June 1972, I was surprised to find that the same male, Shifty, had now transferred to the neighboring Bazaar troop. In 1971, Bazaar troop had contained three adult males, ten subadult and adult females, five infants, and four juveniles. Three of these infants were

now missing. The killing of one had been observed by a local amateur ornithologist who lived beside the bazaar. The three Bazaar troop males remained in the vicinity of their former troop; the second-ranking of these bore a deep wound in his right shoulder.

During 1972, Mug took advantage of Shifty's absence to return to his former troop. At this time Hillside troop consisted of the same six adult females and their four new infants. Two females, an older, one-armed female called Pawless and a very old female named Sol, had no infants. Although Mug was able to return to his troop for extended visits, whenever Shifty left Bazaar troop on reconnaissance to Hillside troop, Mug fled. On at least eight occasions, Mug left the troop abruptly just as the more dominant Shifty arrived, or else the "interloper" was actually chased by Shifty. Typically, Shifty's visits to

Hillside troop were brief, but if one of the Hillside females was in estrus he might remain for as long as eight hours before returning to Bazaar troop.

During the periods Mug was able to spend with his former harem he made repeated attacks on infants that had been born since his loss of control. On at least nine occasions in 1972, Mug actually assaulted the infants he was stalking. Each time one or both childless females intervened to thwart his attack. Despite their heroic intervention, on three occasions the infant was wounded. During this same period, other animals in the troop were never wounded by the male. When the same male, Mug, had been present in the troop in 1971, he had not attacked infants. Similarly, during Shifty's visits to the Hillside troop in 1971, his demeanor toward infants was aloof but never hostile. Whereas Hillside mothers were very

Figure 3. This Hillside infant was conceived in 1971, during the time that Shifty was the troop's resident male, and was later killed by an adult male langur, probably Mug. The age of langur infants can be determined with some precision: between the third and fifth months of life, the all-black natal coat changes to cream color, starting with the top of the head and a little white goatee.

Figure 4. Juveniles and subadults threaten and lunge at two females from another troop. The adult male looks on calmly but does not participate.

restrictive with their infants when Mug was present, gathering them up and moving away whenever he approached, these same mothers were quite casual around Shifty. Infants could be seen clambering about and playing within inches of Shifty without their mothers' taking notice.

In 1973, Mug was joined by a band of five males. Nevertheless, the double usurper Shifty could still chase out all six males whenever he visited the Hillside troop. A daughter born to Pawless during the period when both Shifty and Mug were vying for control of Hillside troop was assaulted on several occasions by the five newer invaders; the infant eventually disappeared and was presumed dead.

By 1974, Mug was once again in sole possession of the Hillside troop and holding his own against Shifty. When the Hillside and Bazaar troops met, Mug remained with his harem. On several occasions, the newly staunch Mug confronted Shifty and in one instance grappled with him briefly before retreating behind females in the Hillside troop. Mug resolutely chased away members of a male band who attempted to enter his troop. By 1975, Mug's star had risen.

When I returned to Abu in March of that year, Shifty was no longer with the Bazaar troop. In his place was Mug. It was not known what had become of the extraordinary old male with the bite out of his left ear. Perhaps he died or moved on to another troop, or perhaps he was at last usurped by his longtime antagonist Mug.

Mug's former position in the Hillside troop was filled by a young adult male called Righty Ear. Righty (with a missing half-moon out of his right ear) was one of the five males who had joined Mug in the Hillside troop two years previously. Since that time, Righty had passed in and out of the troop's range, traveling with other males but not (so far as I knew) attempting to enter the troop. Righty's "waiting game" apparently paid off that March, when he came into sole possession of the Hillside troop. But, as in the case of his predecessors, Hillside troop was only a stepping stone: in April 1975, Righty replaced Mug as the leader of Bazaar troop.

The first indication I had of Righty's arrival in Bazaar troop was a report from local inhabitants that an adult male langur had killed an infant. On the following day when I investigated this report, the young adult male with the unmistakable half-moon out of his right ear was present in Bazaar troop; Mug was nowhere to be found.

An elderly langur mother still carried about the mauled corpse of her infant; by the following day, she had abandoned it. Righty subsequently made more than 50 different assaults on mothers carrying infants. Nevertheless, only one other infant disappeared. Five infants in the Bazaar troop remained unharmed when my observations terminated on June 20.

After Righty switched from Hillside to Bazaar troop, there followed some nine or more weeks during which the Hillside females had no resident male except for brief visits from Righty. Whenever the two troops met at their common border, Hillside females sought out Righty Ear and lingered beside him. These females were fiercely rebuffed by resident females in the Bazaar troop. Hostility of Bazaar troop females toward "trespassers" from Righty's previous harem prevented a merger of the two. The troops were still separate when Harvard biologist James Malcolm visited Abu in October 1975, but the vacuum in Hillside troop had been filled by a new male, christened Slash-neck for the deep gash in his neck.

The evolution of infanticide

Over a period of five years, then, political histories of the Hillside and Bazaar troops were linked by a succession of shared usurpers. First Shifty, then Mug, and finally Righty switched from the small and apparently rather vulnerable Hillside troop to the larger Bazaar troop (Fig. 5). Possibly the shifts were motivated by the greater number of reproductively active females in Bazaar troop. Between 1971 and 1975, at least four different males usurped control of Hillside troop. Infant mortality in this troop between 1971 and 1974 reached 83 percent, and extinction of the troop loomed as a real possibility. In contrast, during the same period, another troop at Abu, the School troop, was exceedingly stable, retaining the same male throughout.

Figure 5. The vicissitudes of male tenure in two troops of langurs are charted during the months the author spent observing the troops at intervals during 1971-75. Observations of infants missing, killed, or assaulted coincided with tenure shifts (as shown in italics in the chart).

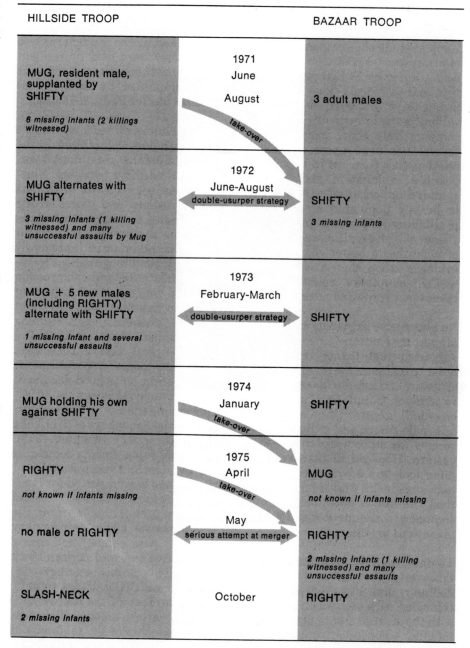

HILLSIDE TROOP		BAZAAR TROOP
MUG, resident male, supplanted by SHIFTY	**1971** June August	3 adult males
6 missing infants (2 killings witnessed)	*take-over*	
MUG alternates with SHIFTY	**1972** June-August double-usurper strategy	SHIFTY
3 missing infants (1 killing witnessed) and many unsuccessful assaults by Mug		*3 missing infants*
MUG + 5 new males (including RIGHTY) alternate with SHIFTY	**1973** February-March double-usurper strategy	SHIFTY
1 missing infant and several unsuccessful assaults		
MUG holding his own against SHIFTY	**1974** January	SHIFTY
	take-over	
RIGHTY	**1975** April	MUG
not known if infants missing	*take-over*	*not known if infants missing*
no male or RIGHTY	May serious attempt at merger	RIGHTY
		2 missing infants (1 killing witnessed) and many unsuccessful assaults
SLASH-NECK	October	RIGHTY
2 missing infants		

Combining all troop studies, the average male tenure at Abu was 27.5 months, a figure astonishingly close to the average tenure of 27 months calculated by Sugiyama for Dharwar (1967).

The short average duration of male tenure among langurs underlies the most crucial counterargument against the social pathology hypothesis: the extent to which adult males appear to gain from killing infants. Given that the tenure of a usurper is likely to be short, he would benefit from telescoping as much of his females' reproductive career as possible into the brief period during which he has access to them. By eliminating infants sired by a competitor, the usurping male hastens the mothers' return to sexual receptivity; on average, a mother whose infant is killed will become sexually receptive again within eight days of the death. In other words, infanticide permits an incoming male to use his short reign more efficiently than if he allowed unweaned infants present in the troop at his entrance to survive, to continue to suckle, and thus to delay the mother's next conception.

In three troops at Dharwar and Abu for which we have reliable information on subsequent births, 70 percent of the females who lost infants gave birth again within 6 to 8 months of the death of their infants, on average—just over one langur gestation period later. In the harsh desert environment of Jodhpur, however, the postinfanticide birth interval was much longer, up to 27 months.

Once infant-killing began, a usurper would be penalized for *not* committing infanticide. If a male failed to kill infants upon taking over a troop, and instead waited for those infants already in the troop to be weaned before he inseminated their mothers, then his infants would still be unweaned and hence vulnerable when the next usurper (presumably an infanticidal male) entered.

Other variations in the social system might likewise be expected to select for changes in male behavior. For example, if the rate of takeovers were speeded up and then held constant over time, male tolerance toward weaned immatures might be drastically altered. With a faster rate of takeover, it would be unlikely that one male could remain in control of a troop long enough for immature females to reach menarche and to give birth to an infant that would in turn grow old enough to survive the next takeover. Immature females, then, would be worth no more to the usurper than young males would be, and they might compete with the productive females of his harem for resources. Under these circumstances, it would behoove a usurper to drive out immatures of both sexes. This is precisely what occurs among a related langur species, *Presbytis senex*, living at very high densities (as high as 215 animals per km[2]) at Horton Plains in Sri Lanka (Rudran 1973). The ousted females travel with former male troopmates in mixed-sex bands.

Up to this point, I have not dealt with the apparent correlation between male takeovers and high population density. At both Dharwar (84–133 langurs per km[2]) and Abu (50 per km[2]), population densities are relatively high. In the desert region near Jodhpur, langurs have vast open areas available to them but tend to cluster about waterholes and garden spots. Infanticide has been reported at all three locations, but it has been recorded for none of the areas with low densities (at Jay's Orcha and Kaukori study sites or at any of three Himalayan sites where langurs have been studied by N. Bishop, the Curtins, and C. Vogel). This finding is inconclusive, however, since observations in the low-density areas were comparatively short, ranging from several months to a year. If the correlation does turn out to be valid, a possible explanation may be the greater numbers of extratroop males in heavily populated areas. If the possibilities for male recruitment are greater at high densities, and if a band of males has a better chance of usurping a troop than a single male would, then there would be more takeovers in crowded areas.

An alternative explanation has been offered by Rudran (1973), who has suggested that takeovers occur in order to maintain the one-male troop structure and infanticide occurs so as to curtail population growth in crowded areas. Unquestionably, one-male troops and reduced infant survival are outcomes of the takeover pattern. However, if takeovers and infanticide are advantageous to the individual males who engage in them, then the above outcomes are only secondary consequences and not explanations for them.

To date, we have information on 15 takeovers, 5 at Dharwar, one at Jodhpur, and 9 at Abu. At least 9 coincided with attacks on infants or with the disappearance of unweaned infants. A conservative estimate of the number of infants who have disappeared at the time of takeovers is 39. The important point (and this is the second line of evidence against the social pathology hypothesis) is that attacks on infants have been observed *only* when males enter the breeding system from outside—even if, as in the case of Mug, they have been only temporarily outside it. Such males are unlikely to be the progenitors of their victims. In contrast to what is considered "pathological" behavior, attacks on infants were highly goal-directed. An important area of future research will be learning exactly what means a langur male has at his disposal for discriminating infants probably his own from those probably sired by some other male. Quite possibly, males are evaluating past consort relations with the

mother (Blaffer Hrdy 1976). Interestingly, infants kidnapped by females from neighboring troops were not attacked by the resident male so long as they were held by resident females from his own troop and were not accompanied by their (alien) mothers (Blaffer Hrdy, 1975 diss.).

The third line of evidence against the social pathology hypothesis is the length of time that conditions favoring infanticide have persisted. Nineteenth-century accounts describing male invasions and fierce fights among males for access to females undermine the position that langur aggression and infanticide are newly acquired traits brought about by recent deforestation and compression of langur ranges. More important (and this constitutes the fourth line of evidence), recent findings concerning other members of the subfamily Colobinae suggest that a time span much longer than a few centuries is at issue. In addition to good documentation for male takeovers and infanticide among the closely related purple-faced leaf-monkeys of Sri Lanka (*Presbytis senex*) (Rudran 1973), adult male replacements coinciding with the death or disappearance of infants have been reported for *Presbytis cristata* of Malaysia (Wolf and Fleagle, in press); *P. potenziani*, the rare Mentawei Island leaf-monkey (R. Tilson, pers. comm.); and among both captive and wild African black and white colobus monkeys (S. Kitchener and J. Oates, pers. comm.). This recurrence of the takeover/infanticide pattern among widely separated members of the subfamily in Africa, India, and Southeast Asia argues strongly for its antiquity. Though the possibility of environmental convergence cannot be ruled out, the case of phylogenetic inheritance of these traits among geographically disparate relatives is a compelling one. Far from being recent responses to crowded conditions, it appears that a predisposition to male takeovers and infanticide has been part of the colobine repertoire since Pliocene times, some ten million or more years ago, when the split between the African and Asian forms occurred.

Beyond the Colobines

But the tale of infanticide does not stop with the Colobines. In what may be the most startling finding by pri-

matologists in recent years, we are discovering that the gentle souls we claim as our near relatives in the animal world are by and large an extraordinarily murderous lot. It is apparent now that the events witnessed at Abu and Dharwar are not aberrations. Increased observation of primates had led to an increase in the number of species in which adult males are known to attack and kill infants—and, occasionally, each other. Although murder is uncommon, cases of adults fighting to the death have been reported for rhesus, pig-tailed, and Japanese macaques, baboons, and chimpanzees, as well as Hanuman langurs.

At the time of this writing, infanticide, either observed or inferred from the disappearance of infants at times when males have usurped new females, has been reported for more than a dozen species of primates. Every major group of primates, including the prosimians, the New and Old World monkeys, apes, and man, is represented.

Not all these reports parallel the pattern of events recorded for Hanuman langurs, but many are disturbingly similar: males attack infants when they come into possession of females who are accompanied by offspring sired by another male. Typically, these are unfamiliar females. Perhaps the clearest illustration of the potential importance of previous acquaintance is provided by an experiment with caged crab-eating macaques (Thompson 1967). Here, infanticide was the unexpected outcome in a cage study on the effects of familiarity or lack of it in relations between male and female *Macaca fascicularis*. When paired with his accustomed companion and her infant, the adult male displayed typical behavior, mounting the female briefly and then casually exploring his surroundings. He entirely ignored the infant. Paired with an unfamiliar mother-infant pair, the male responded quite differently. After a brief attempt at mounting, the male attacked the infant as it lay clutched to its mother's belly. When the mother tried to escape, the male pinned her to the ground and gnawed the infant, making three different punctures in its brain with his canines.

Two suspected cases of infanticide

Figure 6. A female langur holding a newborn infant takes food from a priest of Shiva who lives in one of the sacred caves in the hillsides surrounding Mt. Abu.

among wild hamadryas baboons were occasioned by human manipulation. In the course of capture-and-release experimentation on the process of harem formation among *Papio hamadryas* of Ethiopia, two mothers were switched to new one-male units. In one case the infant was missing a day later; in the other, the infant was seen dead, "its skull pierced and its thighs lacerated by large canine teeth." The witnessed killing of two hamadryas infants at the Zurich Zoo just after their mothers changed "owners" adds plausibility to the inference that the wild infants were similarly murdered (Kummer et al. 1974).

Less contrived perhaps is the following account of chimpanzees from the Gombe Stream Reserve in Tanzania, where a young British researcher, David Bygott, happened to be following a band of five male chimpanzees when they encountered a strange female whom "in hundreds of hours of field observation," Bygott had never seen before. This female and her infant were immediately and intensely attacked by the males. For a few moments, the screaming mass of chimps disappeared from Bygott's view. When he relocated them, the strange female had disappeared and one of the males held a struggling infant. "Its nose was bleeding as though from a blow, and [the male], holding the infant's legs, intermittently beat its head against a branch. After 3 minutes, he began to eat the flesh from the thighs of the infant which stopped struggling and calling" (Bygott 1972). In contrast with normal chimp predation, this cannibalized corpse was nibbled by several males but never consumed.

Dian Fossey's remarkable decade-long study of wild mountain gorillas in central Africa provides what may be the most dramatic instances of adult male invaders mauling infants. For several days, a lone "silverback" (or fully mature) male had been following a harem of gorillas, presumably in quest of females. At last, he made his move, penetrating the group with a "violent charging run." A pri-

miparous female who had given birth to an infant on the previous night countered his charge by running at him. Halting within arm's reach of the male, she stood bipedally to beat her chest. The male struck her ventrally exposed body region where her newly born infant was clinging. Immediately following this blow, a "thin wail" was heard from the dying infant. On two other occasions, Fossey witnessed silverbacks kill infants belonging to primiparous mothers. In the best documented of these cases, the mother subsequently copulated with the male who had killed her infant (Fossey 1974; pers. comm.). To date, of the killings witnessed, only firstborn gorilla infants have been seen to be victims. This could be owing to maternal inexperience, or, as I believe is more probable, to Fossey's finding that in gorilla society only *young* females routinely change social units. Since an older mother would in all

likelihood not join a usurper anyway, he would rarely benefit from killing her infant.

Isolated instances of infanticide by adult males have also been reported for various prosimians: among free-ranging Barbary (*Macaca sylvana*) and rhesus macaques (*M. mulatta*) (Burton 1972; Carpenter 1942) and among wild *Cercopithecus ascanius*, the red-tailed monkeys of Africa (T. Struhsaker, in press). Infanticide is suspected among wild chacma baboons (*Papio ursinus*) (Saayman 1971); wild howler monkeys (*Alouatta*) of South America (Collias and Southwick 1952); and among caged squirrel monkeys (*Saimiri*) (Bowden et al. 1967).

The explanation for infanticide need not be the same in every case, but the parallels with the well-documented langur pattern are striking. According to the explanatory hypothesis offered here for langurs, infant-killing is a reproductive strategy whereby the usurping male increases his own reproductive success at the expense of the former leader (presumably the father of the infant killed), the mother, and the infant. If this model applies, the primatewide phenomenon of infanticide might be viewed as yet another outcome of the process Darwin termed *sexual selection:* any struggle between individuals of one sex (typically males) for reproductive access to the other sex, in which the result is not death to the unsuccessful competitor, but few or no offspring (Trivers 1972). Crucial to the evolution of infanticide are, first, a non-seasonal and flexible female reproductive physiology such that it is both feasible and advantageous for a mother to ovulate again soon after the death of her infant and, second, competition between males such that tenure of access to females is on average short.

Female counterstrategies

Confronted with a population of males competing among themselves, often with adverse consequences for females and their offspring, one would expect natural selection to favor those females most inclined and best able to protect their interests. When an alien langur male invades a troop, he may be chased away and harassed by resident females as well as by the resident male. After a new male takes over, females may form temporary alliances to prevent him from killing their infants (e.g. Sol and Pawless's combined front against the infanticidal Mug).

Females are often able to delay infanticide. Less often are they able to prevent it. Pitted against a male who has the option to try again and again until he finally succeeds, females have poor odds. For this reason, one of the best counterinfanticide tactics may be a peculiar form of female deceit. Almost invariably, langur males have attacked infants sired by some other male; a male who attacked his own offspring would rapidly be selected against. It may be significant then that at Dharwar, Jodhpur, and Abu, pregnant females confronted with a usurper displayed the traditional langur estrous signals: the female presents her rump to the male and

frenetically shudders her head. These females mated with the usurper even though they could not possibly have been ovulating at the time. Postconception estrus in this context may serve to confuse the issue of paternity.

After birth, an infant's survival is best ensured if its mother is able to associate with the father, or at least with a male who "considers" himself the father or who acts like one—in short, a male who tolerates her infant. In at least three instances at Abu, females with unweaned infants left recently usurped troops to spend time in the vicinity of males that on the basis of other evidence I suspected of having fathered their offspring.

If all else fails and her infant is attacked and wounded, a mother may continue to care for it, or abandon it. In several cases at Dharwar and Jodhpur, mothers abandoned their murdered infants soon after or even before death (Sugiyama 1967; Mohnot 1971). Rudran has suggested that the mother abandons her infant for fear of injury to herself and "because an adult female is presumably more valuable than an infant to the troop" (1973). It is far more likely, however, that desertion reflects a practical evaluation of what *this* infant's chances are weighed against the probability that her next infant will survive.

Under some circumstances a mother may opt to abandon an unwounded infant. In a single case from Abu, a female in a recently usurped troop who had been traveling apart from the troop (presumably to avoid the new male's assaults) left her partially weaned infant in the company of another mother and returned to the main body of the troop alone. If this was in fact an attempt to save her infant by deserting it, the ploy failed when the babysitter herself returned to the troop, some time later, bringing both infants with her. Nevertheless, both infants did survive the takeover.

Despite the various tactics that a female may employ to counter males, infanticide was the single greatest source of infant mortality at Abu. The plight of these females raises a perplexing question: How has this situation come about? Langur males contribute little to the rearing of off-

spring; apart from insemination, females have little use for males *except* to protect them from other langur males who might otherwise invade the troop and kill infants. Why then should females tolerate males at all, suffering subjection to the tyranny of warring polygynists? On the vast time scale of evolution, alternatives have been open to the female since the dawn of Colobines. Large body size, muscle mass, and saber-sharp canines might just as well have been selected among females as among males. Why should females weigh only 12 kg, on average, and not the 18 kg that males routinely do? Alternatively, female relatives could ally themselves to a much greater extent than they do. The combined 36 kg of three females operating as a united front against an infanticidal male surely should prevail. Infanticide depends for its evolutionary feasibility on the prior female adaptation of conceiving again as soon as possible after the death of an infant. If females failed to ovulate after a male killed their infants, or if they "refused" to copulate with an infanticide, the trait would be eliminated from the population.

The facts that females do not grow so large as males, that they do not selflessly ally themselves to one another, and that they do not boycott infanticides, suggest that counterselection is at work. Once again, the pitfall is intrasexual competition—this time competition among females themselves for representation in the next generation's gene pool. Whereas head-on competition between males for access to females selects for males who are as big and as strong (or stronger) than their opponents, a female who "opted" for large size in order to fight off males might not be so well-adapted for her dual role of ecological survivor and childbearer. An over-sized female might produce fewer offspring than her smaller cousin. In time, the smaller cousin's progeny would prevail.

Intrasexual competition is mitigated by the close genetic relatedness between female troop members, but it is by no means eliminated. A female in her reproductive prime who altruistically defended her kin, in spite of the cost to herself, might be less fit than her cousin who sat on the sidelines. Finally, if infanticide really is advantageous behavior for males, a female who sexually boycotted infanticides would do so to the detriment of her male progeny. Her sons would suffer in competition with the offspring of nondiscriminating mothers.

For generations langur females have possessed the means to control their own destinies. Caught in an evolutionary trap, they have never been able to use them.

References

Blaffer Hrdy, S. 1974. Male-male competition and infanticide among the langurs (*Presbytis entellus*) of Abu, Rajasthan. *Folia. Primat.* 22:19–58.

———. Male and female strategies of reproduction among the langurs of Abu. 1975 diss., Harvard University.

———. 1976. The care and exploitation of nonhuman primate infants by conspecifics other than the mother. In *Advances in the Study of Behavior 6*, ed. J. Rosenblatt, R. Hinde, C. Beer, and E. Shaw. Academic Press.

———. In press. *The Langurs of Abu*. Harvard Univ. Press.

Bowden, D., P. Winter, and D. Ploog. 1967. Pregnancy and delivery behavior of the squirrel monkey (*Saimiri sciureus*) and other primates. *Folia Primat.* 5:1–42.

Burton, F. 1972. The integration of biology and behavior in the socialization of *Macaca sylvana* of Gibraltar. In *Primate Socialization*, ed. F. Poirier. Random House.

Bygott, D. 1972. Cannibalism among wild chimpanzees. *Nature* 238:410–11.

Calhoun, J. 1962. Population density and social pathology. *Sci. Am.* 206:139–48.

Carpenter, C. R. 1942. Societies of monkeys and apes. *Biol. Symposia* 8:177–204.

Collias, N., and C. H. Southwick. 1952. A field study of population density and social organization in howling monkeys. *Proc. of the Amer. Phil. Soc.* 96:143–56.

Eisenberg, J. F., N. A. Muckenhirn, and R. Rudran. 1972. The relation between ecology and social structure in primates. *Science* 176:863–74.

Fossey, D. 1974. Development of the mountain gorilla (*Gorilla gorilla beringei*) through the first thirty-six months. Paper presented at Berg Wartenstein symposium no. 62, The Behavior of the Great Apes, Wenner-Gren Foundation for Anthropological Research.

Hughes, T. H. 1884. An incident in the habits of *Semnopithecus entellus*, the common Indian Hanuman monkey. *Proc. Asiatic Soc. of Bengal*, pp. 147–50.

Jay, P. The social behavior of the langur monkey. 1963 diss., University of Chicago.

———. 1965. The common langur of North India. In *Primate Behavior*, ed. I. DeVore. Holt, Rinehart and Winston.

Kummer, H., W. Gotz, and W. Angst. 1974. Triadic differentiation: An inhibitory process protecting pairbonds in baboons. *Behaviour* 49:62.

Mohnot, S. M. 1971. Some aspects of social change and infant-killing in the Hanuman langur, *Presbytis entellus* (Primates: Cercopithecinae) in Western India. *Mammalia* 35:175–98.

Rudran, R. 1973. Adult male replacement in one-male troops of purple-faced langurs (*Presbytis senex senex*) and its effects on population structure. *Folia Primat.* 19: 166–92.

Saayman, G. S. 1971. Behaviour of the adult males in a troop of free-ranging chacma baboons (*Papio ursinus*). *Folia Primat.* 15: 36–57.

Struhsaker, T. In press. Infanticide in the redtail monkey (*Cereopithecus ascanius schmidti*). In *Proceedings of the Sixth Congress of International Primatological Society*. Academic Press.

Sugiyama, Y. 1965. On the social change of Hanuman langurs (*Presbytis entellus*) in their natural conditions. *Primates* 6:381–417.

———. 1967. Social organization of Hanuman langurs. In *Social Communication among Primates*, ed. S. Altmann. Univ. of Chicago Press.

Thompson, N. S. 1967. Primate infanticide: A note and request for information. *Laboratory Primate Newsletter* 6(3):18–19.

Trivers, R. L. 1972. Parental investment and sexual selection. In *Sexual Selection and the Descent of Man 1871–1971*, ed. B. Campbell, pp. 136–79. Aldine.

Warren, J. M. 1967. Discussion of social dynamics. In *Social Communication Among Primates*, ed. S. Altmann. Univ. of Chicago Press.

Wolf, K., and J. Fleagle. In press. Adult male replacement in a group of silvered leaf-monkeys (*Presbytis cristata*) at Kuala Selangor, Malaysia. *Primates*.

Richard Curtin
Phyllis Dolhinow

Primate Social Behavior in a Changing World

Human alteration of the environment may be pushing the gray langur monkey of India beyond the limits of its adaptability

The gray langur monkey (*Presbytis entellus*) finds itself at present in the midst of heated controversy. Infant mortality in one of the langur troops at Mt. Abu, in northwest India (Fig. 1), rose to an astonishing 83% over a period of four years, and the causes for this striking figure immediately commanded the attention of students of primate behavior. It was assumed by some researchers that the Abu monkeys, and other langurs living at crowded sites, were typical of the species (Mohnot 1971; Sugiyama 1965, 1966, 1967; Hrdy 1974, 1977), and Hrdy has proposed an evolutionary model that includes males routinely killing infants as an adaptation to competition for females. Our observations of langurs at the less crowded sites of Orcha and Kaukori (Dolhinow 1972) and at Junbesi (Curtin 1975 diss.) have forced us to question this interpretation and to ask instead: How could such levels of infant death arise from the more widespread, generally peaceful pattern of behavior associated with the

Richard Curtin, who has studied langur monkeys in the Nepal Himalaya for twenty months, received his Ph.D. in 1975 from the University of California, Berkeley, where he is currently a Research Associate and Lecturer and does research on captive langurs. During his field work he was supported by an NIH Traineeship.
Phyllis Dolhinow, Professor of Anthropology at Berkeley, spent three and one half years in India studying langur monkeys and rhesus macaques. She has also worked in East Africa. Since 1972 she has done research on langur monkeys in a captive colony at the University of California. Her recent research concentrates on langur development, attachment, caretaking and adoption, and the effects of mother-loss. Address: Department of Anthropology, University of California, Berkeley, CA 94720.

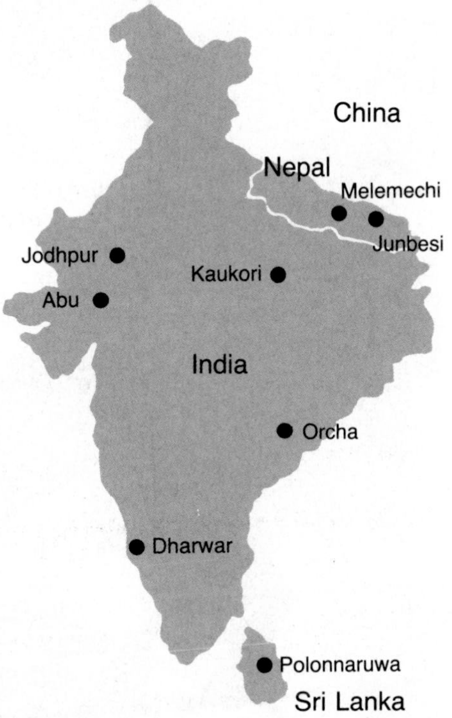

Figure 1. Gray langurs have been studied at a variety of sites in India, Nepal, and Sri Lanka, ranging from heavily populated areas to relatively unspoiled forest.

gray langurs' notable ecological and evolutionary success?

Langurs are members of the subfamily Colobinae, the Old World monkeys that possess digestive tracts which allow them to eat mature leaves. Primate studies have historically focused on the other subfamily—the Cercopithecinae, which includes such animals as baboons and rhesus macaques. Perhaps more important, colobines have been neglected because most of them are forest dwellers, and early studies concentrated on animals that could

be easily seen. However, field researchers of the 1960s and 1970s have been less intimidated by the difficulties of working in forest habitats, and several excellent studies of forest leaf-eaters have been completed or are under way.

Gray langurs live from Nepal to Sri Lanka in an array of habitats ranging from tropical jungle to desert margin, and from undisturbed forest to bustling village. Their behavior varies greatly over this range, but they have now been well studied at a number of sites scattered over the Indian subcontinent, and the limits of their behavioral variation have emerged clearly. The behavior of langurs at Orcha and Junbesi appears to represent a pattern of social organization that probably occurs over the species' whole range and persists under all but the most stressful environmental conditions.

Typical of the gray langur social pattern are multiple infant caretakers and male rivalry that can be dramatic but that does not normally disrupt the orderly rearing of the young. Langur troops include relatively stable cores of adult females plus adult and subadult males whose membership in the group is somewhat less constant. Males can and do move from troop to troop, and although relations among adult males differ from one site to another, competition for females is intense enough in some areas to bring about changes in troop composition. At Orcha, serious fighting among males was never observed during a year's study, but at Junbesi, fighting among males, particularly during the peak mating season, led to animals being driven from the troop, at least temporarily. These populations can be said to have

a "multi-male troop pattern," and are
to be distinguished from populations
in which troops typically have only
one adult male, and a very high percentage of males are not members of
reproductive troops.

The multi-male troop pattern is as
widespread as the species itself. It has
been found in all three continuous
long-term studies of gray langurs
living where human influence on
ecological conditions has been slight:
at Orcha, where leopards and tigers
still prowled the virgin forest; in
minimally disturbed oak forest in
central Nepal (Bishop 1975 diss.); and
at Junbesi, where a marginal habitat
and substantial predation helped
limit population density to a single
langur every square kilometer. The
pattern has also persisted under quite
different circumstances: Ripley
(1967) reported it from Sri Lanka
where langurs lived at a population
density fifty times that at Junbesi,
and the langurs at Kaukori maintained their stable multi-male troop
in an area where 98% of the land was
occupied by villages or under culti-
vation (see Fig. 2). This remarkable
adaptability of langur social organization carries with it a high degree
of behavioral variability, but the
wealth of studies now complete allows
us to describe, in general terms, the
social life a langur is likely to lead.

Normal social life

From the time a newborn presence is
noticed in a group, it becomes a focus
of great interest for young and old
females alike (Fig. 3). The infant is
conspicuous because the bare pink
skin of its hands, feet, face, and disproportionately large and convoluted
ears contrasts sharply with its sparse
dark brown fur. The newborn may be
passed among waiting females more
than fifty times in four hours on the
first day of life. It may be away from
its mother for hours, although normally she can retrieve it when she
desires, and most transfers are calm
and without peril to the infant. Many
females allow the infant to nurse,
even though they are not lactating or
may never have had an infant.

This high frequency of passing the
young infant from female to female
lasts only a few weeks. By the end of
the second month of life the infant is
very active and ably goes where and
when it wishes. In the meanwhile, it

Figure 2. A juvenile gray langur begs a bite of mango from her mother, who ignores her. Although the Kaukori troop, of which they are members, takes a toll of both unripe and ripe fruit from the mango orchards, an abundance of fruit remains on the large trees.

has had many hours of contact with
most of the females in the troop and
has developed strong preferences for
certain of them.

Weaning, at the end of the first year,
is seldom a harsh or traumatic experience. Inconsistent maternal rejection often prolongs weaning, and occasionally an adult with a neonate is
observed being followed closely by her
last infant. At the time of weaning, a
yearling may adopt another adult
female to serve as its mother. Although adoptions seldom have been
witnessed in the field, probably because of difficulties of observation,

they occur frequently in captivity and
may be stimulated by even brief re-
movals of the mother from the group.
This is, of course, assuming that
substitute caretakers are available—as they normally are.

Male and female paths of development diverge soon after the first year
of life. Play is important for both
sexes, but the immature female also
shows interest in neonates. By the
time she is two, she takes an active
part in attempting to take, hold, and
carry new infants. At first, her efforts
are hopelessly clumsy, and her in-
eptness, as reflected in the infant's

struggles and cries, stimulates older females to retrieve it. She gradually acquires skill and becomes a fair or even a competent caretaker well before she becomes a mother. There is a great individual variability in motivation to perform caretaking activities among females of all ages. A woefully inadequate juvenile may become a splendidly able mother!

Furthermore, adult females may treat successive infants in very different manners. The degree of interest in the infants of other females varies greatly among adults, and definite preferences are shown for certain infants by some females. The reasons for these preferences and for changes in caretaking styles of individual females are being investigated, because to the human observer it is not at all apparent why the variations exist. The vast majority of attentions directed toward infants can best be described as adequate and protective. Infants too are far from a homogeneous lot: some are strugglers and squealers, while others seem passive, seldom showing discomfort.

Adult females live an orderly existence. There are social changes as, for example, when males enter and leave the group, but the paramount markers of the passage of time are a succession of infants. A female has her first infant when she is 4 to 5 years old, and her ability to produce viable offspring improves as she ages. In the wild, females usually give birth at intervals of 18 to 24 months. There is variability in peak birth seasons in different areas of the Indian subcontinent, but most areas are characterized by seasonal regularity of births. When a female is in estrus, she solicits and mates with several males if the group is large enough, often showing a preference for one, who may not always be the alpha male. Complete copulations may occur during pregnancy but are by far more frequent during the female's mid-cycle estrous periods.

Precise information on the improvement of reproductive capability with aging is being gathered in a captive colony at Berkeley, where ages are known. By 9 years a female is bearing regularly, and evidence from both field and colony suggests strongly that she continues to do so for many years. No doubt, if she survives long enough, she will eventually experience a diminution of reproductive ability, but very aged females are rarely observed to survive the rigors of feral existence. Estimating ages in the field is exceedingly difficult, and indicators such as worn teeth, wrinkles, bags under the eyes, or generally poor health and emaciation are extremely unreliable.

Adult female social ranks are seldom clearly apparent or linearly ordered, and neither age nor reproductive history is a good predictor of social rank. The older females are by no means usually the highest in social authority, nor do they automatically yield to younger females. The scales of power tip with the events of the reproductive cycle such as sexual receptivity, pregnancy, delivery, and lactation. Fluctuations occur within periods of months as well as over years, and a female may move up or down relative to other females, only to change her position in the hierarchy after some months. Body size does not determine status once females are mature.

Male rivalry

Juvenile males, in contrast to females of the same age, spend most of their time in rough play and rarely show interest in newborn infants. As a male grows he tends to become somewhat peripheralized to the troop because his boisterousness annoys adults. A young male may leave the group when he is sexually mature, and a few do even before that. An exit may be hastened or stimulated by tensions arising between him and the adult males. Such departures are not always permanent, and reentries as well as entries into new troops have been recorded. The adult males in a troop can usually be ranked in a linear hierarchy of status, but there is no clear correlation between size or age and position in the hierarchy.

Rivalry with other males is inevitably an extremely important factor in a male langur's life. Since it is one of the most significant variables in langur behavior, and since it is probably at the root of the infant disappearances reported from Abu and other sites, its expression within the multi-male troop demands close scrutiny.

One end of the male-rivalry spectrum is shown by the relaxed behavior and absence of fighting in the troop at Orcha. Further along the spectrum, rivalry was far more apparent among the Kaukori langurs, where a stable dominance hierarchy was easily discernible among the troop's six adult males. Prolonged harassment with only occasional outbreaks of fighting accompanied the gradual exchange of the two top male positions. Notably, this dominance shift took place *within* the troop; after the exchange was effected, the defeated male assumed a stable, albeit lower, rank within the hierarchy.

In contrast to Kaukori, male rivalry was dramatic among the Junbesi langurs, and its effects were the most striking so far reported within the multi-male troop pattern. Relations among the adult males were tense throughout the 16 months of observation. During the first winter of the study, one of the troop's four original males repeatedly left after skirmishes with the other three males. These departures were temporary and are best described as peripheralizations, since the male did not leave the troop's home range, which was so large that he could stay within it and rarely be in close contact with the troop. He rejoined the troop in late winter but was frequently driven from it during the peak mating season in summer. He was again able to remain within the home range, and as the mating season ended he was gradually able to shorten his enforced distance from the troop. He was one of the two adult males in the troop when the study ended.

The other two males had left the troop permanently after repeated aggressive exchanges and violent fights at the beginning of the summer. One of them was later discovered in a different troop some distance away. A similar gradual process had apparently led to his successful entry, and he was a well-assimilated member of the multi-male troop.

Relations among males within the multi-male troop can thus follow several patterns: males can coexist amicably, with little apparent aggression, as at Orcha. Rivalry can be resolved within a troop by threat, fighting, or temporary peripheralization, or it can lead more dramatically to a male's permanent exclusion from his troop. Males can also enter strange troops or reenter their own with very little fighting. This is a re-

discourse are being continued. And the "measurements" of the last sentence has now become old information, reaching back to the "measured directly" of the preceding sentence. (It also fulfills the promise of the "we have directly measured" with which the paragraph began.) By following our knowledge of reader expectations, we have been able to spot discontinuities, to suggest strategies for bridging gaps, and to rearrange the structure of the prose, thereby increasing the accessibility of the scientific content.

Locating the Action
Our final example adds another major reader expectation to the list.

> Transcription of the 5S RNA genes in the egg extract is TFIIIA-dependent. This is surprising, because the concentration of TFIIIA is the same as in the oocyte nuclear extract. The other transcription factors and RNA polymerase III are presumed to be in excess over available TFIIIA, because tRNA genes are transcribed in the egg extract. The addition of egg extract to the oocyte nuclear extract has two effects on transcription efficiency. First, there is a general inhibition of transcription that can be alleviated in part by supplementation with high concentrations of RNA polymerase III. Second, egg extract destabilizes transcription complexes formed with oocyte but not somatic 5S RNA genes.

The barriers to comprehension in this passage are so many that it may appear difficult to know where to start revising. Fortunately, it does not matter where we start, since attending to any one structural problem eventually leads us to all the others.

We can spot one source of difficulty by looking at the topic positions of the sentences: We cannot tell whose story the passage is. The story's focus (that is, the occupant of the topic position) changes in every sentence. If we search for repeated old information in hope of settling on a good candidate for several of the topic positions, we find all too much of it: egg extract, TFIIIA, oocyte extract, RNA polymerase III, 5S RNA, and transcription. All of these reappear at various points, but none announces itself clearly as our primary focus. It appears that the passage is trying to tell several stories simultaneously, allowing none to dominate.

We are unable to decide among these stories because the author has not told us what to do with all this information. We know who the players are, but we are ignorant of the actions they are presumed to perform. This violates yet another important reader expectation: Readers expect the action of a sentence to be articulated by the verb.

Here is a list of the verbs in the example paragraph:

is
is... is
are presumed to be
are transcribed
has
is... can be alleviated
destabilizes

The list gives us too few clues as to what actions actually take place in the passage. If the actions are not to be found in the verbs, then we as readers have no secondary structural clues for where to locate them. Each of us has to

make a personal interpretive guess; the writer no longer controls the reader's interpretive act.

Worse still, in this passage the important actions never

> *As critical scientific readers, we would like to concentrate our energy on whether the experiments prove the hypotheses.*

---

appear. Based on our best understanding of this material, the verbs that connect these players are "limit" and "inhibit." If we express those actions as verbs and place the most frequently occurring information—"egg extract" and "TFIIIA"—in the topic position whenever possible,* we can generate the following revision:

> In the egg extract, the availability of TFIIIA limits transcription of the 5S RNA genes. This is surprising because the same concentration of TFIIIA does not limit transcription in the oocyte nuclear extract. In the egg extract, transcription is not limited by RNA polymerase or other factors because transcription of tRNA genes indicates that these factors are in excess over available TFIIIA. When added to the nuclear extract, the egg extract affected the efficiency of transcription in two ways. First, it inhibited transcription generally; this inhibition could be alleviated in part by supplementing the mixture with high concentrations of RNA polymerase III. Second, the egg extract destabilized transcription complexes formed by oocyte but not by somatic 5S genes.

As a story about "egg extract," this passage still leaves something to be desired. But at least now we can recognize that the author has not explained the connection between "limit" and "inhibit." This unarticulated connection seems to us to contain both of her hypotheses: First, that the limitation on transcription is caused by an inhibitor of TFIIIA present in the egg extract; and, second, that the action of that inhibitor can be detected by adding the egg extract to the oocyte extract and examining the effects on transcription. As critical scientific readers, we would like to concentrate our energy on whether the experiments prove the hypotheses. We cannot begin to do so if we are left in doubt as to what those hypotheses might be—and if we are using most of our energy to discern the structure of the prose rather than its substance.

Writing and the Scientific Process
We began this article by arguing that complex thoughts expressed in impenetrable prose can be rendered accessible and clear without minimizing any of their complexity. Our

---

*We have chosen these two pieces of old information as the controlling contexts for the passage. That choice was neither arbitrary nor born of logical necessity; it was simply an act of interpretation. All readers make exactly that kind of choice in the reading of every sentence. The fewer the structural clues to interpretation given by the author, the more variable the resulting interpretations will tend to be.

examples of scientific writing have ranged from the merely cloudy to the virtually opaque; yet all of them could be made significantly more comprehensible by observing the following structural principles:

1. Follow a grammatical subject as soon as possible with its verb.

2. Place in the stress position the "new information" you want the reader to emphasize.

3. Place the person or thing whose "story" a sentence is telling at the beginning of the sentence, in the topic position.

4. Place appropriate "old information" (material already stated in the discourse) in the topic position for linkage backward and contextualization forward.

5. Articulate the action of every clause or sentence in its verb.

6. In general, provide context for your reader before asking that reader to consider anything new.

7. In general, try to ensure that the relative emphases of the substance coincide with the relative expectations for emphasis raised by the structure.

None of these reader-expectation principles should be

It may seem obvious that a scientific document is incomplete without the interpretation of the writer; it may not be so obvious that the document cannot "exist" without the interpretation of each reader.

considered "rules." Slavish adherence to them will succeed no better than has slavish adherence to avoiding split infinitives or to using the active voice instead of the passive. There can be no fixed algorithm for good writing, for two reasons. First, too many reader expectations are functioning at any given moment for structural decisions to remain clear and easily activated. Second, any reader expectation can be violated to good effect. Our best stylists turn out to be our most skillful violators; but in order to carry this off, they must fulfill expectations most of the time, causing the violations to be perceived as exceptional moments, worthy of note.

A writer's personal style is the sum of all the structural choices that person tends to make when facing the challenges of creating discourse. Writers who fail to put new information in the stress position of many sentences in one document are likely to repeat that unhelpful structural pattern in all other documents. But for the very reason that writers tend to be consistent in making such choices, they can learn to improve their writing style; they can permanently reverse those habitual structural decisions that mislead or burden readers.

We have argued that the substance of thought and the expression of thought are so inextricably intertwined that changes in either will affect the quality of the other. Note that only the first of our examples (the paragraph about URF's) could be revised on the basis of the methodology to reveal a nearly finished passage. In all the other examples, revision revealed existing conceptual gaps and other problems that had been submerged in the originals by dysfunctional structures. Filling the gaps required the addition of extra material. In revising each of these examples, we arrived at a point where we could proceed no further without either supplying connections between ideas or eliminating some existing material altogether. (Writers who use reader-expectation principles on their own prose will not have to conjecture or infer; they know what the prose is intended to convey.) Having begun by analyzing the structure of the prose, we were led eventually to reinvestigate the substance of the science.

The substance of science comprises more than the discovery and recording of data; it extends crucially to include the act of interpretation. It may seem obvious that a scientific document is incomplete without the interpretation of the writer; it may not be so obvious that the document cannot "exist" without the interpretation of each reader. In other words, writers cannot "merely" record data, even if they try. In any recording or articulation, no matter how haphazard or confused, each word resides in one or more distinct structural locations. The resulting structure, even more than the meanings of individual words, significantly influences the reader during the act of interpretation. The question then becomes whether the structure created by the writer (intentionally or not) helps or hinders the reader in the process of interpreting the scientific writing.

The writing principles we have suggested here make conscious for the writer some of the interpretive clues readers derive from structures. Armed with this awareness, the writer can achieve far greater control (although never complete control) of the reader's interpretive process. As a concomitant function, the principles simultaneously offer the writer a fresh re-entry to the thought process that produced the science. In real and important ways, the structure of the prose becomes the structure of the scientific argument. Improving either one will improve the other.

The methodology described in this article originated in the linguistic work of Joseph M. Williams of the University of Chicago, Gregory G. Colomb of the Georgia Institute of Technology and George D. Gopen. Some of the materials presented here were discussed and developed in faculty writing workshops held at the Duke University Medical School.

Bibliography

Williams, Joseph M. 1988. *Style: Ten Lessons in Clarity and Grace*. Scott, Foresman, & Co.

Colomb, Gregory G., and Joseph M. Williams. 1985. Perceiving structure in professional prose: a multiply determined experience. In *Writing in Non-Academic Settings*, eds. Lee Odell and Dixie Goswami. Guilford Press, pp. 87–128.

Gopen, George D. 1987. Let the buyer in ordinary course of business beware: suggestions for revising the language of the Uniform Commercial Code. *University of Chicago Law Review* 54:1178–1214.

Gopen, George D. 1990. *The Common Sense of Writing: Teaching Writing from the Reader's Perspective*. To be published.

PART II
The Mechanisms of Behavior

The study of animal behavior involves research at different levels of analysis. On the one hand, there are questions about how animals' internal machinery functions to make behavior possible—that is, about the proximate mechanisms that enable individuals to do something. On the other hand, there are questions about why these mechanisms and the behaviors they control exist at all.

Kay Holekamp and Paul Sherman set the stage for the organizational scheme of this book by explaining what is meant by "levels of analysis." They show how the proximate and evolutionary approaches to animal behavior can be subdivided and yet integrated. They do so in the context of a concrete example, the dispersal behavior of Belding's ground squirrels, arguing that all studies of this behavior (or of any other behavior) can be categorized as research into (1) the underlying developmental or (2) physiological mechanisms responsible for the behavior *or* as research on (3) the evolutionary history or (4) fitness consequences of the behavior. These are complementary, not mutually exclusive, approaches: an answer at one level may be correct without eliminating other hypotheses at different levels of analysis. What kinds of academic disagreements would be prevented by adopting such a four-part scheme? What would happen, for example, if the hypothesis that male dispersal is caused by male hormones were pitted against the hypothesis that male dispersal is caused by selection for avoidance of inbreeding?

Not everyone, however, agrees that four fundamental questions underlie the entire field of behavioral research. Some persons have challenged the reasoning used by Holekamp and Sherman when they say that male Belding's ground squirrels disperse because the behavior advances individual reproductive success by reducing the chances of father–daughter and sister–brother incest. These critics state that the fitness consequences of a behavior cannot be invoked as a *cause* of that behavior when they really are an *effect* of the behavior. Do Holekamp and Sherman provide the basis for a suitable response to this criticism?

The other articles in Part II were written by persons interested primarily in the mechanisms underlying behavior, although their analyses at this level do not prevent them from also considering the fitness consequences of behavior. Donald Griffin's article on animal thinking raises the possibility that many animals other than ourselves possess consciousness mechanisms (i.e., physiological systems in the brain) that are fundamentally similar to our own. He encourages researchers to try to get inside the nervous systems of other animals to determine what they are "aware" of. If we could, our understanding of animal behavior would be greatly advanced, according to Griffin.

Griffin's article raises the question of how one would decide whether an animal's actions were based on conscious, as opposed to unconscious, decision making. How would it change your view of "lower animals" if you knew that their decisions were conscious? If crickets (or fish or snakes) could think, would you feel differently about the need to treat them humanely in scientific experiments? Griffin points out that many scientists resist the notion that animals other than ourselves are consciously aware of certain things. He speaks out strongly against this attitude, but readers may wish to ask why so many of Griffin's fellow scientists, after careful consideration of his position, still question the utility of trying to examine consciousness in other animals.

Griffin also discusses the possible fitness consequences of consciousness mechanisms, so his article integrates two different levels of analysis. He proposes that animals might gain reproductive benefits by having conscious thoughts in some situations. But are there any reproductive costs associated with this ability? If so, what determines the spread of the genes underlying the development of the brain systems required to produce consciousness in a killdeer, a sea otter, or any other animal, ourselves included? What sorts of responses are conscious and which are unconscious, and how might knowing this help us understand the costs and benefits of consciousness in ourselves and other creatures?

Although the particular physiological mechanisms that are responsible for human consciousness are not even well defined, let alone thoroughly understood, the genetic and neuronal bases for some elements of human behavior have been identified. Kenneth Blum and his colleagues describe how they and others have tested the hypothesis that genetic differences among people underlie variations in the way their brain cells communicate with one another. In particular, they point to variation in a single gene that codes for a neuronal receptor for the substance dopamine, a neurotransmitter that is released by certain brain cells and detected by others. Differences in this one gene appear to affect the number of dopamine receptors in signal-receiving cells, thereby affecting the cells' response to dopamine. Blum and his fellow researchers make the case that individuals with one form of this key gene are especially prone to develop what they call the "reward deficiency syndrome," which can manifest itself in any of several ways, ranging from alcoholism to cocaine abuse to antisocial personalities.

Can you recreate the scientific process followed by Blum and his colleagues as they reached the conclusion that dopamine and the D_2 receptor are important causes of a variety of addictions and disorders? Try to find at least two predictions and their tests in this article. What does the information presented in their Figure 13 constitute: a hypothesis, a prediction, or a test of a prediction? From an evolutionary perspective, is it surprising that so many millions of Americans are afflicted with addictions and other related disorders? What ideas do you have on why the allele that contributes to the development of these behavioral problems occurs in fully 25% of the general population?

Blum and his colleagues claim that particular behaviors are not absolutely determined by particular genes, yet they trace reward deficiency syndrome to one protein, the D_2 receptor. Haven't they demonstrated a one gene–one behavior relationship? Why or why not?

Blum and his colleagues note that environmental factors, as well as hereditary ones, may contribute to "abnormal" human behaviors, but their focus is largely on the genetic part of the equation. Meredith West and Andrew King, in contrast, devote their article primarily to the environmental factors that influence the "normal" development of a behavior—in this case, singing by starlings. According to modern developmental theory, all behavior is the ontogenetic product of a gene–environment interaction. Thus, the physiological systems that make it possible for a starling to sing are believed to have been produced through the interplay of genetic information and the bird's environment. Developmental theory, however, does not specify what elements of the inanimate or animate environment will affect the development of any given behavior.

West and King wondered how the social environment shaped the development of starling song, since these birds are skillful vocal mimics capable of imitating even human speech. The authors describe how they determined whether and how social cues have developmental effects on communication in the starling, and in so doing, they clear up an entertaining mystery involving Mozart and his pet starling. If West and King are correct, what kinds of bird species should be immune to social influences on the development of their vocalizations? How do you suppose brood-parasitic birds, which are raised individually in the nests of other species, acquire species-typical songs, or indeed, even come to recognize members of their own species? Does the evidence on how starling song develops suggest testable hypotheses about how humans acquire language?

When genes and environmental factors interact, they shape the physiological foundations of behavior, including animal nervous systems and hormones. As illustrated by reward deficiency syndrome, a change in even one protein in the neural networks in one part of the human brain can have profound effects on a person's

behavior. Likewise, in the article by John Wingfield and his coworkers, you will learn how a single hormone helps organize many elements of the behavior of breeding birds. The hormone in question is testosterone, a chemical long implicated in the regulation of sexual and aggressive behavior in various vertebrates.

As Wingfield and his coauthors make clear, it has not been easy to determine precisely what behavior patterns are controlled by testosterone in birds. (Why not?) However, Wingfield's team believes that testosterone's main function is to facilitate male aggression against rival males, something that is essential for males during that part of the breeding season when they are establishing their territories and defending them against intruders. How do these authors use measurements of testosterone levels as a means of testing the "challenge hypothesis" that testosterone primes males to respond to aggressive challenges from others? How do they explain results that contradict this hypothesis? Are they striving to falsify the hypothesis, as some would say was the ethical thing to do, or are they committed to the challenge hypothesis and eager to "explain away" any evidence to the contrary?

Some recent studies have shown that an individual's testosterone levels are extremely labile, changing within minutes in response to events such as being attacked by a rival. If hormone levels can fluctuate quickly and dramatically, does this create any problems for Wingfield and his colleagues, whose conclusions are based on data showing gradual changes in hormone levels over weeks and months?

Testosterone and allied hormones also are examined by Andrew Bass in his article on the mechanisms controlling song in the midshipman. In this remarkable fish, two forms of sexually mature males compete for females: a large type that defends nesting territories under intertidal rocks and produces a loud humming song that females find attractive, and a small, nonterritorial type that can grunt, but not sing, and "sneaks" matings instead of calling females to a nest site. The large singing males can produce so much noise that owners of houseboats floating above the midshipmen's habitat complain of insomnia in the spring when the fish are calling. Bass's article explores a large number of mechanisms that are responsible for the differences between the two types of males, including certain hormonal ones. Thus, the discovery that large and small males differ dramatically in the amount of a particular kind of testosterone in their blood suggests that this hormone is involved in the development of the distinctive song machinery that large males possess.

Bass devotes most of his article to describing how he and others have been able to discover the means by which nerve cells in the brain control the muscles involved in sound production by the large male midshipmen. But in addition, he discusses the possible reasons why two such different kinds of males have evolved in this species. Readers of this article may wish to organize the various hypotheses reviewed by Bass according to the scheme developed by Holekamp and Sherman in their article on ground squirrels. Is their system of classification adequate to deal with the large number of causal explanations presented by Bass? Can each one of Bass's hypotheses be unambiguously assigned to one of the categories outlined by Holekamp and Sherman? If not, should new levels of analysis be created, or should the existing four levels be subdivided?

Mike May uses the level of analysis that focuses on physiological mechanisms as he asks how certain night-flying insects can detect and escape from bats, a topic that has long fascinated biologists. A particularly appealing element of his article is May's reconstruction of his thoughts as he designed and conducted his doctoral research, which enables readers to catch a glimpse of a young researcher's own ontogeny and how one experiment led to another.

In the course of his article, May notes that "Not all useful observations come as a result of premeditated experimental design; sometimes it's useful just to play with the equipment." (The same point is raised by Winston and Slessor in Part V, where they describe their discovery of a way to test how certain chemicals affect the behavior of worker honeybees.) What about this claim? Do you believe that May had no hypothesis at all in mind while he was playing around with a cricket and his experimental equipment? Was May being a Baconian inductivist (see Woodward and Goodstein's article in Part I) during this phase of his work, or was he actually testing specific predictions even though he had not written them down in a clearly formulated manner? Is it possible to make scien-tific progress without using the scientific method? Does the method always require premeditation?

Because some nocturnal insects use bat detection devices and evasive responses, natural selection favors bat predators that can counter their prey's anti-bat tactics. The article by Brock Fenton and James Fullard on bats nicely complements the research by May and others on the insect side of the predator–prey equation. Fenton and Fullard examine the wonderful diversity in foraging behavior exhibited by bat predators as well as the interactions between bats and their potential victims. What level of analysis do Fenton and Fullard employ in their research? Why do different bats use different types of sound production, and how do the bats that don't echolocate at all ever find food? What is the likely evolutionary basis for the immense diversity in bat calls and insect responses? How do you suppose a bat distinguishes the ultrasonic echoes of its own calls from ultrasounds produced by other members of its species, or even by different species of bats?

The articles in Part II show that many different avenues exist for the study of behavioral mechanisms. Moreover, they show how studies of mechanism and evolution are interrelated. An understanding of precisely how a physiological element works helps researchers understand what natural selection designed it to do. And by considering the ultimate, evolutionary aspects of behavior, scientists can better identify what reproduction-enhancing mechanisms to look for in a given species.

Why Male Ground Squirrels Disperse

Kay E. Holekamp
Paul W. Sherman

When they are about two months old, male Belding's ground squirrels (*Spermophilus beldingi*) leave the burrow where they were born, never to return. Their sisters behave quite differently, remaining near home throughout their lives. Why do juvenile males, and only males, disperse? This deceptively simple question, which has intrigued us for more than a decade (*1, 2*), has led us to investigate evolutionary, ecological, ontogenetic, and mechanistic explanations. Only recently have answers begun to emerge.

Dispersal, defined as a complete and permanent emigration from an individual's home range, occurs sometime in the life cycle of nearly all organisms. There are two major types: breeding dispersal, the movement of adults between reproductive episodes, and natal dispersal, the emigration of young from their birthplace (*3, 4*). Natal dispersal occurs in virtually all birds and mammals prior to first reproduction. In most mammals, young males emigrate while their sisters remain near home (the females are said to be philopatric); in birds, the reverse occurs (*4–6*). Although naturalists have long been aware of these patterns, attempts to understand their causal bases have been hindered by both practical and theoretical problems. The former stem from difficulties of monitoring dispersal by free-living animals, and of quantifying the advantages and disadvantages of emigration (*6*). The latter stem from failure to distinguish the two types of dispersal, and from confusion among immediate and long-term explanations for each type.

We begin with a discussion of the latter point and

> *A multilevel analysis helps us to understand why male and not female Belding's ground squirrels leave the area where they were born*

develop the idea that natal dispersal, like other behaviors and phenotypic attributes, can be understood from multiple, complementary perspectives. Separating these levels of analysis helps organize hypotheses about cause and effect in biology (*7*). In the case of natal dispersal, this approach can minimize misunderstandings in terminology and allow for clearer focus on the issues of interest.

Questions of the general form "Why does animal A exhibit trait X?" have always caused confusion among biologists. And even today, the literature is full of examples. The nature-nurture controversy, which arose over the question of whether behaviors are innate or acquired through experience, is a classic case (*8*). After two decades of spirited but inconclusive argument in the nature-nurture debate, it became apparent to Mayr (*9*) and Tinbergen (*10*) that a lack of consensus was caused by the failure to realize that such questions could be analyzed from multiple perspectives.

In 1961, Mayr proposed that causal explanations in biology be grouped into proximate and ultimate categories. Proximate factors operate in the day-to-day lives of individuals, whereas ultimate factors encompass births and deaths of many generations or even entire taxa. Pursuing this theme in 1963, Tinbergen further subdivided each of Mayr's categories. He noted that complete proximate explanations of any behavior involve elucidating both its ontogeny in individuals and its underlying physiological mechanisms. Ultimate explanations require understanding both the evolutionary origins of the behavior and the behavior's effects on reproduction. The former involves inferring the phylogenetic history of the behavior, and the latter requires comparing the fitness consequences of present-day behavioral variants.

There are two key implications of the Mayr-Tinbergen framework. First, competition among alternative hypotheses occurs within and not between the four analytical levels. Second, at least four "correct" answers to any question about causality are possible, because explanations at one level of analysis complement rather than supersede those at another. Deciding which explanations are most interesting or satisfying is largely a matter of training and taste; debating the issue is usually fruitless (*7*).

With the Mayr-Tinbergen framework in mind, let us turn to the question of natal dispersal in ground squir-

Kay Holekamp is a research scientist in the Department of Ornithology and Mammalogy at the California Academy of Sciences. She received a B. A. in psychology in 1973 from Smith College, and a Ph.D. in 1983 from the University of California, Berkeley. From 1983 to 1985 she studied reproductive endocrinology as a postdoctoral fellow at the University of California, Santa Cruz. She is currently observing mother-infant interactions and the development of social behaviors in hyenas in Kenya. Paul Sherman is an associate professor of animal behavior at Cornell. He received a B. A. in biology in 1971 from Stanford, and a Ph.D. in zoology in 1976 from the University of Michigan. Following a postdoctoral appointment at the University of California, Berkeley (1976–78), he joined the psychology faculty there. He moved to his present position at Cornell in 1980, and is currently studying the behaviors of naked mole-rats, Idaho ground squirrels, and wood ducks. Address for Dr. Sherman: Section of Neurobiology and Behavior, Seeley G. Mudd Hall, Cornell University, Ithaca, NY 14853.

Figure 1. A female Belding's ground squirrel *(Spermophilus beldingi)* sits with two of her pups in the central Sierra Nevada of California. The pups are about four weeks old, and have recently emerged above ground. At about six or seven weeks of age, male ground squirrels begin to disperse; young females always remain near home. The causes of male dispersal in ground squirrels and many other mammals are complex, but can be explained by using a multilevel analytical approach in which four categories of causal factors are considered separately. (Photo by Cynthia Kagarise Sherman.)

rels. Following analyses of why natal dispersal occurs from each of the four analytical perspectives, we attempt an integration and a synthesis. Our studies reaffirm the usefulness of levels of analysis in determining biological causality.

From 1974 through 1985 we studied three populations of *S. beldingi* near Yosemite National Park in the Sierra Nevada of California (Figs. 1 and 2). In each population, the animals were above ground for only four or five months during the spring and summer; during the rest of the year they hibernated *(1, 2)*. Females bore a single litter of five to seven young per season, and reared them without assistance from males. Most females began to breed as one-year-olds, but males did not mate until they were at least two. Females lived about twice as long as males, both on average (four versus two years) and at the maximum (thirteen versus seven years) *(11)*.

During each field season ground squirrels were trapped alive, weighed, and examined every two to three weeks. About 5,300 different ground squirrels were handled. The animals were marked individually and observed unobtrusively through binoculars for nearly 6,000 hours. Natal dispersal behavior was measured by a combination of direct observations, livetrapping, radio telemetry, and identification of animals killed on nearby roads *(12)*. The day on which each emigrant was last seen within its mother's home range was defined as its date of dispersal. Only those juveniles that were actually seen after leaving their birthplace were classified as dispersers.

Observations of marked pups revealed that natal dispersal was a gradual process, visually resembling the fissioning of an amoeba (see Fig. 3). Young first emerged from their natal burrow and ceased nursing when they were about four weeks old. Two or three weeks later some youngsters began making daily excursions away from, and evening returns to, the natal burrow. Eventually these young stopped returning, restricting their activities entirely to the new home range; by definition, dispersal had occurred.

As shown in Figure 4, natal dispersal is clearly a sexually dimorphic behavior. In our studies, every one of over 300 surviving males dispersed by the end of its second summer; a large majority (92%) dispersed before their first hibernation, by the age of about 16 weeks. In contrast, only 5% of over 250 females recaptured as two-year-olds had dispersed from their mother's home range. The universality of natal dispersal by males suggested no plasticity in its occurrence; however, there was variation among individuals in the age at which dispersal occurred.

During the summer following their birth, males that

Figure 2. *S. beldingi* in the central Sierra Nevada are found above ground only four or five months of the year, during the spring and summer; they hibernate during the rest of the year. The group above is emerging from an underground burrow. Female adults bear litters of five to seven young each year and rear them in underground burrows without assistance from males. (Photo by George D. Lepp)

had dispersed as juveniles often moved again, always farther from their birthplace (Fig. 4). Yearling males were last found before hibernation an average of 170 m from their natal burrow, whereas yearling females moved on average only 25 m from home in the same time period. As two-year-olds, males mated at locations that were on average ten times farther from their natal burrows than the mating locations of females (13).

By the time they were two years old, male *S. beldingi* had attained adult body size. In the early spring they collected on low ridges beneath which females typically hibernated. As snow melted and females emerged, the males established small mating territories. Only the most physically dominant males—especially the old, heavy ones—retained territories throughout the three-week mating period. Although dominant males usually copulated with multiple females, the majority of males rarely mated. After mating, the most polygynous males again dispersed. They typically settled far from the places where they had mated; indeed, their new home ranges usually did not include their mating territories. Less successful males tended not to move, and they attempted to mate the following season in the same area where they were previously unsuccessful.

Females were all quite sedentary. After mating on a ridge top close to her hibernation burrow, each female dug a new nest burrow or refurbished an old one—sometimes her own natal burrow. There she reared her pups. As a result of philopatry, females spent their lives surrounded by and interacting with female relatives. Close kin cooperated to maintain and defend nesting territories and to warn each other when predators approached (13, 14). Natal philopatry has facilitated similar nepotism, or favoring of kin, among females in many other species of ground-dwelling sciurid rodents (15).

Physiological mechanisms

We began our analysis of natal dispersal in *S. beldingi* by considering physiological mechanisms. Of the two broad categories of such mechanisms, neuronal and hormonal, we were most interested in the latter. Gonadal steroids can influence the development of a specific behavior in two general ways: through organizational effects, which are the result of hormone action, in utero or immediately postpartum, on tissues destined to control the behavior, and through activational effects, which result from the direct actions of hormones on target tissues at the time the behavior is expressed (16). We suspected that gonadal steroids might mediate natal dispersal, and so we tested for organizational versus activational effects of androgens.

Under the activational hypothesis, levels of circulating androgens should be elevated in juvenile males at the time of natal dispersal. Conversely, in the absence of androgens, males should not disperse. To test this, we studied male pups born and reared in the laboratory. Blood samples were drawn every few weeks for four months (17). We also conducted a field experiment: soon after weaning but prior to natal dispersal, a number of juvenile males and females were gonadectomized; sham operations were performed on a smaller sample of each sex. After surgery, these juveniles were released into their natal burrow and subsequent dispersal behavior was monitored.

Castration was found to have little effect on natal dispersal. Although castrated males and those subjected to sham operations dispersed a few days later than untreated males, probably because of the trauma of surgery, castration did not significantly reduce the fraction that dispersed. Likewise, removal of ovaries did not increase the likelihood of dispersal by juvenile females. Finally, radioimmunoassays revealed only traces of testosterone in the blood of lab-reared juvenile males throughout their first four months, and no increase in circulating androgens was detected at the age when natal dispersal typically occurs (7–10 weeks).

Sex and body mass together were the most consistent predictors of dispersal status

---

Under the organizational hypothesis, exposing perinatal or neonatal females to androgens should masculinize subsequent behavior, including natal dispersal. We tested this idea by capturing pregnant females and housing them at a field camp until they gave birth. Soon after parturition, female pups were injected with a small amount of testosterone propionate dissolved in oil; a control group was given oil only. After treatment, the pups and their mothers were taken back to the field, where the mothers found suitable empty burrows and successfully reared their young.

Twelve of the female pups treated with androgens were located when they were at least 60 days old, and

75% of them had dispersed *(17)*. The distances they had traveled and their dispersal paths closely resembled those of juvenile males. By comparison, only 8% of untreated juvenile females in the same study area had dispersed by day 60, whereas 60% of juvenile males from the transplanted litters and 74% of males from unmanipulated litters born in the same area had dispersed by day 60.

It is possible that transplantation and not treatment with androgens caused the juvenile females in our experiment to disperse; unfortunately, we were unable to test this because none of the transplanted females treated with only oil were recovered. However, transplantation did not seem to affect the behavior of the juvenile males in the experiment. Also, other behavioral evidence linked natal dispersal in the females with androgen treatment. For example, treated juvenile females did not differ significantly from untreated juvenile males of the same age, but did differ from control females with respect to several indices of locomotor and social behavior. Androgen treatment masculinized much of the behavior of juvenile females, apparently including the propensity to disperse.

These results, which suggest an organizational role for steroids in sexual differentiation of *S. beldingi*, are consistent with those from studies of many other vertebrates *(18)*. In mammals, females are homogametic (XX) and males are heterogametic (XY), whereas in birds the situation is reversed. In each taxon, natal dispersal occurs primarily in the heterogametic sex. In both birds and mammals, sex-typical adult behavior in the homogametic sex can often be reversed by perinatal exposure to the gonadal steroid normally secreted at a particular developmental stage by the heterogametic sex. These considerations suggest that natal dispersal in mammals and birds has a common underlying mechanism, namely the organizational effects of gonadal steroids on the heterogametic sex.

Ontogenetic processes

Natal dispersal might be triggered during development by changes in either the animal's internal or external environment. We tested two hypotheses about external factors. First, natal dispersal might be caused by aggression directed at juveniles by members of their own species. Under this hypothesis, prior to or at the time of dispersal, the frequency or severity of agonistic behavior between adults and juvenile males should increase. However, observations revealed that adults neither attacked nor chased juvenile males more frequently or vigorously than juvenile females *(19)*, and there was no increase in aggression toward juvenile males at the time of dispersal. Moreover, there were no differences between juvenile males and females in the number and severity of wounds inflicted by other ground squirrels. Thus the data offered no support for the social aggression hypothesis.

A second hypothesis is that natal dispersal occurred because juvenile males attempted to avoid their littermates (current and future competitors) or their mother *(20)*. For a large number of litters, we found no significant relationship between litter size or sex ratio and

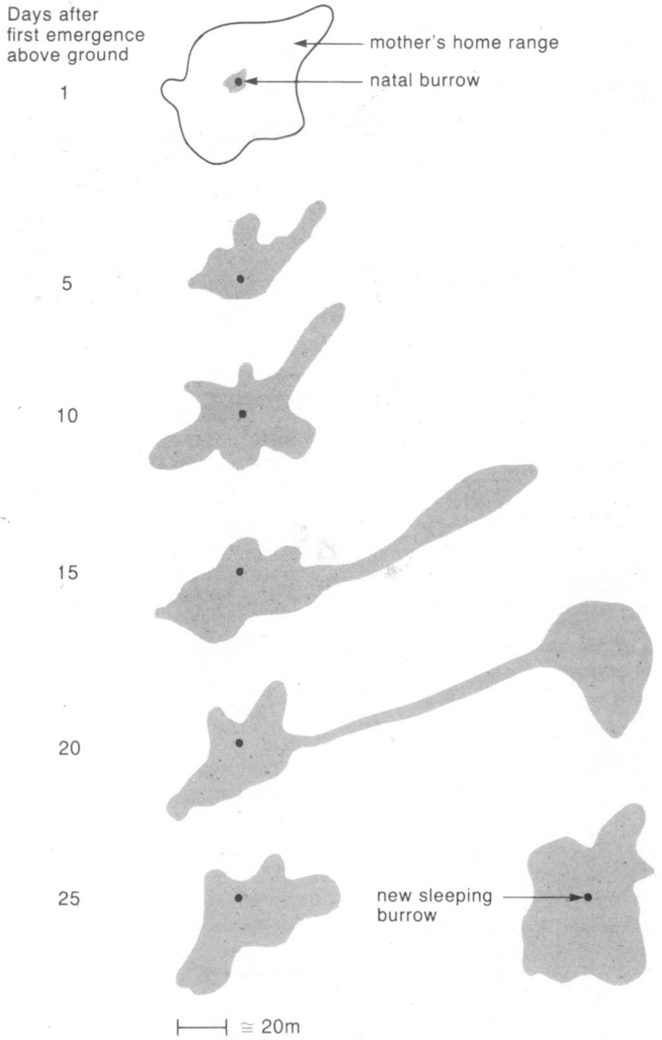

Figure 3. The process by which *S. beldingi* males disperse visually resembles the fissioning of an amoeba. When a male first emerges from the natal burrow, at an age of about four weeks, his daily range of movement is restricted to the immediate vicinity of the burrow. He soon enlarges that range into an amorphous shape, the boundaries of which are established by topographic features or the presence of other animals. By about the 15th day above ground, his range has surpassed the scope of his mother's home range. At this time he may spend long periods far from the natal burrow, yet he will return home at nightfall. Near the 25th day, when he is roughly seven weeks old, he will cease returning at dusk, thereby accomplishing dispersal. (After ref. *35*.)

dispersal behavior *(2, 19)*. Males who dispersed during their natal summer were not from especially large or small litters, or predominantly male or female litters. Also, the timing of juvenile male dispersal depended neither on the mother's age nor on whether the mother was present or deceased. Thus the ontogeny of natal dispersal was apparently not linked to either of the exogenous (external) influences usually invoked to account for it.

In view of these results, we suspected that natal dispersal was triggered by endogenous (internal) factors. In particular, we hypothesized that males might stay home until they attained sufficient size or energy reserves to permit survival during the rigors of emigration. This ontogenetic-switch hypothesis predicts that juvenile

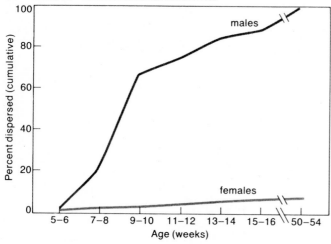

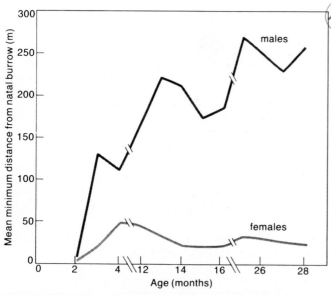

Figure 4. Although a small fraction of female *S. beldingi* disperse, the behavior is very evidently male-biased. The majority of male pups disperse by the 10th week; by about the 54th week, all males have dispersed *(above)*. Although many other mammals exhibit male-biased natal dispersal, *S. beldingi* is unusual in that all males eventually disperse. Males also move considerably farther from their natal burrows than do females, and they continue to move away from home throughout their first three years *(below)*. (After refs. 2 and 12.)

males will disperse when they attain a threshold body mass and that dispersers should be heavier, or exhibit different patterns of weight gain, than predispersal males of equivalent ages.

Our data were consistent with the ontogenetic-switch hypothesis. Emigration dates were correlated with the time at which males reached a minimum body weight of about 125 g, as shown in Figure 5. Emigrant juveniles were significantly heavier than male pups that had not yet dispersed. Most males attained the threshold weight during their natal summer, and dispersed then. Only the smallest males, who did not put on sufficient weight in the first summer, overwintered in their natal area. All these males dispersed the following season once they had become heavy enough.

Sex and body mass together were the most consistent predictors of dispersal status. Occasionally, how-

ever, predispersal and immigrant juvenile males with body weights exceeding the threshold were captured in the same area. This observation suggested that something closely associated with body weight, such as fat stores, may be the actual dispersal trigger.

Behavioral changes also accompanied natal dispersal. The frequencies of movement and distances moved per unit time by juvenile males were found to be greater than those of females, and these behaviors peaked at the time of dispersal. Relative to juvenile females, juvenile males also spent significantly more time climbing and digging and exploring nonfamilial burrows and novel objects—for example, a folding footstool; they also reemerged from a burrow into which they had been frightened much sooner than did females. These observations of spontaneous ontogenetic changes in the behavior of young males reinforced the hypothesis that endogenous factors triggered natal dispersal.

Effects on fitness

Natal dispersal might enable juvenile males to avoid fitness costs associated with life in the natal area and might allow them to obtain benefits elsewhere *(6)*. Possible disadvantages of remaining at home include shortages of food or burrows *(21)*, ectoparasite infestations or diseases, competition with older males for mates *(5, 22)*, and nuclear family incest *(4, 23, 24)*. We examined each of these hypotheses as functional explanations for natal dispersal in *S. beldingi.*

If natal dispersal occurs because of food shortages, then juveniles whose natal burrow is surrounded by abundant food should be more philopatric than those from food-poor areas; immigration to food-rich areas should exceed emigration from them; dispersing individuals should be in poorer condition (perhaps weigh less) than males of the same age residing at home; and, based on the strong sexual dimorphism in natal dispersal, food requirements of young males and females should differ.

Detailed observations revealed that juvenile males and females ate similar amounts of the same plants and at similar rates. Juvenile males spent only slightly more time foraging than did juvenile females. The diets and foraging behaviors of males that had not yet dispersed and males that had immigrated to that same area were indistinguishable. As discussed previously, dispersing males were significantly heavier than predispersal males, a result contrary to that predicted in the scenario of emigration because of lack of food. Finally, juvenile male immigration equaled emigration every year. This is important because preferred foods were unevenly distributed within and among populations *(1, 2)*. Evidence consistently suggested no link between immediate food shortages and natal dispersal.

A second reason for natal dispersal might be to locate a nest burrow. Ground squirrels depend on burrows for safety from predators, as places to spend the night, and as nests in which to hibernate *(25)*. Given the sexual dimorphism in natal dispersal, this hypothesis predicts differences between males and females in the type or location of habitable burrows and implies that dispersers should emigrate from areas of high population density or low burrow quality to areas where unoccupied holes of high quality are available. To test this idea we monitored population density each week and counted burrow entrances in the territories of lactating females. We found that neither the probability of juvenile male dispersal nor its timing was significantly related to population density or burrow availability near home, and that dispersers did not settle in areas of higher burrow density.

The only unusual aspect is that every male eventually leaves home

Another cause of natal dispersal might be ectoparasite infestation. If parasites build up in the natal nest and if juvenile males are more affected by them than are juvenile females, then males in particular might emigrate to avoid them. We examined this hypothesis indirectly, by counting the number of fleas and ticks on every captured juvenile. We found low levels of ectoparasitism throughout the animals' natal summer, and no consistent differences between infestations in males and females prior to or at the time of dispersal.

Do juvenile males disperse to avoid future competition with older males for sexual access to females? Because males always emigrated, it was not possible to determine if dispersers experienced less severe mate competition than hypothetical nondispersers. However, the mate competition hypothesis was examined indirectly by comparing, at sites where males were born and on ridge tops where those males mated two or more years later, three parameters: the density of breeding adult males, the mean number of fights adults engaged in for each successful copulation, and the mean daily ratio of breeding males to receptive females. We found no significant differences in any of these parameters, suggesting that dispersing males did not find better access to females than they would have if they had remained at home.

Do juvenile males disperse to avoid future nuclear family incest? A test of this hypothesis requires comparing the reproductive consequences of various degrees of inbreeding (26, 27). However, of more than 500 copulations observed, none occurred between close kin; therefore we could not directly test this hypothesis. Nonetheless, the nonrandom movements of males away from the natal area clearly resulted in complete avoidance of kin as mates (Fig. 4). Furthermore, during post-breeding dispersal, the highly polygynous males moved farthest. Under this hypothesis, the polygynous males who had sired many female pups in an area would have the most to gain by emigrating. Under the mate-competition hypothesis, successful males would be expected to stay put, while unsuccessful males might gain by dispersing. The observed pattern is thus most consistent with avoiding inbreeding.

Belding's ground squirrels are not unusual in the rarity of close inbreeding. Consanguineous mating is minimized in most mammals and birds (23, 24, 28, 29), often via the mechanism of sex-specific natal dispersal. But why are males the dispersive sex in mammals generally and ground squirrels particularly? The answer probably relates to a sexual asymmetry in the significance

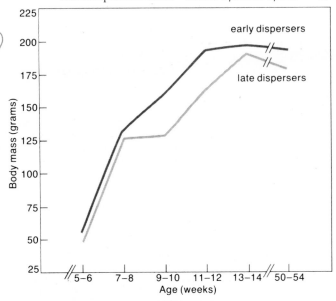

Figure 5. Weight gain among juvenile *S. beldingi* has been positively correlated with the onset of dispersal. Early dispersers (most males) left home at 7–10 weeks of age; late dispersers, in contrast, remained at home until they were 11–14 weeks old. Dispersal seems to occur when a threshold body mass of 125–150 g is attained. (After ref. *19*.)

of the location and quality of burrows for procreation (6, 30). The depth and dryness of nest burrows, their proximity to food, and their degree of protection from both inter- and intraspecific predators are vital to pup survival (31). The significance of the burrow, in turn, favors females who seek out and defend high-quality nest sites and who remain in them from year to year (25). The quality of a nursery burrow is of negligible significance to nonparental males. To avoid predators and inclement weather, and to forage, males can move frequently without jeopardizing the survival of their young. Thus the sexual bias in natal dispersal might occur because inbreeding is harmful to both sexes and males incur lower procreative costs by leaving home.

Sexual selection could reinforce a sex-bias in natal dispersal generated by incest avoidance. If consanguineous mating is indeed harmful, then the philopatric females should prefer to mate with unrelated (unfamiliar) males. A reproductive advantage should therefore accrue to males that seek and locate unfamiliar females (32).

Evolutionary origins

The fourth component of our investigation of natal dispersal was an attempt to infer evolutionary origins. A first hypothesis was that the male bias in natal dispersal arose in an evolutionary ancestor of *S. beldingi* as a developmental error (for example, in the timing of hormone secretion) or as a by-product of natural selection on males for the high levels of activity associated with finding mates and defending mating territories. Alternatively, perhaps natal dispersal was favored directly by selection, for example, as a mechanism to avoid inbreeding, throughout the evolutionary history of *S. beldingi*.

One way to evaluate these alternatives is to consider

Why do juvenile male Belding's ground squirrels disperse? Answers have been found at each of four levels of analysis.

Level of analysis	Summary of findings
Physiological mechanisms	Dispersal by juvenile males is apparently caused by organizational effects of male gonadal steroid hormones. As a result, juvenile males are more curious, less fearful, and more active than juvenile females.
Ontogenetic processes	Dispersal is triggered by attainment of a particular body mass (or amount of stored fat). Attainment of this mass or composition apparently also initiates a suite of locomotory and investigative behaviors among males.
Effects on fitness	Juvenile males probably disperse to reduce chances of nuclear family incest.
Evolutionary origins	Strong male biases in natal dispersal characterize all ground squirrel species, other ground-dwelling sciurid rodents, and mammals in general. The consistency and ubiquity of the behavior suggest that it has been selected for directly across mammalian lineages.

the taxonomic distribution of male-biased natal dispersal. If selection has consistently and directly favored dispersal by juvenile males, then phylogenetic relatives of *S. beldingi* should share this trait to a greater degree than if it were a hormonally mediated side effect or developmental error. This is because any hormonal link between adult male sexual activities and dispersal by juveniles two years previously could presumably be broken by mutation in some species through evolutionary time. This, in turn, would lead to a spotty taxonomic distribution of the behavior if it were neutral for fitness.

Members of the squirrel family first appeared in the fossil record 35 to 40 million years ago; thus they are one of the most ancient of extant rodent families (33). Belding's ground squirrel is one of 32 species in the genus *Spermophilus*; this genus is more closely related to marmots and prairie dogs than to tree squirrels (34). Strongly male-biased natal dispersal occurs in all 12 *Spermophilus* species that have been studied in this regard (5, 15, 35). Male-biased natal dispersal patterns are also the rule in marmots (35) and prairie dogs (36). The dispersal behavior of *S. beldingi* is therefore probably a conservative rather than a derived trait; in other words, it is likely quite ancient.

As far as we know, the only unusual aspect of natal dispersal in *S. beldingi* is that every male eventually leaves home, whereas in a few other species a tiny fraction of males are philopatric. Male-biased natal dispersal is widespread among mammals (4–6, 30, 32, 37), suggesting that this behavior may predate the appearance of the squirrel family. The ubiquity of natal dispersal seems more consistent with the hypothesis that it has been favored directly by natural selection in various

lineages than that it originated as a mistake or a correlated response to selection for some other male attribute and is maintained by phylogenetic inertia rather than adaptive value.

Synthesis

Our data reveal that there are at least four types of answers to the question of why juvenile male Belding's ground squirrels disperse (see the box). These answers complement rather than supersede each other. Clearly, however, the causal variables we have identified within each analytical level do not operate in isolation, and it seems appropriate to consider how they may interrelate.

During embryogenesis, sex chromosomes cause the formation of testes in male *S. beldingi*. The gonads secrete a pulse of androgens before birth, which, we hypothesize, sets up an ontogenetic switch, presumably by modifying the morphology or behavior of neurons or nuclei in the brain. When juvenile males have accumulated sufficient weight or fat stores, the switch turns on. The young males then boldly explore their environment, making increasingly longer forays away from home. The timing of dispersal by each individual may be influenced by any environmental factor that accelerates or delays arrival at the dispersal threshold (for example, food abundance or scarcity). The main cost of natal dispersal is probably mortality during emigration; the main benefits are likely related to reduced inbreeding and optimal outbreeding. Male biases in natal dispersal occur consistently across modern mammalian taxa (37), suggesting an evolutionary history of natural selection favoring such behavior directly, and a taxon-wide consistency of function.

By employing the levels-of-analysis framework for developing and testing hypotheses, we have come to appreciate the complexity of what at first appeared to be a simple behavior. We suspect that our explanations for the proximate and ultimate causes of natal dispersal in *S. beldingi* will be applicable to other species. Perhaps equally important, our study illustrates that there can be multiple correct answers to questions of causality in behavioral biology (38). The usefulness of the levels-of-analysis approach is thereby reemphasized.

References

1. P. W. Sherman. 1976. Natural selection among some group-living organisms. Ph.D. thesis, Univ. of Michigan.
2. K. E. Holekamp. 1983. Proximal mechanisms of natal dispersal in Belding's ground squirrels (*Spermophilus beldingi beldingi*). Ph.D. thesis, Univ. of California.
3. W. Z. Lidicker, Jr. 1975. The role of dispersal in the demography of small mammals. In *Small Mammals: Their Productivity and Population Dynamics*, ed. F. B. Golley, K. Petruscewicz, and C. Ryszkowski, pp. 103–28. Cambridge Univ. Press.
4. P. J. Greenwood. 1980. Mating systems, philopatry and dispersal in birds and mammals. *Animal Behav.* 28:1140–62.
5. F. S. Dobson. 1982. Competition for mates and predominant juvenile male dispersal in mammals. *Animal Behav.* 30:1183–92.
6. A. E. Pusey. 1987. Sex-biased dispersal and inbreeding avoidance in birds and mammals. *Trends in Ecol. and Evol.* 2:295–99.
7. P. W. Sherman. 1988. The levels of analysis. *Animal Behav.* 36: 616–19.
8. D. S. Lehrman. 1970. Semantic and conceptual issues in the

nature-nurture problem. In *Development and Evolution of Behavior*, ed. L. R. Aronson, E. Tobach, D. S. Lehrman, and J. S. Rosenblatt, pp. 17–52. W. H. Freeman.

9. E. Mayr. 1961. Cause and effect in biology. *Science* 134:1501–06.

10. N. Tinbergen. 1963. On aims and methods of ethology. *Zeitschrift für Tierpsychologie* 20:410–33.

11. P. W. Sherman and M. L. Morton. 1984. Demography of Belding's ground squirrels. *Ecology* 65:1617–28.

12. K. E. Holekamp. 1984a. Natal dispersal in Belding's ground squirrels *(Spermophilus beldingi)*. *Behav. Ecol. Sociobiol.* 16:21–30.

13. P. W. Sherman. 1977. Nepotism and the evolution of alarm calls. *Science* 197:1246–53.

14. P. W. Sherman. 1981a. Kinship, demography, and Belding's ground squirrel nepotism. *Behav. Ecol. Sociobiol.* 8:251–59.

15. G. R. Michener. 1983. Kin identification, matriarchies, and the evolution of sociality in ground-dwelling sciurids. In *Recent Advances in the Study of Mammalian Behavior*, ed. J. F. Eisenberg and D. G. Kleiman, pp. 528–72. Am. Soc. Mammal.

16. C. H. Phoenix, R. W. Goy, A. A. Gerall, and W. C. Young. 1959. Organizing action of prenatally administered testosterone propionate on the tissues mediating mating behavior in the female guinea pig. *Endocrinology* 65: 369–82.

17. K. E. Holekamp, L. Smale, H. B. Simpson, and N. A. Holekamp. 1984. Hormonal influences on natal dispersal in free-living Belding's ground squirrels *(Spermophilus beldingi)*. *Hormones and Behavior* 18:465–83.

18. E. Adkins-Regan. 1981. Early organizational effects of hormones: An evolutionary perspective. In *Neuroendocrinology of Reproduction*, ed. N. T. Adler, pp. 159–228. Plenum Press.

19. K. E. Holekamp. 1986. Proximal causes of natal dispersal in Belding's ground squirrels *(Spermophilus beldingi)*. *Ecol. Monogr.* 56: 365–91.

20. S. Pfeifer. 1982. Disappearance and dispersal of *Spermophilus elegans* juveniles in relation to behavior. *Behav. Ecol. Sociobiol.* 10:237–43.

21. F. S. Dobson. 1979. An experimental study of dispersal in the California ground squirrel. *Ecology* 60:1103–09.

22. J. Moore and R. Ali. 1984. Are dispersal and inbreeding avoidance related? *Animal Behav.* 32:94–112.

23. A. E. Pusey and C. Packer. 1987. The evolution of sex-biased dispersal in lions. *Behaviour* 101:275–310.

24. A. Cockburn, M. P. Scott, and D. J. Scotts. 1985. Inbreeding avoidance and male-biased natal dispersal in *Antechinus* spp. (Marsupialia: Dasyuridae). *Animal Behav.* 33:908–15.

25. J. A. King. 1984. Historical ventilations on a prairie dog town. In *The Biology of Ground-dwelling Squirrels*, ed. J. O. Murie and G. R. Michener, pp. 447–56. Univ. of Nebraska Press.

26. W. M. Shields. 1982. *Philopatry, Inbreeding, and the Evolution of Sex*. State Univ. of New York Press.

27. P. J. Greenwood, P. H. Harvey, and C. M. Perrins. 1978. Inbreeding and dispersal in the great tit. *Nature* 271:52–54.

28. J. L. Hoogland. 1982. Prairie dogs avoid extreme inbreeding. *Science* 215:1639–41.

29. K. Ralls, P. H. Harvey, and A. M. Lyles. 1986. Inbreeding in natural populations of birds and mammals. In *Conservation Biology: The Science of Scarcity and Diversity*, ed. M. E. Soulé, pp. 35–56. Sinauer.

30. P. M. Waser and W. T. Jones. 1983. Natal philopatry among solitary mammals. *Q. Rev. Biol.* 58:355–90.

31. P. W. Sherman. 1981b. Reproductive competition and infanticide in Belding's ground squirrels and other animals. In *Natural Selection and Social Behavior*, ed. R. D. Alexander and D. W. Tinkle, pp. 311–31. Chiron Press.

32. A. E. Pusey and C. Packer. 1986. Dispersal and philopatry. In *Primate Societies*, ed. B. B. Smuts, D. L. Cheney, R. M. Seyfarth, R. W. Wrangham, and T. T. Struhsaker, pp. 250–66. Univ. of Chicago Press.

33. W. P. Luckett and L. J. Hartenberger, eds. 1985. *Evolutionary Relationships among Rodents*. Plenum Press.

34. D. J. Hafner. 1984. Evolutionary relationships of the nearctic Sciuridae. In *The Biology of Ground-dwelling Squirrels*, ed. J. O. Murie and G. R. Michener, pp. 3–23. Univ. of Nebraska Press.

35. K. E. Holekamp. 1984b. Dispersal in ground-dwelling sciurids. In *The Biology of Ground-dwelling Squirrels*, ed. J. O. Murie and G. R. Michener, pp. 297–320. Univ. of Nebraska Press.

36. M. G. Garrett and W. L. Franklin. 1988. Behavioral ecology of dispersal in the black-tailed prairie dog. *J. Mammal.* 69:236–50.

37. B. D. Chepko-Sade and Z. T. Halpin, eds. 1987. *Mammalian Dispersal Patterns*. Univ. of Chicago Press.

38. P. W. Sherman. 1989. The clitoris debate and the levels of analysis. *Animal Behav.* 37:697–98.

Animal Thinking

Donald R. Griffin

Ethologists are once again investigating the possibility that animals have conscious awareness

What is it like to be an animal? What do monkeys, dolphins, crows, sunfishes, bees, and ants think about? Or do nonhuman animals experience any thoughts and subjective feelings at all? Aside from Lorenz (1963) and Hediger (1947, 1968, 1980) very few ethologists have discussed animal thoughts and feelings. While seldom denying their existence dogmatically, they emphasize that it is extremely difficult, perhaps impossible, to learn anything at all about the subjective experiences of another species. But the difficulties do not justify a refusal to face up to the issue. As Savory (1959) put the matter, "Of course to interpret the thoughts, or their equivalent, which determine an animal's behaviour is difficult, but this is no reason for not making the attempt to do so. If it were not difficult, there would be very little interest in the study of animal behaviour, and very few books about it" (p. 78).

Most biologists and psychologists tend, explicitly or implicitly, to treat most of the world's animals as mechanisms, complex mechanisms to be sure, but unthinking robots nonetheless. Mechanical devices are usually considered to be incapable of conscious thought or subjective feeling, although it is currently popular to ascribe mental experiences to computer systems. John (in Thatcher and John 1977), among others, has equated consciousness with a sort of internal feedback whereby information about one part of a pattern of information flow acts on another part. This may be a necessary condition for conscious thinking, but it is also an aspect of many physiological processes that operate without any conscious awareness on our part.

Many comparative psychologists seem petrified by the notion of animal consciousness. Historically, the science of psychology has been reacting for fifty years or more against earlier attempts to understand the workings of the human mind by introspective self-examination—trying to learn how we think by thinking about our thoughts. This effort led to confusing and

Donald R. Griffin, a professor at The Rockefeller University, is an authority on animal physiology and behavior, best known for his work on echolocation in bats and other animals. He is the author of numerous articles and books on ethology and comparative physiology, including Listening in the Dark *(1958),* Echoes of Bats and Men *(1959),* Bird Migration *(1964), and* The Question of Animal Awareness *(1976). The present article is adapted by permission of the publisher from* Animal Thinking, *published in April 1984 by Harvard University Press. Address: The Rockefeller University, 1230 York Avenue, New York, NY 10021.*

contradictory results, so in frustration experimental psychologists largely abandoned the effort to understand human consciousness, replacing introspection with objective experiments. While experiments have been very helpful in analyzing learning and other human abilities, the rejection of any concern with consciousness and subjective feelings has gone so far that many psychologists virtually deny their existence or at least their accessibility to scientific analysis.

In one rather extreme form of this denial, Harnad (1982) has argued that only after the functioning of our brains has determined what we will do does an illusion of conscious awareness arise, along with the mistaken belief that we have made a choice or had control over our behavior. The psychologists who thus belittle and ignore human consciousness can scarcely be expected to tell us much about subjective thoughts and feelings of animals. If we cannot gather any verifiable data about our own thoughts and feelings, the argument has run, how can we hope to learn anything about those of other species?

A long-overdue corrective reaction to this extreme antimentalism is well under way. To a wide range of scholars, and indeed to virtually the whole world outside of narrow scientific circles, it has always been self-evident that human thoughts and feelings are real and important (see, for example, MacKenzie 1977 and Whiteley 1973). This is not to underestimate the difficulties that arise when one attempts to gather objective evidence about other people's feelings and thoughts, even those one knows best. But it really is absurd to deny the existence and importance of mental experiences just because they are difficult to study.

Why do so many psychologists appear to ignore a central area of their subject matter when most other branches of science refrain from such self-inflicted paralysis? The usual contemporary answer to such a question is that a relatively new sort of cognitive psychology has developed during the past twenty or thirty years, based in large part on the analysis of human and animal behavior in terms of information-processing (reviewed in Norman 1981). Analogies to computer programs play a large part in this approach, and many cognitive psychologists draw their inspiration from the success of computer systems, feeling that certain types of programs can serve as instructive models of human thinking. Words that used to be reserved for conscious human beings are now commonly used to describe the impressive accomplishments of computers. Despite the

Figure 1. A chimpanzee, having selected a suitably shaped small branch, strips it of twigs and leaves, transforming it into a serviceable tool for capturing termites. The chimpanzee will then walk to a termite nest, often at a considerable distance, and probe it with the stick for termites. Such behavior, which differs radically from other activities of chimpanzees, seems difficult to understand unless the animal is consciously thinking about gathering termites while preparing the probe. (Photograph by Linda Koebner, Bruce Coleman Inc.)

an empirical test, but the extent and the complexity of information-processing in our brains is so great that available procedures can detect only a tiny fraction of it, and even if it could be monitored in full detail, we do not know whether any computer system could duplicate it.

The difference between conscious and nonconscious states is a significant one, yet most scientists concerned with animal behavior have felt that looking for consciousness in animals would be a futile anachronism. This defeatist attitude is based in part on convincing evidence that we do a great deal of problem-solving, decision-making, and other kinds of information-processing without any consciousness of what is going on. Harnad (1982) bases his belief that human consciousness is merely an illusion on the fact that we are conscious of only the tip of the iceberg of information-processing in our brains. Indeed the ratio of conscious to unconscious brain activity is probably even smaller than the density ratio of ice to water. The intellectual excitement of this discovery has obscured the obvious fact that we are conscious some of the time, and we certainly do experience many sorts of thoughts and feelings that are very important to us and our companions. If the choice were open, would anyone prefer a lifelong state of sleepwalking?

optimism of computer enthusiasts, however, it is highly unlikely that any computer system can spontaneously generate subjective mental experience (Boden 1977; Dreyfus 1979; Baker 1981).

Conspicuously absent from most of contemporary cognitive psychology is any serious attention to conscious thoughts or subjective feelings. For example, Wasserman (1983) defends cognitive psychology to his fellow behaviorists by arguing that it is not subjective and mentalistic. Analyzing people as though they were computers may be useful as an initial, limited approach, just as physiologists began their analysis of the functioning of hearts by drawing analogies to mechanical pumps. But it is important to recognize the limitations inherent in this approach; it suffers from the danger of leading us into what Savory (1959) called by the apt but unfortunately tongue-twisting name of "the synechdochaic fallacy." This means the confusion of a part of something with the whole, or as Savory put it, "the error of nothing but." Information-processing is doubtless a necessary condition for mental experience, but is it sufficient? Human minds do more than process information; they think and feel. We experience beliefs, desires, fears, expectations, and many other subjective mental states.

Many cognitive psychologists imply that a computer system that could process information exactly as the human brain does would duplicate all essential elements of thinking and feeling; others simply feel that subjective experience is beyond the reach of scientific investigation. Perhaps the issue will someday be put to

What behavior suggests conscious thinking?

Just what is it about some kinds of behavior that leads us to feel that it is accompanied by conscious thinking? Comparative psychologists and biologists worried about this question extensively around the turn of this century. No clear and generally accepted answers emerged from their thoughtful efforts, and this is one reason why the behavioristic movement came to dominate psychology.

Complexity is often taken as evidence that some behavior is guided by conscious thinking. But complexity is a slippery attribute. One might think that simply running away from a frightening stimulus was a rather simple response, yet if we make a detailed description of every muscle contraction during turning and running away, the behavior becomes extremely complex. But, one might object, this complexity involves the physiology of locomotion; what is simple is the direction in which the animal moves. If we then ask what sensory and central nervous mechanisms cause the animal to move in this direction, the matter again becomes complex. Does the animal continuously listen to the danger signal and push more or less hard with its right or left legs in order to keep the signal directly behind it? Or does it head directly toward some landmark? If the latter,

Figure 2. The assassin bug at the top has camouflaged itself chemically and tactilely by gluing bits of a termite nest all over its body. In this way it is able to capture a termite at the opening of the nest without alarming the soldier termites. After sucking out the termite's semifluid organs, the assassin bug jiggles the empty exoskeleton in front of the nest opening in order to attract another termite worker, which will normally attempt to consume or dispose of the corpse. When a second termite worker seizes the first, it is then captured and consumed itself, as shown in the photograph below, and the process may be repeated continuously many times by the same assassin bug. The extraordinary complexity and coordination of these actions strongly suggest conscious thought, even though the assassin bug's central nervous system is very small. (Photographs by Raymond A. Mendez.)

how does it coordinate vision and locomotion? Again one might say that the direction of motion is simple, and it is irrelevant to worry about the complexities of the physiological mechanisms involved.

But how is this simple direction "away from the danger" represented within the animal's central nervous system? Does the animal employ the concepts of *away from* and *danger*? If so, how are such concepts established? Even though we cannot answer the question in neurophysiological terms, it is clear that running away from something is a far simpler behavior than, say, the

construction of a bird's nest. Conversely, even the locomotor motions of a caterpillar that will move toward a light with a machine-like consistency hour after hour are not simple when examined in detail. What is simple is the abstract notion of *toward* or *away*, but the mechanistic interpretation of animal behavior tends to deny that the animal could think in terms of even such a simple abstraction.

One very important attribute of animal behavior that seems intuitively to suggest conscious thinking is its adaptability to changing circumstances. If an animal repeats some action in the same way regardless of the results, we assume that a rigid physiological mechanism is at work, especially if the behavior is ineffective or harmful to the animal. When a moth flies again and again at a bright light or burns itself in an open flame, it is difficult to imagine that the moth is thinking, although one can suppose that it is acting on some thoughtful but misguided scheme. When members of our own species do things that are self-damaging or even suicidal, we do not conclude that their behavior is the result of a mechanical reflex. But to explain the moth flying into the flame as thoughtful but misguided seems far less plausible than the usual interpretation that such insects automatically fly toward a bright light, which leads them to their death in the special situation where the brightest light is an open flame.

Conversely, if an animal manages to obtain food by a complex series of actions that it has never performed before, intentional thinking seems more plausible than rigid automatism. For example, Japanese macaques learned a new way to separate grain from inedible material by throwing the mixture into the water; the kernels of grain would float while the inorganic sand and other particles tended to sink (Kawai 1965). These new types of food handling were first devised by a few monkeys, then were gradually acquired by other members of their social group through observational learning.

Behavioral versatility came into play in a spectacular fashion in the 1930s when two species of tits discovered that milk bottles delivered to British doorsteps could be a source of food (Fisher and Hinde 1949; Hinde and Fisher 1951). At that time milk-bottle tops were made of soft metal foil, and the milk was not homogenized, so the cream rose to the top of the bottle. One or more birds discovered that the same type of behavior used to get at insects hidden under tree bark could also be used to get cream from milk bottles. The people whose milk was disturbed immediately noticed it, and careful studies were made of the gradual spread of this behavior throughout much of England. A change in the technology of covering milk bottles eventually ended the whole business, but meanwhile thousands of birds had learned, almost certainly through observation, to exploit a newly available food source.

Connected patterns of behavior

Another criterion upon which we tend to rely in inferring conscious thinking is the element of interactive steps in a relatively long sequence of appropriate behavior patterns. Effective and versatile behavior often entails many steps, each one modified according to the results of the previous actions. In such a complex sequence the animal must pay attention not only to the immediate stimuli, but also to information obtained in the past. Psychologists once postulated that complex behavior can be understood as a chain of rigid reflexes, the outcome of one serving as stimulus for the next. Students of insect behavior have generally accepted this explanation for such complex activities as the construction of elaborate shelters or prey-catching devices, ranging from the underwater nets spun by certain caddis-fly larvae to the magnificent webs of spiders. But the steps an animal takes often vary, depending on the results of the previous behavior and on many influences from the near or distant past. The choice of *which* past events to attend to may be facilitated by conscious selection from a broad spectrum of memories.

An outstanding example of such sequences of interactive behaviors is the use of probes by chimpanzees to gather termites from their mounds (Goodall 1968, 1971). The chimpanzee prepares a probe by selecting a suitable branch, pulling off its leaves and side branches, breaking the stick to the right length, carrying it—often for several minutes—to a termite mound, and then probing into the openings used by the termites (see Fig. 1). If the hole yields nothing, the chimpanzee moves to another one. Even after the tool has been prepared, its use is far from stereotyped. When curious scientists try to imitate the chimpanzees' techniques, they find it rather difficult and seldom gather as many termites. It is especially interesting that the young chimpanzees seem to learn this use of tools by watching their mothers or other members of their social group. Youngsters have been observed making crude and relatively ineffectual attempts to prepare and use their own termite probes; the termite "fishing" of chimpanzees gives every evidence of being learned.

Examples can be found among insects of long, complex sequences of behaviors that are as suggestive of thought as the chimpanzee's use of a termite probe. McMahan (1982) has discovered that in tropical rain forests one species of assassin bug, a predatory insect, uses two effective tricks, illustrated in Figure 2, to capture the workers of termite colonies. The bug glues small bits of the outer layers of a termite nest to its head, back, and sides. Then it stands near an opening to the termite colony. The bits of termite nest on the assassin bug apparently smell and perhaps feel familiar to the termites, so no alarm signals are emitted, which otherwise would attract the well-armed members of the soldier caste that attack intruders. Although the assassin bug's actions often attract soldier termites, its camouflage seems to prevent them from recognizing it as an intruder, and they return to the nest. This chemical and tactile camouflage allows the assassin bug to reach into the opening and capture a termite worker, which it kills and con-

Figure 3. With an apparent deliberateness that suggests intentional thinking, nesting killdeer will conspicuously lead a potential predator away from their nest or young, adjusting the speed of their abnormally awkward movements away from the nest so as to remain within sight but out of reach of the intruder. Quite often the killdeer acts as if it is injured, as shown here, a behavior that would make it more attractive to a predator. The killdeer will sometimes employ a very different tactic in response to approaching cattle, which may trample on eggs or nestlings but will not eat them; the birds will then stand close to the nest and spread their wings in a conspicuous display that usually causes the cattle to step aside. (Photograph by Noble S. Proctor.)

sumes by sucking out all the semifluid internal organs, leaving only the exoskeleton.

Such camouflage-assisted prey capture is remarkable enough, but the next step is even more thought-provoking. The assassin bug pushes the empty exoskeleton of its victim into the nest opening and jiggles it gently. Another termite worker seizes the corpse as part of a normal behavior pattern of devouring the body of a dead sibling or carrying the corpse away for disposal. The assassin bug pulls the exoskeleton of the first victim out with the second worker attached. This one is eaten and its empty exoskeleton used in another "fishing" effort. In one case an assassin bug was observed to thus devour thirty-one termites before moving away with a fully distended abdomen.

When chimpanzees fashion sticks to probe for termites, their actions are considered among the most convincing cases of intentional behavior yet described for nonhuman animals. When McMahan discovers assassin bugs carrying out an almost equally elaborate feeding behavior, must we assume that the insect is only a genetically programmed robot incapable of understanding what it does? Perhaps we should be ready to infer conscious thinking whenever any animal shows such ingenious behavior, regardless of its taxonomic group and our preconceived notions about limitations of animal consciousness.

An example even more suggestive of thoughtful behavior among insects is the so-called "dance language" of honeybees, which was discovered by the remarkably brilliant and original experiments of von Frisch (reviewed by von Frisch 1967, 1972). One significant reaction to von Frisch's discovery was that of Jung (1973). Late in his life he wrote that although he had believed insects were merely reflex automata:

this view has recently been challenged by the researches of Karl von Frisch. . . . Bees not only tell their comrades, by means of a peculiar sort of dance, that they have found a feeding-place, but they also indicate its direction and distance, thus enabling beginners to fly to it directly. This kind of message is no different in principle from information conveyed by a human being. In the latter case we would certainly regard such behavior as a conscious and intentional act and can hardly imagine how anyone could prove in a court of law that it had taken place unconsciously. . . . We are . . . faced with the fact that the ganglionic system apparently achieves exactly the same result as our cerebral cortex. Nor is there any proof that bees are unconscious.

In many cases the networks of informative events that suggest animal thinking are sufficiently complicated that we are not sure what the animal is doing even when we know most of the relevant facts. Consider, for example, how certain ground-nesting birds, such as the killdeer or piping plovers, lead predators away from their nests or young. At a considerable distance, long before an approaching large intruder, such as a person or other mammal, can see the cryptically colored bird or its eggs, the plover may stand up and walk slowly to a point a few meters from the nest. Then the bird may flutter slowly but conspicuously away from the nest, staying relatively close to the intruder. It almost always makes loud piping or peeping sounds similar to those a bird makes when disturbed or mildly irritated.

It is common for the bird to hold its tail or wing in an abnormal position as it moves. Often the tail almost drags on the ground, and the wings are slightly extended, sometimes one more than the other, strongly suggesting some weakness or injury. After running a few meters, the bird may flop about on the ground, extending one or both wings, as if injured. This behavior, shown in Figure 3, is often called the "broken-wing display," and it requires considerable effort for an observer to believe that the bird is really quite healthy. Predators are extremely sensitive to minor differences in the gait and demeanor of potential prey and are much more likely to attack animals that are behaving abnormally.

Throughout most of this predator-distraction behavior, the bird watches the intruder. Typically it does not move in a straight line and stops from time to time. If the intruder approaches, the bird moves farther ahead. If not, the bird usually flies back closer to the intruder and repeats the behavior. The bird will allow the intruder to approach quite close, sometimes within two meters, but it always moves just fast enough and far enough to avoid capture. Typically the bird continues the injury simulation while leading the intruder some distance away from the nest or young. Finally, however, it flies away rapidly, usually in the same direction, then circling back to the general vicinity, though seldom to the exact spot, where the eggs or young are located. One Wilson's plover, a close relative of the killdeer and piping plover, led me more than 300 meters along a sandy beach before flying off.

Killdeer have been observed to use very different tactics when their nests are approached by cattle, which may trample on eggs or nestlings but will not eat them. Rather than moving away from the nest and fluttering as though injured, the birds stand close to the nest and spread their wings in a conspicuous display that usually causes the cattle to step aside (Skutch 1976).

Adaptations to novelty

One further consideration can help refine the criteria for determining the presence of conscious thought. We can easily change back and forth between thinking consciously about our own behavior and not doing so. When we are learning some new task such as swimming, riding a bicycle, driving an automobile, flying an airplane, operating a vacuum cleaner, caring for our teeth by some new technique recommended by a dentist, or any of the large number of actions we did not formerly know how to do, we think about it in considerable detail. But once the behavior is thoroughly mastered, we give no conscious thought to the details that once required close attention.

This change can also be reversed, as when we make the effort to think consciously about some commonplace and customary activity we have been carrying out for some time. For example, suppose you are asked about the pattern of your breathing, to which you normally give no thought whatsoever. But you can easily take the trouble to keep track of how often you inhale and exhale, how deeply, and what other activities accompany different patterns of breathing. You can find out that it is extremely difficult to speak while inhaling, so talking continuously requires rapid inhalation and slower exhalation. This and other examples that will readily come to mind if one asks the appropriate questions show that we can bring into conscious focus activities that usually go on quite unconsciously.

The fact that our own consciousness can be turned on and off with respect to particular activities tells us that in at least one species it is not true that certain behavior patterns are always carried out consciously while others never are. It is reasonable to guess that this is true also for other species. Well-learned behavior patterns may not require the same degree of conscious attention as those the animal is learning how to perform. This in turn means that conscious awareness is more likely when the activity is novel and challenging; striking and unexpected events are more likely to produce conscious awareness.

Thus it seems likely that a widely applicable, if not all-inclusive, criterion of conscious awareness in animals is *versatile adaptability of behavior to changing circumstances and challenges*. If the animal does much the same things regardless of the state of its environment or the behavior of other animals nearby, we are less inclined to judge that it is thinking about its circumstances or what it is doing. Consciously motivated behavior is more plausibly inferred when an animal behaves appropriately in a novel and perhaps surprising situation that requires specific actions not called for under ordinary circumstances. This is a special case of versatility, of course, but the rarity of the challenge combined with the appropriateness and effectiveness of the response are important indicators of thoughtful actions.

For example, Janes (1976) observed nesting ravens make an enterprising use of rocks. He had been closely observing ten raven nests in Oregon, eight of which were near the top of rocky cliffs. At one of these nests two ravens flew in and out of a vertical crack that extended from top to bottom of a twenty-meter cliff. Janes and a companion climbed up the crevice and inspected the six nearly fledged nestlings. As they started down,

Figure 4. When sea otters cannot open shellfish with their claws or teeth, they often employ a stone for the purpose, smashing it against the shells. The apparently conscious, intentional nature of the behavior is further indicated by the fact that an otter carefully selects a stone of suitable size and weight and often carries the stone under its armpit for considerable periods. (Photograph by William F. Bryan.)

both parents flew at them repeatedly, calling loudly, then landed at the top of the cliff, still calling. One of the ravens then picked up small rocks in its bill and dropped them at the human intruders. Several of the rocks showed markings where they had been partly buried in the soil, so the birds presumably had pried them loose. Only seven rocks were dropped, but the raven seemed to be seeking other loose ones and apparently stopped only because no more suitable rocks were available.

While many birds make vigorous efforts to defend their nests and young from intruders, often flying at people who come too close, regurgitating or defecating on them, and occasionally striking them with their bills, rock throwing is most unusual. Nor do ravens pry out rocks and drop them in other situations. It is difficult to avoid the inference that this quite intelligent and adaptable bird was anxious to chase the human intruders away from its nest and decided that dropping rocks might be effective.

There are limits to the amount of novelty with which a species can cope successfully, and this range of versatility is one of the most significant measures of mental adaptability. This discussion of adaptable versatility as a criterion of consciousness implies that conscious thinking occurs only during learned behavior, but we should be cautious in accepting this belief as a rigid doctrine.

Another aspect of conscious thinking is anticipation and intentional planning of an action with conscious awareness of its likely results. An impressive example is the use of small stones by sea otters to detach and open shellfish (Kenyon 1969). These intelligent aquatic carnivores feed mostly on sea urchins and mollusks. The sea otter must dive to the bottom and pry the mollusk loose with claws or teeth, but some shells, especially abalones, are tightly attached to the rocks and have shells that are too tough to be loosened in this fashion. The otter will search for a suitable stone, which it carries while diving, then uses the stone to hammer the shellfish loose, holding its breath all the while.

The otter usually eats while floating on its back, as shown in Figure 4. If it cannot get at the fleshy animal inside the shell, it will hold the shell against its chest with one paw and pound it with the stone. The otter often tucks a good stone under an armpit as it swims or dives. Although otters do not alter the shapes of the stones, they do select ones of suitable size and weight and often keep them for considerable periods. The otters use tools only in areas where sufficient food cannot be obtained by other methods. In some areas only the young and very old sea otters use stones; vigorous adults can dislodge the shellfish with their unaided claws or teeth. Thus it is far from a simple stereotyped behavior pattern, but one that is used only when it is helpful. Sea otters sometimes use floating beer bottles to hammer open shells. Since the bottles float, they need not be stored under the otter's armpit.

Anticipation and planning are of course impossible to observe directly in another person or animal, but indications of their likelihood are often observable. As early as the 1930s Lorenz studied the intention movements of birds (Lorenz 1971), and other ethologists have noted that these movements, small-scale preliminaries to major actions such as flying, often serve as signals to others of the same species. Although Lorenz interpreted the movements as indications that the bird was planning and preparing to fly, the term *intention movement* has been quietly dropped from ethology in recent years. I suspect this is because the behavioristic ethologists fear that the term has mentalistic implications. Earlier ethologists such as Daanje (1951) described a wide variety of intention movements in many kinds of animals, but their interest was in whether the movements had gradually become specialized communication signals in the course of behavioral evolution. The possibility that intention movements indicate the animal's conscious intention has been totally neglected by ethologists during their behavioristic phase, but we may hope that the revival of scientific interest in animal thinking will lead cognitive ethologists to study whether such movements are accompanied by conscious intentions.

Animal communication

The very fact that intention movements so often evolve into communicative signals may reflect a close linkage between thinking and the intentional communication of thoughts from one conscious animal to another. These considerations lead us directly to a recognition that because communicative behavior, especially among social animals, often seems to convey thoughts and feelings from one animal to another, it can tell us something about animal thinking: it can be an important "window" on the minds of animals.

Human communication is hardly limited to formal language; nonverbal communication of mood or intentions also plays a large and increasingly recognized role in human affairs. We make inferences about people's feelings and thoughts, especially those of very young children, from many kinds of communication, verbal and nonverbal; we should similarly use all available evidence in exploring the possibility of thoughts or feelings in other species. When animals live in a group and depend on each other for food, shelter, warning of dangers, or help in raising the young, they need to be able to judge correctly the moods and intentions of their companions. This extends to animals of other species as well, especially predators or prey. It is important for the animal to know whether a predator is likely to attack or whether the prey is so alert and likely to escape that a chase is not worth the effort. Communication may either inform or misinform, but in either case it can reveal something about the conscious thinking of the communicator.

Vervet monkeys, for example, have at least three different categories of alarm calls, which were described by Struhsaker (1967) after extensive periods of observation. He found that when a leopard or other large carnivorous mammal approached, the monkeys gave one type of alarm call; quite a different call was used at the sight of a martial eagle, one of the few flying predators that captures vervet monkeys. A third type of alarm call was given when a large snake approached the group. This degree of differentiation of alarm calls is not unique, although it has been described in only a few kinds of animals. For example, ground squirrels of western North America use different types of calls when frightened by a ground predator or by a predatory bird such as a hawk (Owings and Leger 1980).

The question is whether the vervet monkey's three types of alarm calls convey to other monkeys information about the type of predator. Such information is important because the animal's defensive tactics are different in the three cases. When a leopard or other large carnivore approaches, the monkeys climb into trees. But leopards are good climbers, so the monkeys can escape them only by climbing out onto the smallest branches, which are too weak to support a leopard. When the monkeys see a martial eagle, they move into thick vegetation close to a tree trunk or at ground level. Thus the tactics that help them escape from a leopard make them highly vulnerable to a martial eagle, and vice versa. In response to the threat of a large snake they stand on their hind legs and look around to locate the snake, then simply move away from it, either along the ground or by climbing into a tree.

To answer this question, Seyfarth, Cheney, and Marler (1980a, b) conducted some carefully controlled playback experiments under natural conditions in East Africa. From a concealed loudspeaker, they played tape recordings of vervet alarm calls and found that the playbacks of the three calls did indeed elicit the appropriate responses. The monkeys responded to the leopard alarm call by climbing into the nearest tree; the martial eagle alarm caused them to dive into thick vegetation; and the python alarm produced the typical behavior of standing on the hind legs and looking all around for the nonexistent snake.

Inclusive behaviorists—that is, psychologists interested only in contingencies of reinforcement during an individual's lifetime, and ethologists or behavioral ecologists solely concerned with the effects of natural selection on behavior—insist on limiting themselves to stating that an animal benefits from accurate information about what the other animal will probably do. But within a mutually interdependent social group, an individual can often anticipate a companion's behavior most easily by empathic appreciation of his mental state. The inclusive behaviorists will object that all we need postulate is behavior appropriately matched to the probabilities of the companions *behaving* in this way or that—all based on contingencies of reinforcement learned from previous situations or transmitted genetically.

But empathy may well be a more efficient way to gauge a companion's disposition than elaborate formulas describing the contingencies of reinforcement. All the animal may need to know is that another is aggressive, affectionate, desirous of companionship, or in some other common emotional state. Judging that he is aggressive may suffice to predict, economically and parsimoniously, a wide range of behavior patterns depending on the circumstances. Neo-Skinnerian inclusive behaviorists may be correct in saying that this empathy came about by learning, for example, the signals that mean a companion is aggressive. But our focus is on the animal's possible thoughts and feelings, and for this purpose the immediate situation is just as important as the history of its origin.

Humphrey (1976) has extended an earlier suggestion by Jolly (1966) that consciousness arose in primate evolution when societies developed to the stage where it became crucially important for each member of the group to understand the feelings, intentions, and thoughts of others. When animals live in complex social groupings, where each one is crucially dependent on cooperative interactions with the others, they need to be "natural psychologists," as Humphrey puts it. They need to have internal models of the behavior of their companions, to feel with them, and thus to think consciously about what the other one must be thinking or feeling.

Following this line of thought, we might distinguish between the animals' interactions with some feature of the physical environment or with plants, and their interactions with other reacting animals, usually their own species, but also predators and prey. Although Humphrey has so far restricted his criterion of consciousness to our own ancestors within the past few million years, it could apply with equal or even greater force to other animals that live in mutually interdependent social groups.

All this adds up to the simple idea that when animals communicate to one another they may be conveying something about their thoughts or feelings. If so, eavesdropping on the communicative signals they exchange may provide us with a practicable source of data about their mental experiences. When animals devote elaborate and specifically adjusted activities to communication, each animal responding to messages from its companion, it seems rather likely that both sender and receiver are consciously aware of the content of these messages.

The adaptive economy of conscious thinking

The natural world often presents animals with complex challenges best met by behavior that can be rapidly adapted to changing circumstances. Environmental conditions vary so much that for an animal's brain to have programmed specifications for optimal behavior in all situations would require an impossibly lengthy instruction book. Whether such instructions stem from the animal's DNA or from learning and environmental influences within its own lifetime, providing for all likely contingencies would require a wasteful volume of specific directions. Concepts and generalizations, on the other hand, are compact and efficient. An instructive analogy is provided by the hundreds of pages of official rules for a familiar game such as baseball. Once the general principles of the game are understood, however, quite simple thinking suffices to tell even a small boy approximately what each player should do in most game situations.

Of course, simply thinking about various alternative actions is not enough; successful coping with the challenges of life requires that thinking be relatively rapid and that it lead both to reasonably accurate decisions and to their effective execution. Thinking may be economical without being easy or simple, but consideration of the likely results of doing this or that is far more efficient than blindly trying every alternative. If an animal thinks about what it might do, even in very simple terms, it can choose the actions that promise to have desirable consequences. If it can anticipate probable events, even if only a little way into the future, it can avoid wasted effort. More important still is being able to avoid dangerous mistakes. To paraphrase Popper (1972), a foolish impulse can die in the animal's mind rather than lead it to needless suicide.

I have suggested that conscious thinking is economical, but many contemporary scientists counter that the problems mentioned above can be solved equally well by unconscious information-processing. It is quite true that skilled motor behavior often involves complex, rapid, and efficient reactions. Walking over rough ground or through thick vegetation entails numerous adjustments of the balanced contraction and relaxation of several sets of opposed muscles. Our brains and spinal cords modulate the action of our muscles according to whether the ground is high or low or whether the vegetation resists bending as we clamber over it. Little, if any, of this process involves conscious thought, and yet it is far more complex than a direct reaction to any single stimulus.

We perform innumerable complex actions rapidly, skillfully, and efficiently without conscious thought. From this evidence many have argued that an animal does not need to think consciously to weigh the costs and benefits of various activities. Yet when we acquire a new skill, we have to pay careful conscious attention to details not yet mastered. Insofar as this analogy to our own situation is valid, it seems plausible that when an animal faces new and difficult challenges, and when the stakes are high—often literally a matter of life and death—conscious evaluation may have real advantages.

Inclusive behaviorists often find it more plausible to suppose that an animal's behavior is more efficient if it is automatic and uncomplicated by conscious thinking. It has been argued that the vacillation and uncertainty involved in conscious comparison of alternatives would slow an animal's reactions in a maladaptive fashion. But when the spectrum of possible challenges is broad, with a large number of environmental or social factors to be considered, conscious mental imagery, explicit anticipation of likely outcomes, and simple thoughts about them are likely to achieve better results than thoughtless reaction. Of course, this is one of the many areas where we have no certain guides on which to rely. And yet, as a working hypothesis, it is attractive to suppose that if an animal can consciously anticipate and choose the most promising of various alternatives, it is likely to succeed more often than an animal that cannot or does not think about what it is doing.

References

Baker, L. R. 1981. Why computers can't act. *Am. Philos. Q.* 18:157–63.

Boden, M. A., ed. 1977. *Artificial Intelligence and Natural Man.* Basic Books.

Daanje, A. 1951. On the locomotory movements in birds and the intention movements derived from them. *Behaviour* 3:48–98.

Dreyfus, H. L. 1979. *What Computers Can't Do: The Limits of Artificial Intelligence*, rev. ed. Harper and Row.

Fisher, J., and R. A. Hinde. 1949. The opening of milk-bottles by birds. *Brit. Birds* 42:347–57.

Frisch, K. von. 1967. *The Dance Language and Orientation of Bees*. Harvard Univ. Press.

———. 1972. *Bees, Their Vision, Chemical Senses and Language*, 2nd ed. Cornell Univ. Press.

Goodall, J. van Lawick. 1968. Behaviour of free-living chimpanzees of the Gombe Stream area. *Anim. Beh. Monogr.* 1:165–311.

———. 1971. *In the Shadow of Man*. Houghton Mifflin.

Harnad, S. 1982. Consciousness: An afterthought. *Cog. Brain Theory* 5:29–47.

Hediger, H. 1947. Ist das tierliche Bewusstsein unerforschbar? *Behaviour* 1:130–37.

———. 1968. *The Psychology of Animals in Zoos and Circuses*. Dover.

———. 1980. *Tiere verstehen, Erkenntnisse eines Tierpsychologien*. Munich: Kindler.

Hinde, R. A., and J. Fisher. 1951. Further observations on the opening of milk bottles by birds. *Brit. Birds*. 44:393–96.

Humphrey, N. K. 1976. The social function of intellect. In *Growing Points in Ethology*, ed. P. P. G. Bateson and R. A. Hinde. Cambridge Univ. Press.

Janes, S. W. 1976. The apparent use of rocks by a raven in nest defense. *Condor* 78:409.

Jolly, A. 1966. Lemur social behavior and primate intelligence. *Science* 153:501–06.

Jung, C. G. 1973. *Synchronicity, a Causal Connecting Principle*. Princeton Univ. Press.

Kawai, M. 1965. Newly acquired pre-cultural behavior of the natural troop of Japanese monkeys on Koshima Islet. *Primates* 6:1–30.

Kenyon, K. W. 1969. *The Sea Otter in the Eastern Pacific Ocean*. North American Fauna, no. 68. US Bureau of Sport Fisheries and Wildlife.

Lorenz, K. 1963. Haben Tiere ein subjectives Erleben? *Jahr. Techn. Hochs. München.* Eng. trans., Do animals undergo subjective experience? In *Studies in Animal and Human Behavior*, vol. 2. Harvard Univ. Press.

———. 1971. *Studies in Animal and Human Behavior*, vol. 2. Harvard Univ. Press.

MacKenzie, B. D. 1977. *Behaviorism and the Limits of Scientific Method*. London: Routledge and Kegan Paul.

McMahan, E. A. 1982. Bait-and-capture strategy of termite-eating assassin bug. *Insectes Sociaux* 29:346–51.

Norman, D. A., ed. 1981. *Perspectives on Cognitive Science*. Hillsdale, N. J.: Erlbaum.

Owings, D. H., and D. W. Leger. 1980. Chatter vocalizations of California ground squirrels: Predator- and social-role specificity. *Z. Tierpsychol.* 54:163–84.

Popper, K. R. 1972. *Objective Knowledge*. Oxford Univ. Press.

Savory, T. H. 1959. *Instinctive Living, a Study of Invertebrate Behaviour*. London: Pergamon.

Seyfarth, R. M., D. L. Cheney, and P. Marler. 1980a. Monkey responses to three different alarm calls: Evidence for predator classification and semantic communication. *Science* 210:801–03.

———. 1980b. Vervet monkey alarm calls: Semantic communication in a free-ranging primate. *Anim. Beh.* 28:1070–94.

Skutch, A. F. 1976. *Parent Birds and Their Young*. Univ. of Texas Press.

Struhsaker, T. T. 1967. *The Red Colobus Monkey*. Univ. of Chicago Press.

Thatcher, R. W., and E. R. John. 1977. *Foundations of Cognitive Processes*. Hillsdale, N. J.: Erlbaum.

Wasserman, E. A. 1983. Is cognitive psychology behavioral? *Psychol. Record.* 33:6–11.

Whiteley, C. H. 1973. *Mind in Action, an Essay in Philosophical Psychology*. Oxford Univ. Press.

Reward Deficiency Syndrome

Addictive, impulsive and compulsive disorders—including alcoholism, attention-deficit disorder, drug abuse and food bingeing—may have a common genetic basis

Kenneth Blum, John G. Cull, Eric R. Braverman and David E. Comings

In 1990 one of us published with his colleagues a paper suggesting that a specific genetic anomaly was linked to alcoholism (Blum *et al.* 1990). Unfortunately it was often erroneously reported that they had found the "alcoholism gene," implying that there is a one-to-one relation between a gene and a specific behavior. Such misinterpretations are common—readers may recall accounts of an "obesity gene," or a "personality gene." Needless to say, there is no such thing as a specific gene for alcoholism, obesity or a particular type of personality. However, it would be naive to assert the opposite, that these aspects of human behavior are not associated with any particular genes. Rather the issue at hand is to understand how certain genes and behavioral traits are connected.

Kenneth Blum is a professor of pharmacology at the University of Texas Health Science Center in San Antonio. He has been working in the field of addictive behavior and genetics for over three decades. He has chaired two Gordon Research Conferences on alcoholism and is currently developing genetic-based diagnostic tests for addictive disorders and related "reward" behaviors. John G. Cull received his Ph.D. from Texas Tech University and is co-founder of the Neurodevelopmental Institutes of North America/NeuRecovery International. He was an assistant commissioner of the Virginia State Department of Rehabilitation and an independent clinical psychologist for more than 12 years. Eric R. Braverman is on the staff of the Department of Psychiatry at New York University Medical School, where he received his M.D. In addition to his clinical practice he conducts research on neurological illnesses and heart disease in Princeton, New Jersey. David E. Comings is the director of the Department of Medical Genetics at the City of Hope National Medical Center in Duarte, California. He is a former editor of the American Journal of Human Genetics *and past president of the American Society of Human Genetics. Blum's address: Department of Pharmacology, University of Texas Health Center, San Antonio, TX 78284.*

In the past five years we have pursued the association between certain genes and various behavioral disorders. In molecular genetics, an *association* refers to a statistically significant incidence of a genetic variant (an allele) among genetically unrelated individuals with a particular disease or condition, compared to a control population. In the course of our work we discovered that the genetic anomaly previously found to be associated with alcoholism is also found with increased frequency among people with other addictive, compulsive or impulsive disorders. The list is long and remarkable—it comprises alcoholism, substance abuse, smoking, compulsive overeating and obesity, attention-deficit disorder, Tourette's syndrome and pathological gambling.

We believe that these disorders are linked by a common biological substrate, a "hard-wired" system in the brain (consisting of cells and signaling molecules) that provides pleasure in the process of rewarding certain behavior. Consider how people respond positively to safety, warmth and a full stomach. If these needs are threatened or are not being met, we experience discomfort and anxiety. An inborn chemical imbalance that alters the intercellular signaling in the brain's reward process could supplant an individual's feeling of well being with anxiety, anger or a craving for a substance that can alleviate the negative emotions. This chemical imbalance manifests itself as one or more behavioral disorders for which one of us (Blum) has coined the term "reward deficiency syndrome."

This syndrome involves a form of sensory deprivation of the brain's pleasure mechanisms. It can be manifested in relatively mild or severe forms that follow as a consequence of an individual's biochemical inability to derive reward from ordinary, everyday activities. We believe that we have discovered at least one genetic aberration that leads to an alteration in the reward pathways of the brain. It is a variant form of the gene for the dopamine D_2 receptor, called the A_1 allele. This is the same genetic variant that we previously found to be associated with alcoholism. In this review we shall look at evidence suggesting that the A_1 allele is also associated with a spectrum of impulsive, compulsive and addictive behaviors. The concept of a reward deficiency syndrome unites these disorders and may explain how simple genetic anomalies give rise to complex aberrant behavior.

The Biology of Reward

The pleasure and reward system in the brain was discovered by accident in 1954. The American psychologist James Olds was studying the rat brain's alerting process, when he mistakenly placed the electrodes in a part of the limbic system, a group of structures deep within the brain that are generally believed to play a role in emotions. When the brain was wired so that the animal could stimulate this area by pressing a lever, Olds found that the rats would press the lever almost nonstop, as many as 5,000 times an hour. The animals would stimulate themselves to the exclusion of everything else except sleep. They would even endure tremendous pain and hardship for an opportunity to press the lever. Olds had clearly found an area in the limbic system that provided a powerful reward for these animals.

Research on human subjects revealed that the electrical stimulation of some areas of the brain (the medial hypothalamus) produced a feeling of quasi-or-

Figure 1. Alcoholism is partly a consequence of a genetically based deficiency that affects the pleasure and reward areas of the brain, according to the authors. The deficient genes code for receptors and transporters for the neurotransmitter dopamine. Recent studies suggest that the same genetic variants are associated with a spectrum of disorders involving compulsive, impulsive and addictive behavior, which comprise a reward deficiency syndrome. The authors propose a neurobiological mechanism to account for the manifestation of the syndrome and suggest possibilities for treatment.

gasmic sexual arousal (Olds and Olds 1969). If certain other areas of the brain were stimulated, an individual experienced a type of light-headedness that banished negative thoughts. These discoveries demonstrated that pleasure is a distinct neurological function that is linked to a complex reward and reinforcement system (Hall, Bloom and Olds 1977).

During the past several decades research on the biological basis of chemical dependency has been able to establish some of the brain regions and neurotransmitters involved in reward. In particular it appears that the dependence on alcohol, opiates and cocaine relies on a common set of biochemical mechanisms (Cloninger 1983, Blum *et al.* 1989). A neuronal circuit deep in the brain in-

volving the limbic system and two regions called the nucleus accumbens and the globus pallidus appears to be critical in the expression of reward for people taking these drugs (Wise and Bozarth 1984). Although each substance of abuse appears to act on different parts of this circuit, the end result is the same: Dopamine is released in the nucleus accumbens and the hippocampus (Koob and Bloom 1988). Dopamine appears to be the primary neurotransmitter of reward at these reinforcement sites.

Although the system of neurotransmitters involved in the biology of reward is complex, at least three other neurotransmitters are known to be involved at several sites in the brain: serotonin in the hypothalamus, the enkephalins (opioid peptides) in the

ventral tegmental area and the nucleus accumbens, and the inhibitory neurotransmitter GABA in the ventral tegmental area and the nucleus accumbens (Stein and Belluzi 1986, Blum 1989). Interestingly, the glucose receptor is an important link between the serotonergic system and the opioid peptides in the hypothalamus. An alternative reward pathway involves the release of norepinephrine in the hippocampus from neuronal fibers that originate in the locus coeruleus.

In a normal person, these neurotransmitters work together in a cascade of excitation or inhibition—between complex stimuli and complex responses—leading to a feeling of well being, the ultimate reward (Cloninger 1983, Stein and Belluzi 1986, Blum and Koslowski 1990). In the

Figure 2. Rat in a Skinner box is a typical laboratory scenario used in the investigation of reward-seeking behavior. Early studies on reward-seeking behavior assumed that an animal's response to pleasurable stimuli was largely learned. Since the 1950s, however, it has become evident that identifiable structures deep within the brain modulate the animal's experience of pleasure in response to stimuli associated with food, sex and thirst. Here a rat can directly stimulate the pleasure regions of the brain by pressing a lever that activates an electrode in its head. Such animals will stimulate themselves as many as 5,000 times an hour. (Photograph courtesy of the authors.)

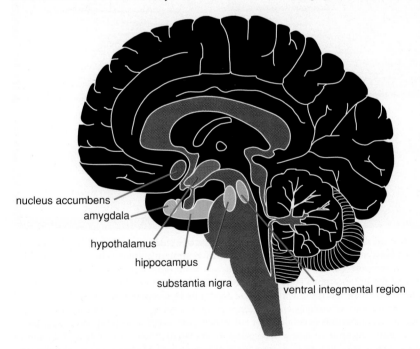

nucleus accumbens

amygdala

hypothalamus

hippocampus

substantia nigra

ventral integmental region

Figure 3. Structures deep within the limbic system play a crucial role in the expression of emotions and the activity of the reward system of the brain. The experience of pleasure and the modulation of reward is based on a reward "cascade," a chain of neurons within the limbic system that interact through various signaling molecules, or neurotransmitters. The authors propose that a biochemical deficiency in one or more of these neurons or signaling molecules can supplant an individual's feeling of well being with anxiety, anger or a craving for a substance that can alleviate the negative emotions.

cascade theory of reward, a disruption of these intercellular interactions results in anxiety, anger and other "bad feelings" or in a craving for a substance that alleviates these negative emotions. Alcohol, for example, is known to activate the norepinephrine system in the limbic circuitry through an intercellular cascade that includes serotonin, opioid peptides and dopamine. Alcohol may also act directly through the production of neuroamines that interact with opioid receptors or with dopaminergic systems (Alvaksinen *et al.* 1984; Blum and Kozlowski 1990). In the cascade theory of reward, genetic anomalies, prolonged stress or long-term abuse of alcohol can lead to a self-sustaining pattern of abnormal cravings in both animals and human beings.

Support for the cascade theory can be derived from a series of experiments on strains of rats that prefer alcohol to water. Compared to normal rats, the alcohol-preferring rats have fewer serotonin neurons in the hypothalamus, higher levels of enkephalin in the hypothalamus (because less is released), more GABA neurons in the nucleus accumbens (which inhibit the release of dopamine), a reduced supply of dopamine in the nucleus accumbens and a lower density of dopamine D_2 receptors in certain areas of the limbic system (Russell, Lanin and Taljaard 1988; McBride *et al.* 1990; Zhou *et al.* 1990; McBride *et al.* 1993).

These studies suggest a four-part cascade in which there is a reduction in the amount of dopamine released in a key reward area in the alcohol-preferring rats. The administration of substances that increase the supply of serotonin at the synapse or that directly stimulate dopamine D_2 receptors reduce craving for alcohol (McBride *et al.* 1993). For example, D_2 receptor agonists reduce the intake of alcohol among rats that prefer alcohol, whereas D_2 dopamine-receptor antagonist increase the drinking of alcohol in these inbred animals (Dyr *et al.* 1993).

Support for the cascade theory of alcoholism in human beings is found in a series of clinical trials. When amino-acid precursors of certain neurotransmitters (serotonin and dopamine) and a drug that promotes enkephalin activity were given to alcoholic subjects, the individuals experienced fewer cravings for alcohol, a reduced incidence of stress, an increased likelihood of recovery and a reduction in relapse rates (Brown *et al.* 1990; Blum and Tractenberg 1988; Blum,

Briggs and Tractenberg 1989). Furthermore, the notion that dopamine is the "final common pathway" for drugs such as cocaine, morphine and alcohol is supported by recent studies by Jordi Ortiz and his associates at Yale University School of Medicine and the University of Connecticut Health Services Center. These authors demonstrated that the chronic use of cocaine, morphine or alcohol results in several biochemical adaptations in the limbic dopamine system. They suggest that these adaptations may result in changes in the structural and functional properties of the dopaminergic system.

We believe that the biological substrates of reward that underlie the addiction to alcohol and other drugs are also the basis for impulsive, compulsive and addictive disorders comprising the reward deficiency syndrome.

Alcoholism and Genes

An alteration in any of the genes that are involved in the expression of the molecules in the reward cascade might predispose an individual to alcoholism. Indeed, the evidence for a genetic basis to alcoholism has accumulated steadily over the past five decades. The earliest report comes from studies of laboratory mice by the American psychologist L. Mirone in 1952. Mirone found that, given a choice, certain mice preferred alcohol to water. Gerald McLearn at the University of California at Berkeley took this a step farther by producing an inbred mouse (the C57 strain) that had a marked preference for alcohol. The alcohol-preferring C57 strain bred true through successive generations—it was the first clear indication that alcoholism has a genetic basis (McLearn and Rodgers 1959).

The first evidence that alcoholism has a genetic basis in human beings came in 1972 when scientists at the Washington University School of Medicine in St. Louis found that adopted children whose biological parents were alcoholics were more likely to have a drinking problem than those born to nonalcoholic parents (Schuckit, Goodwin and Winokur 1972). In 1973 Goodwin and Winokur, working at the Psykologisk Institut in Copenhagen, studied 5,483 men in Denmark who had been adopted in early childhood. They found that the sons born to alcoholic fathers were three times more likely to become alcoholic than the sons of nonalcoholic fathers.

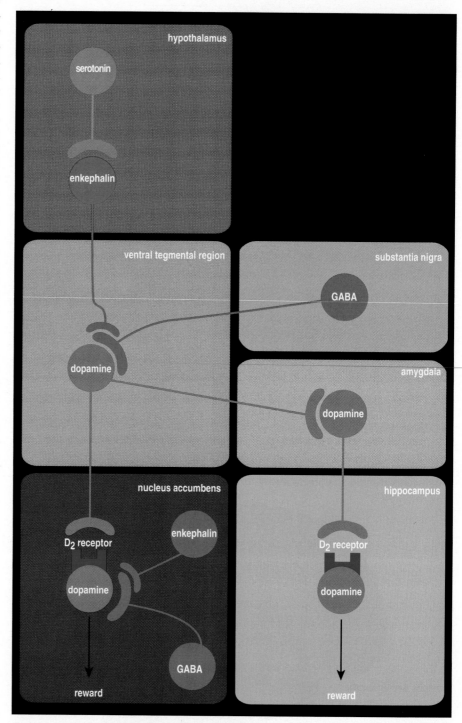

Figure 4. Reward cascade in the limbic system consists of excitatory *(blue)* and inhibitory *(red)* connections between neurons that are modulated by neurotransmitters. The activation of the dopamine D_2 receptor *(green)* by dopamine on the cell membranes of neurons in the nucleus accumbens and the hippocampus is hypothesized by the authors to be the "final common pathway" of the reward cascade. If the activity of the dopamine D_2 receptor is deficient, the activity of neurons in the nucleus accumbens and the hippocampus is decreased, and the individual experiences unpleasant emotions or cravings for substances that can provide temporary relief by releasing dopamine. Alcohol, cocaine and nicotine are known to promote the release of dopamine in the brain. A simplified version of the cascade is presented here. Disorders of the cells and molecules in the "upstream" part of the cascade may also disrupt the normal activity of the reward system. The cascade begins with the excitatory activity of serotonin-releasing neurons in the hypothalamus. This causes the release of the opioid peptide met-enkephalin in the ventral tegmental area, which inhibits the activity of neurons that release the inhibitory neurotransmitter gamma-aminobutyric acid (GABA). The disinhibition of dopamine-containing neurons in the ventral tegmental area allows them to release dopamine in the nucleus accumbens and in certain parts of the hippocampus, permitting the completion of the cascade.

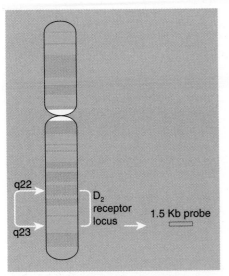

Figure 5. Human chromosome number 11 carries the gene that codes for the dopamine D$_2$ receptor, one of six known dopamine receptors. The gene is located on the long (*q*) arm of the chromosome and was cloned and sequenced in 1990, providing investigators with the opportunity to test for genetic variations in the population. A 1.5-kilobase sequence taken from one end of the gene is used as probe to search for variants.

In the late 1980s research on the inheritance of alcoholism suggested that there might be important genetic differences between alcoholics and nonalcoholics (Cloninger, Bohman and Sigvardsson 1981; Goodwin 1979). One of us (Blum) and his colleagues suspected that the activity of the chemical signaling molecules in the reward pathways of the brain might be involved. Over the course of two years we compared eight genetic markers associated with various neurotransmitters (including serotonin, endogenous opioids, GABA, transferrin, acetylcholine, alcohol dehydrogenase and aldehyde dehydrogenase). In each instance we failed to find a direct association between the genetic markers and alcoholism.

The opportunity to investigate a ninth genetic marker arose after Olivier Civelli of the Vollum Institute at Oregon University cloned and sequenced the gene for one form of the dopamine D$_2$ receptor. The D$_2$ receptor is one of at least five physiologically distinct dopamine receptors (D$_1$, D$_2$, D$_3$, D$_4$ and D$_5$) found on the synaptic membranes of neurons in the brain (Sibley and Monsma 1992). Previous studies had established that D$_2$ receptors are expressed in neurons within the cerebral cortex and the limbic system, including the nucleus accumbens, the amygdala and the hippocampus. Because these are the same areas of the brain (with the exception of the cortex) that are believed to be involved in the reward cascade, Civelli's work provided the opportunity to investigate an important molecular candidate for genetic aberrations among alcoholics.

The technique we used to distinguish between the D$_2$ receptor genes of alcoholics and those of nonalcoholics relies on the detection of restriction-fragment-length polymorphisms (RFLPs). This approach involves the use of DNA-cutting enzymes (restriction endonucleases) that cleave the DNA molecule at specific nucleotide sequences. If there are genetic differences between two individuals such that a restriction enzyme cuts their DNA along different points in (or near) a gene, the resulting fragments of their genes will be of different lengths. These differing fragments, or polymorphisms, are recognized by the use of a radioactively labeled DNA probe—in this case a short sequence of the D$_2$ receptor gene—that binds to a complementary DNA sequence on the fragments. Radiolabeled fragments of different lengths signify a difference in the cleavage sequence recognized by the restriction enzyme (Grandy *et al.* 1989).

The restriction enzyme (*Taq* 1) cuts the nucleotide sequence at a site just outside the coding region for the D$_2$ receptor gene. This produces the *Taq 1A* polymorphisms. To date there are four *Taq 1A* alleles known, the A$_1$, A$_2$, A$_3$ and A$_4$ alleles. The A$_3$ and A$_4$ alleles are rare, whereas the A$_2$ allele is found in nearly 75 percent of the general population and the A$_1$ allele in about 25 percent of the population.

In 1990 we used the *Taq* I enzyme to search for *Taq* IA polymorphisms in the DNA extracted from the brains of deceased alcoholics and a control population of nonalcoholics. The results were striking: In our sample of 35 alcoholics we found that 69 percent had the A$_1$ allele and 31 percent had the A$_2$ allele. In 35 nonalcoholics we found that 20 percent had the A$_1$ allele and 80 percent had the A$_2$ allele.

Since our 1990 study, some laboratories have failed to find a connection between the A$_1$ allele and alcoholism. However, a review of their work shows that their samples were not limited to *severe* forms of alcoholism, which we believe to be an important distinguishing criterion. In our original study, over 70 percent of the alcoholics had cirrhosis of the liver, a disease suggestive of severe and chronic alcoholism. Moreover, the negative studies failed to adequately assess controls to eliminate alcoholism, drug abuse and other related "reward behaviors." In this regard, Katherine Neiswanger and Shirley Hill of the University of Pittsburgh recently found a strong association of the A$_1$ allele and alcoholism and suggested that early failures were the result of poor assessment of a true phenotype in the controls (Neiswanger, Kaplan and Hill 1995). To date, 14 independent laboratories have supported the finding that the A$_1$ allele

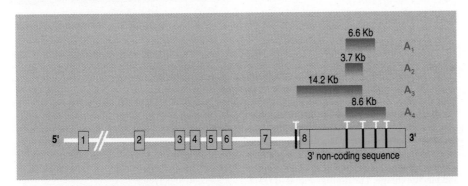

Figure 6. Map of the dopamine D$_2$ receptor gene shows the location of the genetic variants in the *A* region (*peach*) at the 3' end of the gene. The coding region of the gene contains eight exons (*blue*), containing nucleotide sequences that code for the structure of the gene, which are interrupted by introns (*gray*), which do not code for the gene. The *A* region contains nucleotide sequences that do not code for the structure of the gene, but may have an important role in regulating the expression of the gene. The restriction enzyme *Taq* I cuts the *A* region of the gene at several sites (*T*), which vary according to differences in the sequence of the nucleotide base pairs. Four genetic variations in the *A* region (the A$_1$, A$_2$, A$_3$ and A$_4$ polymorphisms or alleles) are produced by the *Taq* I enzyme. The A$_3$ and A$_4$ alleles are rare, whereas the A$_1$ allele is present in about 25 percent of the general population and the A$_2$ allele in nearly 75 percent of the population. The A$_1$ allele is associated with the addictive-impulsive-compulsive disorders of the reward deficiency syndrome.

is a causative factor in severe forms of alcoholism, though perhaps not in milder forms (Blum and Noble 1994). These findings do not prove that the A_1 allele of the dopamine D_2 receptor gene is the *only* cause of severe alcoholism, but they are a powerful indication that the A_1 allele is involved with alcoholism.

Further evidence for the role of biology in alcoholism comes from efforts to find electrophysiological markers that might indicate a predisposition to the addictive disorder. One such marker is the latency and the magnitude of the positive 300-millisecond (P300) wave, an indicator of the general electrical activity of the brain that is evoked by a specific stimulus such as a tone. It turns out that abnormalities in the electrical activity of the brain are evident in the young sons of alcoholic fathers. Their P300 waves are markedly reduced in amplitude compared to the P300 waves of the sons of nonalcoholic fathers. These results raised the question as to whether this deficit had been transferred from father to son and whether this deficit would predispose the son to substance abuse in the future (Begleiter, Porjexa, Bihari and Kissin 1984).

Experiments carried out since then have answered both questions. The alcoholic fathers had the same P300-wave deficit seen in their sons, and the sons showed increased drug-seeking behaviors (including alcohol and nicotine) compared to the sons of nonalcoholic fathers. Moreover, the sons of alcoholic fathers had an atypical neurocognitive profile (Whipple, Parker and Noble 1988). It now appears that children with P300 abnormalities are more likely to abuse drugs and tobacco in later years (Berman, Whipple, Fitch and Noble 1993).

Remarkably, Noble and his colleagues found an association between the A_1 allele and a prolonged latency of the P300 wave in children of alcoholics (Noble *et al.* 1994). Two of us (Blum and Braverman) extended this work and observed a similar correlation between the A_1 allele and a prolonged P300 latency in a neuropsychiatric population. Subjects who are homozygous for the A_1 allele showed significantly prolonged P300 latency compared to A_1/A_2 and A_2/A_2 carriers.

Drug Addiction and Smoking

Cocaine can bring intense, but temporary, pleasure to the user. The aftermath is addiction and severe psychological and physiological harm. Various psycho-social theories have been advanced to account for the abuse of cocaine and other illicit drugs. In contrast to alcoholism, where growing empirical evidence is implicating hereditary factors, relatively little has been known about the genetics of human cocaine dependence. However, some recent studies have suggested that hereditary factors are involved in the use and abuse of cocaine and other illicit drugs.

Studies of adopted children, for example, show that a biological background of alcohol problems in the parents predicts an increased tendency toward illicit drug abuse in the children (Cadoret, Froughton, O'Gorman and Heywood 1986). Similarly, family studies of cocaine addicts show a high percentage of first- or second-degree relatives who have been diagnosed as alcoholics (Miller, Gold, Belkin and Klaher 1989; Wallace 1990).

Behavioral anomalies such as conduct disorder (in which children violate social norms and the rights of others) and antisocial personality (the adult equivalent of conduct disorder) are often found to be associated with alcohol and drug problems. Several investigators have noted that sociopathic behavior in children predicts a tendency toward antisocial personality behavior, alcohol abuse and drug problems later in life. An analysis of 40 studies showed a strong positive correlation between alcoholism and drug abuse, between alcoholism and antisocial personality, and between drug abuse and antisocial personality (Schubert *et al.* 1988).

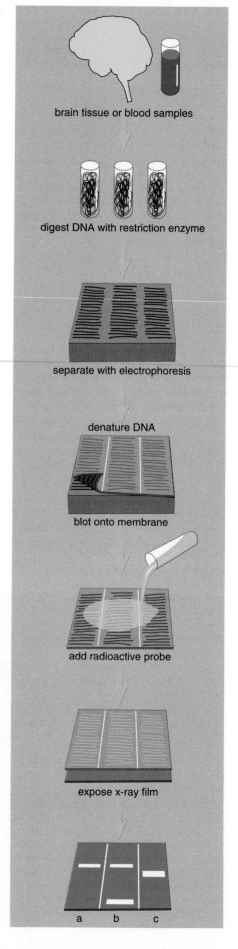

Figure 7. Method of recognizing genetic variations in the dopamine D_2 receptor gene relies on the detection of restriction-fragment-length polymorphisms (RFLPs). DNA is extracted from brain tissue, blood samples or other body tissues and then cut into many fragments with a restriction enzyme (such as *Taq* I), which cleaves the genetic material at a specific nucleotide sequence. The fragments are separated from each other in a gelatin solution by an electric current that carries the fragments different distances according to their lengths and electric charge. The double-stranded DNA molecules are denatured into single strands before being blotted onto a membrane to allow further processing. A radioactive probe (a 1.5-kilobase sequence of the dopamine D_2 receptor gene extracted from a known source) binds to complementary sequences of the single-stranded fragments. The fragments of the single-stranded DNA that bind the probe are visualized when x-ray film is exposed to the membrane. Here the genetic variations are revealed in the x-ray film, producing a DNA fingerprint for each of three individuals (*a, b, c*).

Caption within figure:
brain tissue or blood samples

digest DNA with restriction enzyme

separate with electrophoresis

denature DNA

blot onto membrane

add radioactive probe

expose x-ray film

a   b   c

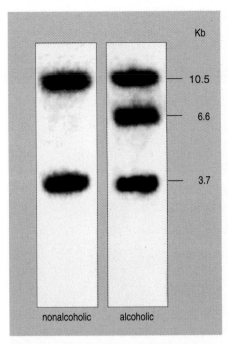

Figure 8. DNA fingerprint of an individual carrying the A_1 allele of the dopamine D_2 receptor gene *(right)* contains an extra fragment that is about 6.6 kilobases in length. The other DNA fingerprint *(left)* shows that another individual carries the A_2 allele rather than the A_1 allele, since the 6.6-kilobase fragment is absent. The individual with the A_1 allele was an alcoholic, the individual with the A_2 allele was not an alcoholic. Many studies (discussed in the text) now reveal that people who carry the A_1 allele are at risk for developing various disorders associated with the reward deficiency syndrome.

Although there is little known about the genetics of cocaine dependence, extensive scientific data are available on the effects of cocaine on brain chemistry. The current view is that the system that uses dopamine in the brain plays an important role in the pleasurable effects of cocaine. In animals, for example, the principal location where cocaine takes effect is the dopamine D_2 receptor gene on chromosome 11 (Koob and Bloom 1988). Recently George Koob and his colleagues of the Scripps Research Institute in La Jolla, California, found evidence suggesting that the dopamine D_3 receptor gene is a primary site of cocaine effects. The exact effect of cocaine on gene expression is unknown. However, we do know that D_2 receptors are decreased by chronic cocaine administration, and this may induce severe craving for cocaine and possibly cocaine dreams (Volkow *et al.* 1993).

A recent study by Ernest Noble of the University of California at Los Angeles and Blum found that about 52 percent of cocaine addicts have the A_1 allele of the dopamine D_2 receptor gene, compared to only 21 percent of nonaddicts. The prevalence of the A_1 allele increases significantly with three risk factors: parental alcoholism and drug abuse; the potency of the cocaine used by the addict (intranasal versus "crack" cocaine); and early-childhood deviant behavior,

such as conduct disorder. In fact, if the cocaine addict has three of these risk factors, the prevalence of the A_1 allele rises to 87 percent. These findings suggest that childhood behavioral disorders may signal a genetic predisposition to drug or alcohol addiction (Noble *et al.* 1993).

A recent survey by the National Institute of Drug Abuse of five independent studies showed that the A_1 allele is also associated with polysubstance dependence (Uhl, Blum, Noble and Smith 1993). The A_1 allele is also associated with an increase in the amount of money spent for drugs by polysubstance-dependent people (Comings *et al.* 1994).

Although not viewed in the same light as the use of cocaine and other illicit drugs, cigarette smoking is another form of chemical addiction. Most attempts to stop smoking are associated with withdrawal symptoms typical of the other chemical addictions. Although environmental factors may be important determinants of cigarette use, there is strong evidence that the acquisition of the smoking habit and its persistence are strongly influenced by hereditary factors.

Of particular significance are studies of identical twins, which show that when one twin smokes, the other tends to smoke. This is not the case in nonidentical twins. In one twin study, Dorit Carmelli of the Stanford Research Insti-

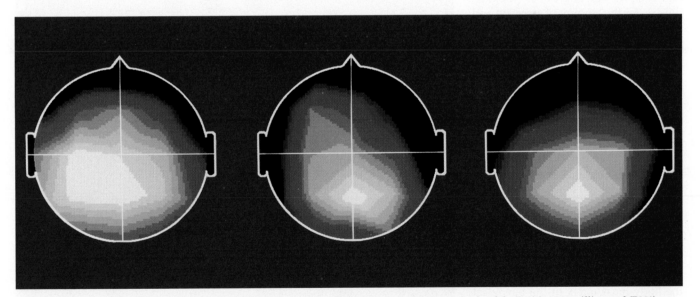

Figure 9. Differences in the electrical activity of the brain are evident in the latency and the magnitude of the Positive 300 millisecond (P300) wave of evoked potentials in individuals who carry the A_1 allele for the dopamine D_2 receptor gene. In normal individuals *(left)* the P300 wave typically occurs between 300 and 330 milliseconds and has a high amplitude (about 10 microvolts). In certain brain disorders associated with the neurotransmitters dopamine and acetylcholine the latency of the P300 wave increases and its magnitude diminishes. Here an obese patient *(middle)* with a A_1/A_2 heterozygous genotype displays a normal latency but a decreased magnitude (about seven microvolts) of the P300 wave. An alcoholic patient *(right)* with a homozygous A_1/A_1 genotype displays an abnormally long latency (364 milliseconds) and a decreased magnitude (about six microvolts). Previous studies have shown that the young sons of alcoholic fathers have an abnormally delayed latency of the P300 wave, which is a predictor of adolescent substance abuse. (Photograph courtesy of the authors.)

tute and her associates examined a national sample of male twins who were veterans of World War II. A unique aspect of this study was that the twins were surveyed twice, once in 1967–68 and again 16 years later. This allowed an examination of genetic factors in all aspects of smoking—initiation, maintenance and quitting. In general, whatever happened to one identical twin happened to the other—including the long-term pattern of not smoking, smoking and then quitting smoking. The absence of these similarities in a control population of nonidentical twins suggests a strong biogenetic component in smoking behavior (Swan *et al.* 1990).

Animal studies have suggested that the dopaminergic pathways of the brain may be involved. For example, the administration of nicotine to rodents disturbs dopamine metabolism in the reward centers of the brain to a greater extent than does the administration of alcohol.

With this in mind, one of us (Comings) and his colleagues investigated the incidence of the A_1 allele in a population of Caucasian smokers. These smokers did not abuse alcohol or other drugs, but had made at least one unsuccessful attempt to stop smoking. It turned out that 48 percent of the smokers carried the A_1 allele. The higher the prevalence of the A_1 allele, the earlier had been the age of onset of smoking, the greater the amount of smoking and the greater the difficulty experienced in attempting to stop smoking. In another sample of Caucasian smokers and nonsmokers, Noble and his colleagues found that the prevalence of the A_1 allele was highest in current smokers, lower in those who had stopped smoking and lowest in those who had never smoked (Noble *et al.* 1994).

Compulsive Bingeing and Gambling
Obesity is a disease that comes in many forms. Once thought to be primarily environmental, it is now considered to have both genetic and environmental components. In a Swedish adoption study, for example, the weight of the adult adoptees was strongly related to the body-mass index of the biological parents *and* to the body-mass index of the adoptive parents. The links to both genetic and environmental factors were dramatic. Other studies of adoptees and twins suggest that heredity is an important contributor to the development of obesity, whereas childhood environ-

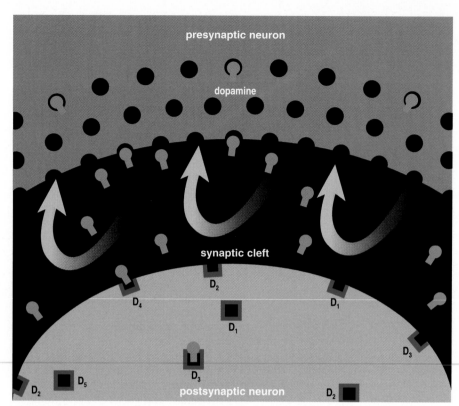

A_1 allele with one-third normal number of receptors

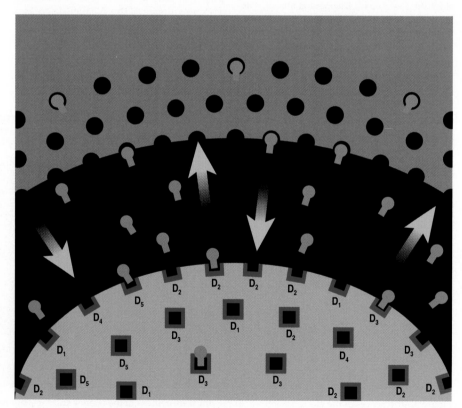

A_2 allele with normal number of receptors

Figure 10. Individuals who carry the A_1 allele *(top)* **of the dopamine D_2 receptor gene have a lower density of dopamine D_2 receptors** *(green)* **compared to individuals who carry the A_2 allele** *(bottom)*. **The authors propose that a decreased number of dopamine D_2 receptors in the reward pathways of the brain results in anger, anxiety and a craving for substances, such as cocaine, alcohol or nicotine, that increase the release of the neurotransmitter dopamine in the brain.**

ment has little or no influence. Moreover, the distribution of fat around the body has also been found to have heritable elements. The inheritance of subcutaneous fat distribution is genetically separable from body fat stored in other compartments (among the viscera in the abdomen, for example). It has been suggested that there is evidence for both single and multiple gene anomalies (Bouchard 1995).

Given the complex array of metabolic systems that contribute to overeating and obesity, it is not surprising that a number of neurochemical defects have been implicated. Indeed at least three such genes have been found: one associated with cholesterol production, one with fat transport and one related to insulin production (Bouchard 1995). The *ob* gene and its product the leptin protein have also been implicated in regulating long-term eating behavior (Zhang *et al.* 1994). Most recently another protein, glucagon-like peptide 1 (GLP-1) has been found to be involved in the regulation of short-term eating behavior (Turton *et al.* 1996). The relationship between leptin and GLP-1 is not known. The *ob* gene may be involved in the animal's selection of fat, but perhaps not in the ingestion of carbohydrates, which appears to be regulated by the dopaminergic system. It may be that the *ob* gene is functionally linked to the opioid peptodergic systems involved in reward.

Whatever the relation between these systems, the complexity of compulsive eating disorders suggests that more than one defective gene is involved. Indeed, the relation between compulsive overeating and drug and alcohol addiction is well documented (Krahn 1991, Newman and Gold 1992). Neurochemical studies show that pleasure-seeking behavior is a common denominator of addiction to alcohol, drugs and carbohydrates (Blum *et al.* 1990). Alcohol, drugs and carbohydrates all cause the release of dopamine in the primary reward area of the brain, the nucleus accumbens. Although the precise localization and specificity of the *pleasure-inducing* properties of alcohol, drugs and food are still debated, there is general agreement that they work through the dopaminergic pathways of the brain. Other studies suggest the involvement of at least three other neurotransmitters serotonin, GABA and the opioid peptides.

Variants of the dopamine D_2 receptor gene appear to be risk factors in obesity. The A_1 allele was present in 45 percent of obese subjects as compared to 19 percent of nonobese subjects (Noble, Noble and Ritchie 1994). Furthermore, the A_1 allele was not associated with a number of other metabolic and cardiovascular risks, including elevated levels of cholesterol and high blood pressure. In contrast, when the subject's profile included factors such as parental obesity, a later onset of obesity and carbohydrate preference, the prevalence of the A_1 allele rose to 85 per-

cent. More recently another study found a significant association between genetic variants of the D_2 receptor and obese subjects (Comings *et al.* 1993).

There is also an increased prevalence of the A_1 allele in obese subjects who have severe alcohol and drug dependence (Blum *et al.* 1996a). When obesity, alcoholism and drug addiction were found in a patient, the incidence of the A_1 allele rose to 82 percent. In contrast, the allele had an incidence of zero percent in nonobese patients who were also not substance abusers and did not have a family history of substance abuse. The presence of the dopamine D_2 receptor gene variants increases the risk of obesity and related behaviors.

Pathological gambling—in which an individual becomes obsessed with the act of risking money or possessions for greater "payoffs"—occurs at a rate of less than two percent in the general population. Although it is the most socially acceptable of the behavioral addictions, pathological gambling has many affinities to alcohol and drug abuse. Clinicians have remarked on the similarity between the aroused euphoric state of the gambler and the "high" of the cocaine addict or substance abuser. Pathological gamblers express a distinct craving for the "feel" of gambling; they develop tolerance in that they need to take greater risks and make larger bets to reach a desired level of excitement, and they experience withdrawal-like symptoms (anxiety and irritability) when no "action" is available (Volberg and Steadman 1988). Indeed, there is a typical course of progression through four stages of the compulsive-gambling syndrome: winning, losing, desperation and hopelessness—a series not uncommon to other addictive behaviors.

Might the dopamine pathways in the brain be involved with pathological gambling? A recent study of Caucasian pathological gamblers found that 50.9 percent carried the A_1 allele of the dopamine D_2 receptor (Comings *et al.* 1996b). The more severe the gambling problem, the more likely it was that the individual was a carrier of the A_1 allele. Finally, in a population of males with drug problems who were also pathological gamblers the incidence of the A_1 allele rose to 76 percent.

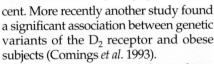

Figure 11. Likelihood of carrying the A_1 allele increases as the number of risk factors increases among cocaine-dependent people. Three risk factors are especially significant: parental alcoholism and drug abuse, the potency of the cocaine used by the addict (intranasal versus "crack" cocaine), and early childhood deviant behavior, such as conduct disorder. The study included 49 subjects (Noble, Blum and Khalsa 1993).

Attention-Deficit Disorder

This disorder is most commonly found among school-age boys, who are at least four times more likely to express the

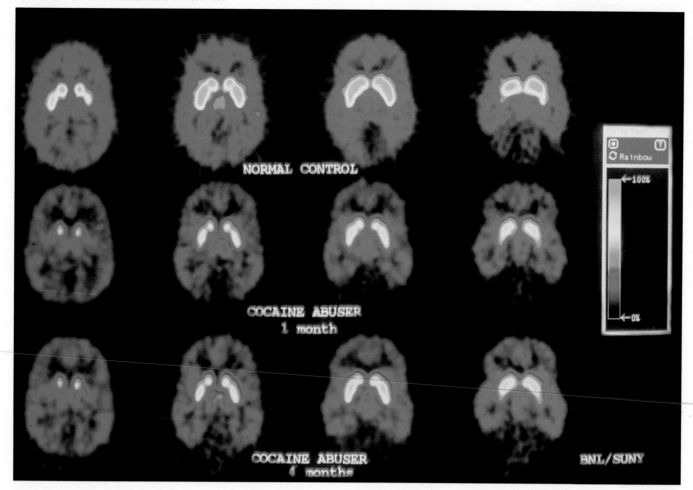

Figure 12. Differences in the density of dopamine D_2 receptors in the brain of a normal subject *(top row)* and the brain of a cocaine abuser one month *(middle row)* and 4 months *(bottom row)* after withdrawal are revealed by the binding of a radioactive tracer in these positron-emission-tomography (PET) scans. The long-term decrease of dopamine D_2 receptors in the cocaine abuser suggests that the lower activity of these receptors is a chronic condition. The decrease in the dopamine D_2 receptors may contribute to the individual's craving for cocaine. Here the density of the dopamine D_2 receptors is shown at four different levels in the basal ganglia. (Photograph courtesy of Nora Volkow, Brookhaven National Laboratory. Used with permission from *Synapse* © 1993, Wiley-Liss, Inc.)

symptoms than are young girls. These children have difficulty applying themselves to tasks that require a sustained mental effort, they can be easily distracted, they may have difficulty remaining seated without fidgeting and they may impulsively blurt out answers in the classroom or fail to wait their turn. Although normal children occasionally display these symptoms, attention-deficit disorder is diagnosed when the behavior's persistence and severity impedes the child's social development and education.

Early speculation about the causes of attention-deficit disorder focused on potential sources of stress within the child's family, including marital discord, poor parenting, psychiatric illness, alcoholism or drug abuse. It has become progressively clear, however, that stress within the family cannot explain the incidence of the disorder.

There is now little doubt that the disorder has a genetic basis.

Evidence in support of this notion comes from patterns of inheritance in the families of children with the disorder and from studies of identical twins. For example, consider instances in which full siblings and half-siblings (who have only half of the genetic identity of full siblings) are both raised in the same family environment. If the behavioral symptoms of attention-deficit disorder were "learned" in the family, then the incidence of the disorder should be the same for full siblings as it is for half-siblings. In fact, half-siblings of children with attention-deficit disorder have a significantly lower frequency of the disorder than full siblings (Lopez 1965). In another study, investigators found that if one identical twin had attention-deficit disorder, there was a 100 percent probability that the other

also had the disorder. In contrast, the incidence of concordance among nonidentical twins was only 17 percent. This result has been supported by two other independent studies of identical twins (Willerman 1973). Finally, one of us (Comings) and his coworkers found that the A_1 allele of the dopamine D_2 receptor gene was present in 49 percent of the children with attention-deficit disorder compared to only 27 percent of the controls (Comings *et al.* 1991).

Some other recent work has linked attention-deficit disorder with another impulsive disorder: Tourette syndrome. More than 100 years ago the French neurologist Giles de la Tourette described a condition that was characterized by compulsive swearing, multiple muscle tics and loud noises. He found that the disorder usually appeared in children between 7 to 10 years old, with boys more likely to be affected than

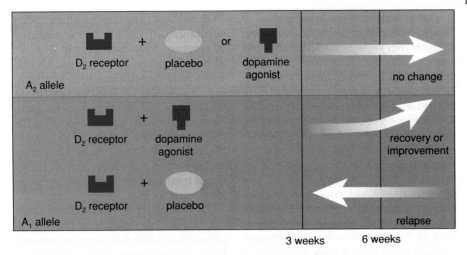

Figure 13. Effectiveness of dopamine agonists *(brown)* in the treatment of certain forms of alcoholism may depend on the individual's genotype for the dopamine D$_2$ receptor gene. The authors propose that alcoholics who carry the A$_1$ allele are more likely to respond positively to treatment with a dopamine agonist (such as bromocriptine). However, if such individuals are treated with a placebo *(beige)* they are more likely to relapse into alcoholism. Alcoholics with the A$_2$/A$_2$ genotype do not respond to dopamine agonists (or to a placebo) because their alcoholism is not associated with the dopamine D$_2$ receptor. The authors suggest that the use of dopamine agonists to treat alcoholics with the A$_1$ allele initiates a feedback system that produces more dopamine receptors after a period of about six weeks.

girls. Tourette suggested that the condition might be inherited.

In the early 1980s one of us (Comings) and his colleagues studied 246 families in which at least one member of the family had Tourette disorder. The study indicated that virtually all cases of Tourette syndrome are genetic (Comings *et al.* 1991). Subsequent studies also found that there was a high incidence of impulsive, compulsive, addictive, mood and anxiety disorders on both sides of the affected individual's family (Comings and Comings 1987). The A$_1$ allele was implicated in a recent report showing that nearly 45 percent of the people diagnosed with Tourette disorder carried the aberrant gene (Comings *et al.* 1991). Moreover, the A$_1$ allele had the highest incidence among people who had the severest manifestations of the disorder.

As mentioned earlier, Tourette syndrome appears to be tightly coupled to attention-deficit disorder. In studies of the two disorders, it was found that 50 to 80 percent of the people with Tourette syndrome also had attention-deficit disorder. Furthermore, an increased number of relatives of individuals with Tourette disorder also had attention-deficit/hyperactivity disorder (Knell and Comings 1993). It now appears that Tourette syndrome is a complex illness that may include attention-deficit disorder, conduct disorder, obsessive, compulsive and addictive disorders and

other related disorders. The close coupling between these disorders has led one of us (Comings) to propose that Tourette syndrome is a severe form of attention-deficit disorder (Comings and Comings 1989; Comings 1995).

The high frequency of the A$_1$ allele among people with Tourette syndrome and attention-deficit disorder raises the question of whether other genes affecting dopaminergic function might also be involved in these disorders. Two others that have been considered are the gene for the enzyme dopamine B-hydroxylase, which converts dopamine to norepinephrine, and the gene for the dopamine transporter, which takes dopamine back into the presynaptic terminal after it is released into the synapse. In both cases, variant forms of these genes are associated with Tourette syndrome (Comings *et al.* 1996c). The anomalous dopamine B-hydroxylase gene (the "DBH *Taq* B1" allele) was further associated with learning disabilities, conduct disorder and substance abuse, whereas the variant of the dopamine transporter (the "10 repeat" allele) was also associated with alcohol abuse, depression and obsessive-compulsive disorder. This observation was supported by other work showing that the 10 repeat allele for the dopamine transporter gene was associated with attention-deficit/hyperactivity disorder (Cook *et al.* 1995). Moreover, elevated levels of the dopamine transporter mol-

ecule have been found in the brains of patients with Tourette syndrome (Malison *et al.* 1995).

If these dopamine-related molecules are indeed associated with various behavioral disorders, it might be expected that having more than one variant would increase the severity or the likelihood of having a disorder. Indeed, this is the case: The severity of attention-deficit disorder, conduct disorder, substance abuse and mood disorders progressively increased from individuals carrying none of the genes to those who carried all three genes (Comings *et al.* 1996c).

Given the widespread prevalence of attention-deficit disorder among children, and its frequent association with alcoholism, drug dependence and other behavioral disorders, it may be that childhood attention-deficit disorder is a predisposing cause to various disorders among adults. For example, there is a significant correlation between attention-deficit hyperactivity disorder and adult drug abuse (Gittleman, Mannuzza, Shenker and Bonagura 1985).

The Dopamine D$_2$ Receptor

The A$_1$ allele carries a behavioral risk factor that shows up not only in substance addiction and attention-deficit disorder, but also in antisocial behavior, conduct disorder and violent or aggressive behavior. In a recent study the A$_1$ allele was present in 60 percent of a sample population of young adolescents between 12 and 18 years old who were diagnosed as "pathologically violent" subjects (Blum Unpublished). A variant of the dopamine transporter gene (VENT 10 repeat) was present in 100 percent of the adolescents. Of these 70 percent had the so-called 10/10 form whereas 30 percent carried the 10/9 allelic form. Another study found that 59 percent of Vietnam veterans with posttraumatic stress disorder also carried the A$_1$ allele, compared to only 5 percent of veterans who were exposed to similar stress but did not develop the disorder (Comings, Muhleman and Gysin 1996).

Why would carriers of the A$_1$ allele be predisposed to the spectrum of disorders associated with the reward deficiency syndrome? Individuals having the A$_1$ allele have approximately 30 percent fewer D$_2$ receptors than those with the A$_2$ allele (Noble *et al.* 1991). Since the D$_2$ receptor gene controls the production of these receptors, the finding suggests that the A$_1$ allele is responsible for the reduction in receptors. In some way that

cycles. The kinds of vocalizations produced by a species can differ considerably throughout the year, with the most "interesting" sounds in the form of territorial or mating signals occurring for only a few months each year. In sum, songbirds are a handful.

Mimetic species add another layer of difficulty by including sounds made by other birds, other animals, and even machines. Thus, in addition to exploring how members of a mimetic species develop species-typical calls and songs—that is, vocalizations with many shared acoustic properties within a population—investigators routinely encounter individual idiosyncracies. Why does one starling mimic a goat and another a cat? Given the abundance of sounds in the world, what processes account for the selection of models?

Baylis (6) advocated studying just part of the mimic's repertoire as a first step, suggesting the example of mockingbirds frequently mimicking cardinals (Cardinalis cardinalis). Although mockingbirds mimic many

> Hier ruht ein lieber Narr,
> Ein Vogel Staar.
> Noch in den besten Jahren
> Musst er erfahren
> Des Todes bittern Schmerz.
> Mir blut't das Herz,
> Wenn ich daran gedenke.
> O Leser! schenke
> Auch du ein Thränchen ihm.
> Er war nicht schlimm;
> Nur war er etwas munter,
> Doch auch mitunter
> Ein lieber loser Schalk,
> Und drum kein Dalk.
> Ich wett', er ist schon oben,
> Um mich zu loben
> Für diesen Freundschaftsdienst
> Ohne Gewinnst.
> Denn wie er unvermuthet
> Sich hat verblutet,
> Dacht er nicht an den Mann,
> Der so schön reimen kann.
>
> Den 4ten Juni 1787.

> A little fool lies here
> Whom I held dear—
> A starling in the prime
> Of his brief time,
> Whose doom it was to drain
> Death's bitter pain.
> Thinking of this, my heart
> Is riven apart.
> Oh reader! Shed a tear,
> You also, here.
> He was not naughty, quite,
> But gay and bright,
> And under all his brag
> A foolish wag.
> This no one can gainsay
> And I will lay
> That he is now on high,
> And from the sky,
> Praises me without pay
> In his friendly way.
> Yet unaware that death
> Has choked his breath,
> And thoughtless of the one
> Whose rime is thus well done.

species, cardinals are a favorite. Why? What consequences accrue for mimic or model? By focusing on one model-mimic system, scientists might answer a number of questions surrounding the nature and function of mimicry. Further control of the model-mimic system can be gained by exposing birds to human speech, a vocal code with a more favorable "signal-to-noise" ratio. This heightens the probability that investigators can detect mimicry and makes it easier to identify the origin of mimicked sounds and the environmental conditions facilitating or inhibiting interspecific mimicry (7). Here, the use of human language is not comparable to efforts with apes or dolphins aimed at uncovering possible analogues to human language. Rather, the use of speech sounds is more properly compared to the use of a radioactive isotope to trace physiological pathways. Thus, when a captive starling utters, "Does Hammacher Schlemmer have a toll-free number?" it is easier to trace the phrase's origin and how often it has been said than to trace the history of the bird's production of "breep, beezus, breep, beeten, beesix."

Over the past decade, we have studied nine starlings, each hand-reared from a few days of age (8). We have also collected information on the behavior of five other starlings (Fig. 1), raised under similar conditions by individuals unaware of our work and unaware of starlings' mimicking abilities when their relationship with the birds began (9). Although many questions remain about the species's vocal capacities, the findings shed light on Mozart's response to his starling's death.

The 14 starlings experienced different social relationships with humans. Eight birds lived individually in what is called interactive contact with the humans who hand-reared them. Their cages were placed in busy parts of the home, and the birds had considerable freedom to associate with their caregivers in diverse ways: feeding from hands; perching on fingers, shoulders, or heads; exploring caregivers' possessions; and inserting themselves into activities such as meal preparation, piano lessons, baths, showers, and telephone conversations (Fig. 2). The humans spontaneously talked to the birds, whistled to them, and gestured by kissing, snapping fingers, and waving good-bye.

Explicit procedures to teach human words using methods prescribed for other mimicking species were not used. Six of the eight caregivers did not know that such training would have an effect until the birds themselves demonstrated their mimicking ability, and two refrained because they were instructed by us to do so. The birds could obtain food and water (and avian companionship in five of eight cases) without interacting with humans.

Three other starlings lived under conditions of limited contact with humans. After 30 days of hand rearing by us, they were individually placed in new homes, along with a cowbird (Molothrus ater). They lived in cages, rarely flew free, and were passively exposed to humans. They heard speech but were not "spoken to" because they did not engage in the kinds of social interactions described for the first group. The final three starlings lived together in auditory contact with humans. They were housed in an aviary on a screened porch of the caregivers raising one of the freely interacting birds. As a result, their auditory environment was loosely yoked to that of the other bird.

The information gathered on the starling's mimicry

differed by setting and caregiver. Extensive audio taping was carried out for the nine subjects studied under our supervision. For three of the remaining birds involved in interactive contact, we used repertoires available in published works, supplemented by personal inquiries. For the last two we obtained verbal reports from caregivers.

Social transmission of the spoken word

The starlings' mimetic repertoires varied consistently by social context: only the birds in interactive contact mimicked sounds with a clearly human origin. None of the other subjects imitated such sounds, although all mimicked their cowbird companions, each other, wild birds, and mechanical noises. For the purposes of this article, we have elected to focus solely on the actions of the birds in interactive contact.

All of these birds mimicked human sounds—including clear words, sounds immediately recognizable as speech but largely unintelligible, and whistled versions of songs identified as originating from a human source—and mechanical sounds whose source could be identified within the households. For the three audiotaped birds, roughly two-thirds of their vocalizations were related to the words or actions of caregivers. The same categories applied to the remaining five birds, who mimicked speech, whistles, and human-derived or mechanical sounds (Table 1).

Many of the more impressive properties of the starlings' vocal capacities defy simple categorization. The most striking feature was their tendency to mimic con-

---

Fewer than 200 starlings were released in Central Park in the 1890s; population estimates in the 1980s hovered around 200,000,000 birds, a millionfold increase

---

nected discourse, imitating phrases rather than single words. Words most often mimicked alone included the birds' names and words associated with humans' arrivals and departures, such as "hi" or "good-bye." All phrases were frequently recombined, sometimes giving the illusion of a different meaning. One bird, for example, frequently repeated, "We'll see you later," and "I'll see you soon." The phrase was often shortened to "We'll see," sounding more like a parental ploy than an abbreviated farewell. Another bird often mimicked the phrase "basic research" but mixed it with other phrases, as in "Basic research, it's true, I guess that's right."

The audiotapes and caregivers' reports made clear, however, that nonsensical combinations (from a human speaker's point of view) were as frequent as seemingly sensible ones: the only difference was that the latter were more memorable and more often repeated to the birds. Sometimes, the speech utterances occurred in highly incongruous settings: the bird mentioned above blasted his owners with "Basic research!" as he struggled frantically with his head caught in string; another screeched, "I have a question!" as she squirmed while being held to have her feet treated for an infection. The tendency for the birds to produce comical or endearing combinations did much to facilitate attention from humans. It was difficult to ignore a bird landing on your shoulder announcing, "Hello," "Give me a kiss," or "I think you're right."

The birds devoted most of their singing time to rambling tunes composed of songs originally sung or whistled to them intermingled with whistles of unknown origin and starling sounds. Rarely did they preserve a melody as it had been presented, even if caregivers repeatedly whistled the "correct" tune. The tendency to sing off-key and to fracture the phrasing of the music at unexpected points (from a human perspective) was reported for seven birds (no information on the eighth). Thus, one bird whistled the notes associated with the words "Way down upon the Swa-," never adding "-nee River," even after thousands of promptings. The phrase was often followed by a whistle of his own creation, then a fragment of "The Star-spangled Banner," with frequent interpositions of squeaking noises. Another bird whistled the first line of "I've Been Working on the Railroad" quite accurately but then placed unexpectedly large accents on the notes associated with the second line, as if shouting, "All the livelong day!" Yet another routinely linked the energetically paced *William Tell* Overture to "Rockaby Baby."

One category of whistles escaped improvisation. Seven of the eight caregivers used a so-called contact whistle to call the birds, typically a short theme (e.g., "da da da dum" from Beethoven's Fifth Symphony). This fragment of melody escaped acoustic improvisation in all cases, although the whistles were inserted into other melodies as well. One bird, however, often mimicked her contact whistle several times in succession, with each version louder than the preceding one (perhaps a quite accurate representation of the sound becoming louder as her caregiver approached her).

All the birds in interactive contact showed an interest in whistling and music when it was performed. They often assumed an "attentive" stance, as shown in Figures 1 and 2: they stood very quietly, arching their necks and moving their heads back and forth. The birds did not vocalize while in this orientation. Records for all eight subjects contained verbal or pictorial reports of the posture.

Clear mimicry of speech was relatively infrequent, due in large part to the birds' tendency to improvise on the sounds, making them less intelligible although definitely still speechlike. Other aspects of their speech imitations were also significant. First, the birds would mimic the same phrase, such as "see you soon" or "come here," but with different intonation patterns. At times, the mimetic version sounded like a human speaking in a pleasant tone of voice, and at other times in an irritated tone. Second, when the birds repeated speech sounds, they frequently mimicked the sounds that accompany speaking, including air being inhaled, lips smacking, and throats being cleared. One bird routinely preceded his rendition of "hi" with the sound of a human sniffing, a combination easily traced to his caregiver being allergic to birds. Finally, the quality of the mimicry of the human voice was surprisingly high. Many visitors who heard the mimicry "live" looked for an unseen human. Those listening to tapes asked which sounds were the starlings' and which the humans', when the only voices were the birds'.

The particular phrases that were mimicked varied, although a majority fell into the broad semantic category of socially expressive speech used by humans as greetings or farewells, compliments, or playful responses to children and pets (see Table 1). Several of the starlings used phrases of greeting or farewell when they heard the sound of keys or saw someone putting on a coat or approaching a door. Several mimicked household events such as doors opening and closing, keys rattling, and dishes clinking together. One bird acquired the word "mizu" (Japanese for water), which she routinely used after flying to the kitchen faucet. Another chanted "Defense!" when the television was on, a sound that she apparently had acquired as she observed humans responding to basketball games.

Figure 1. Kuro is a starling who was hand-reared in captivity. Living in daily close contact with the Iizuka family, she has spontaneously developed, like other starlings in similar circumstances studied by the authors, a rich repertoire of imitations of human speech, songs, and household sounds. Here Kuro listens to whistling. (Photo by Birgitte Nielsen; reprinted by permission of Nelson Canada from *Kuro the Starling*, by Keigo Iizuka and family.)

Caregivers reported that it took anywhere from a few days to a few months for new items to appear in the birds' repertoires. Acquisition time may have depended on the kind of material: one of the birds in limited contact, housed with a new cowbird, learned its companion's vocalization in three days, while one bird in interactive contact took 21 days to mimic his cowbird companion. The latter bird, however, repeated verbatim the question, "Does Hammacher Schlemmer have a toll-free number?" a day after hearing it said only once.

Starlings copy the sounds of other birds and animals, weaving these mimicked themes into long soliloquies that, in captive birds, can contain fragments of human speech

---

Some whistled renditions of human songs also appeared after intervals of only one or two days. An important variable in explaining rate of acquisition and amount of human mimicry may be the birds' differential exposure to other birds. The three birds without avian cage mates appeared to have more extensive repertoires, but they were also older than the other subjects.

The birds did not engage much in mutual vocal exchanges with their caregivers—that is, a vocalization directed to a bird did not bring about an immediate vocal response, although it often elicited bodily orientation

and attention. Thus, the mimicry lacked the "conversational" qualities that have been sought after in work with other animals (10). As no systematic attempt had been made to elicit immediate responding by means of food or social rewards, reciprocal exchanges may nevertheless be possible. Ongoing human conversation not involving the starlings, however, was a potent stimulus for simultaneous vocalizing. The birds chattered frequently and excitedly while humans were talking to each other in person or on the telephone.

The starlings' lively interest and ability to participate in the activities of their caregivers created an atmosphere of mutual companionship, a condition that may be essential in motivating birds to mimic particular models, as indicated by the findings with the birds in limited and auditory contact. The capacity of starlings to learn the sounds of their neighbors fits with what is known about their learning of starling calls, especially whistles, in nature. They learn new whistles as adults by means of social interactions, an ability that is quite important when they move into new colonies or flocks (11). Analyses of social interactions between wild starling parents and their young also indicate the use, early in ontogeny, of vocal exchanges between parent and young and between siblings (12). Thus, the capacities identified in the mimicry of human speech and their dependence on social context seem relevant to the starling's ecology.

Other mimics and songsters

Studies of another mimic, the African gray parrot (*Psittacus erithacus*), also indicate linkages between mimicry and social interaction (13). This species mimics human speech when stimulated to do so by an "interactive modeling technique" in which a parrot must compete for the attention of two humans engaged in conversation. Extrinsic rewards such as food are avoided. The reinforcement is physical acquisition of the object being talked about and responses from human caregivers. Such procedures lead to articulate imitation and often highly appropriate use of speech sounds. Pepperberg reports that one bird's earliest "words" referred to objects he could use: "paper," "wood," "hide" (from rawhide chips), "peg wood," "corn," "nut," and "pasta" *(14)*. The parrot also employed these mimicked sounds during exchanges with caregivers in which he answered questions about the names of objects and used labels identifying shape and color in appropriate ways. The parrot's use of "no" and "want" also suggested the ability to form functional relationships between speech and context, a capacity perhaps facilitated by the trainer's explicit attempts to arrange training sessions meaningful for the student.

Explanations of mimicry of human sounds in this and other species originate in the idea that hand-reared birds perceive their human companions in terms of the social roles that naturally exist among wild birds. Lorenz and von Uexküll elaborated on the kinds of relationships between and among avian parents, offspring, siblings, mates, and rivals *(15)*. In the case of captive birds, humans become the companion for all seasons, with the nature of the relationship shifting with the changing developmental and hormonal cycles in a bird's life.

Mimics are not the only birds to show clear evidence of the effects of companions on vocal capacities. Two examples from nonmimetic species are relevant. In the white-crowned sparrow *(Zonotrichia leucophrys)*, the capacity to learn the songs of other males differs according to the tutoring procedure used. For example, young males learn songs from tape recordings until they are 50 days of age but not afterward. They do acquire songs well after 50 days from live avian tutors with whom they can interact, copying the song of another species, even if

---

Explanations of mimicry of human sounds originate in the idea that hand-reared birds perceive their human companions in terms of the social roles that naturally exist among wild birds

---

they can hear conspecifics in the background. The potency of social tutors has led to a comprehensive reinterpretation of the nature of vocal ontogeny in this species *(16)*. We tried tutoring nine of the starlings using tapes of the caregiver's voice singing songs and reciting prose. There was no evidence of mimicry, except that one bird learned the sound of tape hiss. And thus, if we had relied on tape tutoring, as has been done with many species to assess vocal capacity, we would have vastly underestimated the starlings' skills.

What are the characteristics of live tutors that make them so effective? The studies of white-crowned sparrows suggest that it is not the quality of the tutor's voice, but the opportunity for interaction. Indeed, we have studied a case where voice could not be a cue at all because the "tutor" could not sing. In cowbirds, as in many songbirds, only males sing. Females are frequently the recipients of songs and display a finely tuned perceptual sensitivity to conspecific songs *(17)*. We have documented that acoustically naive males produce distinct themes when housed with female cowbirds possessing different song preferences. We have also identified one important element in the interaction. When males sang certain themes, females responded with distinctive wing movements. The males responded in turn to such behavior by repeating the songs that elicited the females' wing movements. Such data show that singers attend to visual, as well as acoustic, cues and that tutors can be salient influences even when silent. In this species, the social, as distinct from the vocal, conduct of a male's audience is of consequence.

Figure 2. Kuro adopts a listening posture during a music lesson, with neck arched and head moving back and forth. (Photo by Birgitte Nielsen; reprinted by permission of Nelson Canada from *Kuro the Starling*, by Keigo Iizuka and family.)

Studies of another avian group, domestic fowl (*Gallus gallus*), also direct attention to the importance of a signaler's audience (18). In this species, male cockerels produce different calls in the presence of different social companions. Emitting a food call in the presence of food is not an obligatory response but one modulated by the signaler's observations of his audience. Similar findings with cockerel alarm calls indicate the need to consider the multiple determinants of vocal production. Taken as a whole, the findings reveal that, for many birds, acoustic communication is as much visual as vocal experience.

Mozart as birdcatcher

Mozart knew how to look at, as well as listen to, audiences, especially when one of his compositions was the object of their attention. After observing several audiences watching *The Magic Flute*, he wrote to his wife, "I have at this moment returned from the opera, which was as full as ever. . . . But what always gives me most pleasure is the *silent approval!* You can see how this opera is becoming more and more esteemed" (19). Mozart's enjoyment of the less obvious reactions of his audience suggests that, like a bird, he too was motivated not only by auditory but by visual stimuli. The German word he used can be translated "applause" as well as "approval," suggesting his search for rewards more meaningful than the expected clapping of hands. We now turn to the case of Mozart's starling and to the kinds of social and vocal rewards offered to him by his choice of an avian audience.

Mozart recorded the purchase of his starling in a diary of expenses, along with a transcription of a melody whistled by the bird and a compliment (Fig. 3). He had begun the diary at about the same time that he began a catalogue of his musical compositions. The latter effort was more successful, with entries from 1784 to 1791, the year of his death. His book of expenditures, however, lapsed within a year, with later entries devoted to practice writing in English (20). The theme whistled by the starling must have fascinated Mozart for several reasons. The tune was certainly familiar, as it closely resembles a theme that occurs in the final movement of the Piano Concerto in G Major, K. 453 (see Fig. 3). Mozart recorded the completion of this work in his catalogue on 12 April in the same year. As far as we know, just a few people had heard the concerto by 27 May, perhaps only the pupil for whom it was written, who performed it in public for the first time at a concert on 13 June. Mozart had expressed deep concern that the score of this and three other concertos might be stolen by unscrupulous copyists in Vienna. Thus, he sent the music to his father in Salzburg, emphasizing that the only way it could "fall into other hands is by that kind of cheating" (21). The letter to his father is dated 26 May 1784, one day before the entry in his diary about the starling.

Mozart's relationship with the starling thus begins on a tantalizing note. How did the bird acquire Mozart's music? Our research suggests that the melody was certainly within the bird's capabilities, but how had it been transmitted? Given our observation that whistled tunes are altered and incorporated into mixed themes, we assume that the melody was new to the bird because

Table 1. Sounds mimicked by starlings

Greetings and farewells

hi	hey there	I'll (we'll) *see you* soon
good morning	c'mon, c'mere	breakfast
hello	go to your cage	it's time
hey buddy	night night	

Attributions

you're a crazy bird	nutty bird	you're gorgeous
good girl	rascal	see you soon baboon
pretty bird	you're kidding	baby
silly bird		

Conversational fragments

it's true	OK	have the kids called
I suggest	I have a question	*whatcha doing*
that's right	defense	what's going on
basic research	thank you	all right you guys
because	*right*	this is Mrs. Suthers
I guess	who is coming	calling

Human sounds

sighing	sniffing	kissing
coughing	lip smacking	wolf whistle
throat clearing	laughing	

Household sounds

door squeaking	alarm clock	dishes clinking
cat meowing	telephone beep	gun shots
dog barking	keys rattling	

Categories refer to social contexts in which humans produced the sounds, not necessarily the ones in which starlings repeated them. Italicized entries were imitated by four or more birds.

it was so close a copy of the original. Thus, we entertain the possibility that Mozart, like other animal lovers, had already visited the shop and interacted with the starling before 27 May. Mozart was known to hum and whistle a good deal. Why should he refrain in the presence of a bird that seems to elicit such behavior so easily?

A starling in May would be either quite young, given typical spring hatching times, or at most a year old, still young enough to acquire new material but already an accomplished whistler. Because it seems unlikely to us that a very young bird could imitate a melody so precisely, we envision the older bird. The theme in question from K. 453 has often been likened to a German folk tune and may have been similar to other popular tunes already known to the starling, analogous to the highly familiar tunes our caregivers used. But to be whistled to by Mozart! Surely the bird would have adopted its listening posture, thereby rewarding the potential buyer with "silent applause."

Given that whistles were learned quite rapidly by the starlings we studied, it is not implausible that the Vienna starling could have performed the melody shortly after hearing it for the first time. Of course, we cannot rule out a role for a shopkeeper, who could have repeated Mozart's tune from its creator or from the starling. In any case, we imagine that Mozart returned to the shop and purchased the bird, recording the expense

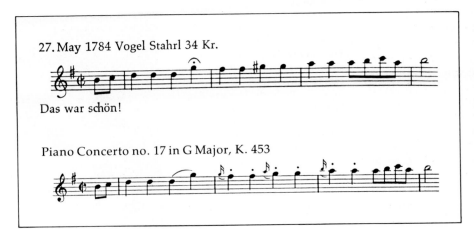

27. May 1784 Vogel Stahrl 34 Kr.

Das war schön!

Piano Concerto no. 17 in G Major, K. 453

Figure 3. Wolfgang Amadeus Mozart was also the delighted owner of a pet starling. He recorded the purchase of the starling in an expense book, noting the date, price, and a musical fragment the bird was whistling. The pleasure he expressed at hearing the starling's song—"Das war schön!" (that was beautiful!)—is all the more understandable when one compares the beginning of the last movement of his Piano Concerto in G Major, K. 453, which was written about the same time. Somehow the bird had learned the theme from Mozart's concerto. It did however sing G sharp where Mozart had written G natural, giving its rendition a characteristically off-key sound.

out of appreciation for the bird's mimicry. Some biographers suggest an opposite course of transmission—from the starling to Mozart to the concerto—but the completion date of K. 453 on 12 April makes this an unlikely, although not impossible, sequence of events.

Given the sociable nature of the captive starlings we studied, we can imagine that some of the experiences that followed Mozart's purchase must have been quite agreeable. Mozart had at least one canary as a child and another after the death of the starling, suggesting that it would not be hard for him to become attached to so inventive a housemate. Moreover, he shared several behavioral characteristics with captive starlings. He was fond of mocking the music of others, often in quite irreverent ways. He also kept late hours, composing well into the night (22). The caregivers of the starlings we

The mimicry of vocal acts such as lip noises, sniffs, and throat clearing brought to the attention of caregivers routine dimensions of their own behavior that they rarely took notice of

studied uniformly reported—and sometimes complained about—the tendency of their birds to indulge in more than a little night music.

The text of Mozart's poem on the bird's death suggests other perceptions shared with the caregivers. Mozart dubbed his pet a "fool"—the German word could also be translated as "clown" or "jester"—an attribution in keeping with the modern starlings' vocal productions of "crazy bird," "rascal," "silly bird," and "nutty bird" and the even more frequent use of such terms in the written description of life with starlings. Mozart gets to the heart of the starling's character when he states that the bird was "not naughty quite, / But gay and bright, / And under all his brag, / A foolish wag." And thus, when we contemplate Mozart's emotions at the bird's death, we see no reason to invoke attributions of displaced grief. We regard Mozart's sense of loss as genuine, his epitaph as an apt gesture.

No other written records of Mozart's relationship with his pet are known. He may have said more, given his prolific letter writing, but much of his correspondence during this period has been lost. The lack of other accounts, however, cannot be considered to indicate a lack of interest in his starling. We are inclined to believe that other observations by Mozart on the starling do exist but have not been recognized as such. Our case rests in part on recent technical analyses of the original (autograph) scores of Mozart's compositions, investigations describing changes in handwriting, inks, and paper. Employing new techniques to date paper by analyzing the watermarks pressed into it at the time of its manufacture, Tyson (23) has established that the dates and places assigned to some of Mozart's compositions can be questioned, reaching the general conclusion that many pieces were written over an extended period of time and not recorded in his catalogue until the time of completion. The establishment of an accurate chronology of Mozart's compositions is obviously essential to those attempting to understand the development of his musical genius. It also serves our purposes in reconstructing events after the starling's funeral.

One composition examined by Tyson is a score entered in Mozart's catalogue on 12 June 1787, the first to appear after the deaths of his father and the starling. The piece is entitled *A Musical Joke* (K. 522). Consider the following description of it from a record jacket: "In the first movement we hear the awkward, unproportioned, and illogical piecing together of uninspired material . . . [later] the andante cantabile contains a grotesque cadenza which goes on far too long and pretentiously and ends with a comical deep pizzicato note . . . and by the concluding presto, our 'amateur composer' has lost all control of his incongruous mixture" (24). Is the piece a musical joke? Perhaps. Does it bear the vocal autograph of a starling? To our ears, yes. The "illogical piecing together" is in keeping with the starlings' intertwining of whistled tunes. The "awkwardness" could be due to the starlings' tendencies to whistle off-key or to fracture musical phrases at unexpected points. The presence of drawn-out, wandering phrases of uncertain structure also is characteristic of starling soliloquies. Finally, the abrupt end, as if the instruments had simply ceased to work, has the signature of starlings written all over it.

Tyson's analysis of the original score of K. 522

indicates that it was not written during June 1787, but composed in fragments between 1784 and 1787, including an excerpt from K. 453. This period coincides with Mozart's relationship with the starling. A common interpretation is that *A Musical Joke* was meant to caricature the kinds of music popular in Mozart's day. Writing such music, a course of action urged on him by his father, might have earned Mozart more money. And thus, the composition has also been interpreted in regard to the father/son relationship *(25)*. Tyson disputes this view on the basis of the physical nature of the autograph score, as much of it was written before Leopold's death, and the lack of solid evidence that Mozart's relationship with his father was bitter enough to cause him to commemorate his first and foremost teacher with a parody.

Although we do not presume to explain all the layers of compositional complexity contained in K. 522, we propose that some of its starling-like qualities are pertinent to understanding Mozart's intentions in writing it. Given the propensities of the starlings we studied and the character and habits of Mozart, it is hard to avoid the conclusion that some of the fragments of K. 522 originated in Mozart's interactions with the starling during its three-year tenure. The completion of the work eight days after the bird's death might then have been motivated by Mozart's desire to fashion an appropriate musical farewell, a requiem of sorts for his avian friend.

Last words

We have offered these observations on starlings and on Mozart for two reasons. First, to give music scholars new insights with which to evaluate one of the world's most studied composers. The analyses of the autograph scores and recent reinterpretations of Mozart's illnesses and death demonstrate the power of present-day knowledge to inform our understanding of the past. We have provided the profile of captive starlings as another way to gain perspective on Mozart's genius.

Second, we hope to spark further interest in the analysis of the social stimulation of vocal learning. Although the role of social companions in motivating avian vocal learning is now well established, the mechanisms by which social influence exerts its effects have only begun to be articulated *(26)*. Part of the problem is defining the nature of social contexts. To say birds interact is to say something quite vague. Interact how? By fighting? By feeding? By flocking? By sitting next to one another? Measuring sound waves is easy compared to calibrating degrees of social influence. Moreover, social signals are multi-modal. The species described here make much use of visual, as well as vocal, stimulation. By what means do they link sights and sounds? Why are only certain linkages made? Answering these questions is the next challenge for students of communication.

One of the founders of the study of bird song, W. H. Thorpe, speculated that birds' imitation of sounds represents a quite simple cognitive process: "The essence of the point may be summed up by saying that while it is very difficult for a human being (and perhaps impossible for an animal) to see himself as others see him, it is much less difficult for him to hear himself as others hear him"

Figure 4. Relationships between starlings and human beings appear to reflect the behavior of birds in the wild. Hand-reared starlings interact with their human companions in terms of the social roles of wild birds. In particular, they learn by observing vocal and other responses to their own expressive efforts. (Photos by Birgitte Nielsen.)

(27). Although we recognize the law of parsimony in Thorpe's remark, we are led by the evidence to seek a phylogenetic middle ground between self-awareness and vocal matching. We propose that some birds use acoustic probes to test the contingent properties of their environment, an interpretation largely in keeping with concepts of communication as processes of social negotiation and manipulation *(28)*. An analogy with the capacities of echo-locating animals may be appropriate. Like bats or dolphins emitting sounds to estimate distance, some birds may bounce sounds off the animate environment, using behavioral reverberations to gauge the effects of their vocal efforts. They are not using Thorpe's behavioral mirror, necessary for self-reflection, but instead a social sounding board with which to shape functional repertoires.

In the case of our starlings, we also conclude that social sonar works two ways: human caregivers cast many 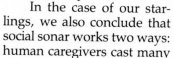 sounds in the direction of their starlings and were often educated by the messages returned. The mimicry of vocal acts such as lip noises, sniffs, and throat clearing brought to the attention of caregivers routine dimensions of their own behavior that they rarely took notice of. The birds' echoing of greetings, farewells, and words of affection conveyed a sense of shared environment with another species, a sensation hard to forget (Fig. 4). The caregivers' sadness in response to the illnesses, absence, or death of their avian companions also suggests that they had been beguiled by the chance to glimpse a bird's-eye view of the world. Most found themselves at a loss for words. And thus we turn to Mozart for fitting emotional expressions—his poem, his *Musical Joke,* and his appropriately grand burial for a "starling bird."

References

1. G. Nottebohm. 1880. *Mozartiana.* Breitkopf and Härtel.
   O. E. Deutsch. 1965. *Mozart: A Documentary Biography.* Stanford Univ. Press.
2. O. Jahn. 1970. *Life of Mozart,* trans. P. D. Townsend. Cooper Square.
   B. Brophy. 1971. In *W. A. Mozart. Die Zauberflöte.* Universe Opera Guides.
   W. Hildesheimer. 1983. *Mozart,* trans. M. Faber. Vintage.
   P. J. Davies. 1989. *Mozart in Person: His Character and Health.* Greenwood.
3. F. M. Chapman. 1934. *Handbook of Birds of Eastern North America.* Appleton.
   E. W. Teale. 1948. *Days without Time.* Dodd, Mead.
4. E. A. Armstrong. 1963. *A Study of Bird Song.* Oxford Univ. Press.
   C. Feare. 1984. *The Starling.* Oxford Univ. Press.
5. R. Dyer-Bennet, trans. 1967. Impatience. In *The Lovely Milleress (Die schöne Müllerin).* Schirmer.
6. J. R. Baylis. 1982. Avian vocal mimicry: Its function and evolution. In *Acoustic Communication in Birds,* vol. 2, ed. D. E. Kroodsma and E. H. Miller, pp. 51–84. Academic Press.
7. D. Todt. 1975. Social learning of vocal patterns and models of their application in grey parrots. *Zeitschrift für Tierpsychologie* 39:178–88.
   I. M. Pepperberg. 1981. Functional vocalizations by an African Grey Parrot (*Psittacus erithacus*). *Zeitschrift für Tierpsychologie* 55:139–60.
8. M. J. West, A. N. Stroud, and A. P. King. 1983. Mimicry of the human voice by European starlings: The role of social interactions. *Wilson Bull.* 95:635–40.
9. H. B. Suthers. 1982. Starling mimics human speech. *Birdwatcher's Digest* 2:37–39.
   M. S. Corbo and D. M. Barras. 1983. *Arnie the Darling Starling.* Houghton Mifflin.
   K. Iizuka. 1988. *Kuro the Starling.* Nelson.
   M. S. Corbo and D. M. Barras. 1989. *Arnie and a House Full of Company.* Fawcett Crest.
   A. DeMotos, pers. com.
   W. R. Fox, unpubl. data.
   A. Peterson and T. Peterson, pers. com.
10. I. M. Pepperberg. 1986. Acquisition of anomalous communicatory systems: Implication for studies on interspecies communication. In *Dolphin Behavior and Cognition: Comparative and Ethological Aspects,* ed. R. J. T. Schusterman and F. Wood, pp. 289–302. Erlbaum.
11. M. Adret-Hausberger. 1982. Temporal dynamics of the whistled songs of sedentary starlings. *Ethology* 71:140–52.
    ———. 1986. Species specificity and dialects in starlings' whistles. In *Acta 19th Congr. Intl. Ornithol.,* vol. 2, pp. 1585–97.
12. M. Chaiken. 1986. Vocal communication among starlings at the nest: Function, individual distinctiveness, and development of calls. Ph.D. diss., Rutgers Univ.
13. I. M. Pepperberg. 1988. An interactive modeling technique for acquisition of communication skills: Separation of "labeling" and "requesting" in a psittacine subject. *App. Psycholing.* 9:59–76.
14. Pepperberg. Ref. 7.
15. K. Lorenz. 1957. Companionship in bird life. In *Instinctive Behavior: The Development of a Modern Concept,* ed. C. H. Schiller, pp. 83–128. International Universities Press.
    J. von Uexküll. 1957. A stroll through the world of animals and men. In *Instinctive Behavior: The Development of a Modern Concept,* ed. C. H. Schiller, pp. 5–82. International Universities Press.
16. L. F. Baptista and L. Petrinovich. 1984. Social interaction, sensitive periods, and the song template hypothesis in the white-crowned sparrow. *Animal Behav.* 36:1753–64.
    L. Petrinovich. 1989. Avian song development: Methodological and conceptual issues. In *Contemporary Issues in Comparative Psychology,* ed. D. A. Dewsbury, pp. 340–59. Sinauer.
17. A. P. King and M. J. West. 1988. Searching for the functional origins of song in eastern brown-headed cowbirds, *Molthrus ater ater. Animal Behav.* 36:1575–88.
    M. J. West and A. P. King. 1988. Female visual displays affect the development of male song in the cowbird. *Nature* 334:244–46.
18. P. Marler, A. Dufty, and R. Pickert. 1986. Vocal communication in the domestic chicken. II. Is a sender sensitive to the presence of a receiver? *Animal Behav.* 34:194–98.
    S. J. Karakashian, M. Gyger, and P. Marler. 1988. Audience effects on alarm calling in chickens (*Gallus gallus*). *J. Comp. Psychol.* 102:129–35.
19. E. Anderson, ed. 1989. *The Letters of Mozart and His Family,* p. 907. Norton.
20. Jahn. Ref. 2.
21. Anderson. Ref. *19,* p. 877.
22. F. Niemtschek. 1956. *Life of Mozart,* trans. H. Mautner. Leonard Hyman.
    Jahn. Ref. 2.
    Davies. Ref. 2.
23. A. Tyson. 1987. *Mozart: Studies of the Autograph Scores.* Harvard Univ. Press.
24. W. A. Mozart. *A Musical Joke.* Liner notes by P. Cohen. Deutsche Grammophon. 400 065–2.
25. Ref. 2.
26. Ref. *16.*
27. W. H. Thorpe. 1961. *Bird-Song,* p. 79. Cambridge Univ. Press.
28. D. W. Owings and D. F. Hennessy. 1984. The importance of variation in sciurid visual and vocal communication. In *The Biology of Ground-dwelling Squirrels: Annual Cycles, Behavioral Ecology, and Sociality,* ed. J. O. Murie and G. R. Michener, pp. 167–200. Univ. Nebraska Press.

Testosterone and Aggression in Birds

*John C. Wingfield, Gregory F. Ball, Alfred M. Dufty, Jr.,
Robert E. Hegner, Marilyn Ramenofsky*

The familiar spring sound of birdsongs heralds the onset of territory formation and a complex sequence of interrelated events that make up the breeding period. Such songs are an integral part of the repertoire of aggressive behaviors that males use to advertise and defend territorial boundaries and to attract mates (Fig. 1). It is well established that hormones, particularly testosterone, have stimulatory effects on aggression in reproductive contexts. The prevailing "challenge" hypothesis asserts that testosterone and aggression correlate only during periods of heightened interactions between males. Under more stable social conditions, according to the hypothesis, relationships among males are maintained by other factors such as social inertia, individual recognition of status, and territorial boundaries, and testosterone levels remain low. Recent research has suggested ways in which the hypothesis should be modified or extended. In this article we will consider the complexities of aggressive behaviors and their regulation, focusing specifically on species differences in territorial behavior of male birds as models for the multiple interactions of hormones, environment, and behavior.

The secretion of testosterone by interstitial cells in the testis is controlled primarily by a glycoprotein, luteinizing hormone, secreted from the anterior pituitary gland (Fig. 2). Testosterone stimulates the development of secondary sex characteristics such as wattles, combs, spurs, the cloacal protuberance (a copulatory organ), and

> *Testosterone may not trigger aggressive behavior but may facilitate responses to it*

in some species bright-colored skin and nuptial plumage. These characteristics are used extensively in sexual and aggressive displays (Witschi 1961).

Testosterone is also transported in the blood to the brain, where it influences the expression of reproductive behaviors. Classical experiments conducted on a variety of vertebrates, including birds, showed that if the testes are removed, there is a decline in the frequency and intensity of aggressive and sexual behaviors such as singing (or equivalent vocalizations), threat postures, and actual fights. If exogenous testosterone is given to these castrates, the frequency of aggressive behaviors increases again (for reviews on birds see Harding 1981; Balthazart 1983).

The extent to which aggressive behaviors decline after castration or increase after administration of exogenous testosterone varies greatly from species to species, in part because of the different ways in which testosterone can influence behavior. Two mechanisms have been proposed involving organizational and activational effects. Organizational effects of testosterone occur early in development, often immediately after hatching, and once adulthood is reached the neurons involved can operate independently. Activational effects require the immediate presence of testosterone for the sensitive neurons to function normally. Whether organizational or activational effects predominate depends on context and stage in the breeding period. However, in birds it appears that testosterone may have important activational effects regulating short-term changes in territorial aggression within the breeding season.

Over the past 15 years, radioimmunoassay has been used to determine circulating levels of testosterone. If testosterone does activate aggressive behavior, plasma levels should correlate with the behavior in reproductive contexts. Recent work on rodents (Schuurman 1980; Brain 1983; Sachser and Pröve 1984) and primates (Eaton and Resko 1974; Dixson 1980; Phoenix 1980; Bernstein et al. 1983; Sapolsky 1984) suggests that there are such correlations, but that they depend to a great extent on taxonomic class, age, experience, social context, and other environmental influences. The mechanisms underlying such variation are still largely unknown.

In birds, the evidence for correlations of testosterone and aggression is more convincing, although not completely so. Once again, social context must be taken

John C. Wingfield is an associate professor at the University of Washington. He has combined laboratory techniques in comparative endocrinology with field investigations to study the responses of birds to their social and physical environment. He obtained his Ph.D. from the University College of North Wales in 1973 and was on the faculty of The Rockefeller University before moving to the University of Washington. Gregory F. Ball obtained his Ph.D. at the Institute of Animal Behavior, Rutgers University, and is now assistant professor at The Rockefeller University; Alfred M. Dufty performed his doctoral work at SUNY Binghamton and is a post-doctoral fellow at The Rockefeller University; Robert E. Hegner graduated with a Ph.D. from Cornell and completed postdoctoral work at Oxford, The Rockefeller University, and the University of Washington; and Marilyn Ramenofsky was awarded a Ph.D. from the University of Washington, was visiting assistant professor at Vassar College, and is now a research associate at the University of Washington. Address for Professor Wingfield: Department of Zoology, NJ-15, University of Washington, Seattle, WA 98195.

Figure 1. As part of the annual ritual of establishing territories and attracting mates, male birds engage in a variety of aggressive behaviors. In the photograph on the left, a male song sparrow (*Melospiza melodia*) assumes the posture that heralds an attack on an intruder, in this case a decoy in a cage. Recent research has shown how the steroid hormone testosterone stimulates aggression in response to such perceived threats. Mist nets stretched between aluminum poles are used to catch birds in the field. After removing a bird from the net (*right*), the scientist collects a blood sample from a wing vein. The bird is then released unharmed. (Photographs by J. C. Wingfield.)

into account, as well as environmental influences such as length of day, presence of a mate, and nest sites (Wingfield and Ramenofsky 1985). At least some of this confusion can be eliminated by bringing a comparative approach to bear on a variety of avian species. Birds are ideal for this kind of research because there is much diversity in social systems across species. They are also relatively easy to study under free-living conditions, enabling us to conduct parallel field and laboratory investigations.

Seasonal changes

If testosterone is as intimately involved with territorial aggression in birds as is usually presumed, testosterone levels in the blood should parallel the expression of seasonal territoriality. This relationship has been investigated in several species of birds under free-living conditions, thus reducing possible artifacts of captivity (see Wingfield and Farner 1976).

It is crucial when analyzing these kinds of data to determine the precise stage in the reproductive period at which each individual is sampled. This point is illustrated in Figure 3, which depicts plasma levels of luteinizing hormone and testosterone in free-living house sparrows (*Passer domesticus*). If plasma levels of a number of individuals are organized by calendar date, several stages of reproductive activity (prelaying, laying, incubating, renesting) are averaged out on any given date, and the result is a pair of curves, with luteinizing hormone and testosterone rising in spring, remaining relatively high during the breeding season, and then declining to basal as reproduction ends in August and September. If the data are reorganized according to the phase of the breeding cycle, the true pattern of hormone variation is revealed, making allowance for the average

time it takes a pair to progress through each stage (about 4 to 6 days to lay the first egg, 5 days to produce a clutch, and 11 to 14 days to incubate).

Figure 4 compares levels of testosterone in several monogamous species sampled in free-living conditions. Typically, testosterone is highest when territories are first established and aggressive interactions among males are most frequent. For the song sparrow (*Melospiza melodia*), there are two peaks of testosterone, the first associated with the establishment of territory and the second with the egg-laying period for the first clutch, when the male guards his sexually receptive mate. Plasma levels of testosterone decline markedly just prior to or during the parental phase (incubation) and gradually diminish to basal concentrations by the end of the breeding season.

There is no increase in plasma levels of testosterone during the egg-laying period of the second brood for many of the species with open-cup nests, such as the song sparrow and the European blackbird (*Turdus merula*), because there are virtually unlimited sites for these nests, and competition focuses on maintaining territorial boundaries and guarding mates. However, species such as the house sparrow and the European starling (*Sturnus vulgaris*) that nest in holes, a limited resource for which there often is intense competition (in addition to guarding mates), do show an increase in testosterone level with each egg-laying period (see Figs. 3 and 4).

An interesting contrast is provided by the western gull (*Larus occidentalis wymani*). Individuals of this species are long-lived, may breed for 20 years or more, usually mate for life, and return to the same breeding territory year after year. Furthermore, there is an excess of females and no shortage of nest sites at one of the breeding colonies, Santa Barbara Island (Hunt et al. 1980). As a result, competition between males is minimal, and it is not surprising, given the low level of aggression, that the cycle of plasma testosterone in male western gulls is of very low amplitude (Wingfield et al. 1982).

As Figure 5 shows, the same relationship of testosterone levels and aggression can be found in polygamous and promiscuous species, but males of these species have high levels of testosterone for longer periods than do monogamous species. For example, male red-winged blackbirds (*Agelaius phoeniceus*) generally do not feed young but rather display at one another throughout the breeding season in an attempt to maintain territorial boundaries and retain females.

Both monogamous and polygynous males are found within populations of the pied flycatcher (*Ficedula hypoleuca*). Monogamous males have testosterone levels similar to those of monogamous males in other species, but polygynous males maintain high levels of testosterone until the second female has begun incubating. Only then do levels decline rapidly, followed by a return of the male to his first mate, whose young he helps to feed (see Silverin and Wingfield 1982).

Male brown-headed cowbirds (*Molothrus ater*) are unusual in that they do not defend a territory but form dominance hierarchies for access to females. They are brood parasites, showing no parental care. Males spend the entire breeding season guarding females from the attentions of other males. Accordingly, we see prolonged high levels of testosterone that decline only gradually during the season (Dufty and Wingfield 1986a).

Laboratory tests of the challenge hypothesis

As we have seen, field investigations of free-living birds suggest that testosterone is elevated during periods of elevated competition between males, and that parental behavior in males is preceded by a decline in testosterone. Only in species in which males do not feed young or are exposed to intense competition do plasma levels of testosterone remain elevated. These results have led to the challenge hypothesis.

What is the experimental evidence in support of the hypothesis? A positive correlation of aggressive displays with plasma testosterone was found when Japanese quail (*Coturnix coturnix*) were paired in a tournament lasting several days, but the correlation was apparent only immediately prior to the first fighting day and during the following three days (Ramenofsky 1984). From the fifth day onward, levels of testosterone in quail that won fights were indistinguishable from levels in those that lost. By that time, dominance relationships had been established. This may explain why Balthazart and his colleagues (1979) and Tsutsui and Ishii (1981) could find no correlation of plasma testosterone level

and dominance status in groups of male quail with well-established social relationships.

Other experiments confirmed these findings. Captive flocks of house sparrows formed social hierarchies in which dominant individuals had higher plasma levels of testosterone than subordinates only during the first week after the birds were grouped. Before grouping, and more than one week after, there were no correlations of

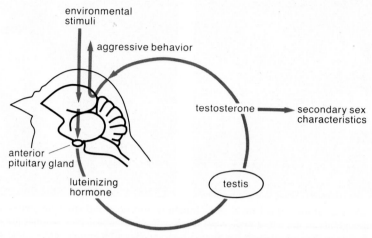

Figure 2. The system through which testosterone influences aggressive behavior begins with the secretion, in response to environmental stimuli, of the glycoprotein luteinizing hormone from the anterior pituitary gland at the base of the brain. Luteinizing hormone in turn stimulates secretion of testosterone by interstitial cells in the testis, where testosterone is produced. In addition to arousing aggressive behavior, testosterone contributes to the development of secondary sex characteristics, such as combs, spurs, and bright plumage.

testosterone level and social status (Hegner and Wingfield 1986). This is consistent with the challenge hypothesis, since testosterone levels were elevated only for a short period as relationships were established. Similarly in the brown-headed cowbird, three males grouped with a single female formed social relationships, and the dominant male gained access to the female. Plasma levels of testosterone in dominant males were elevated one day after grouping, but not before or one week after (Dufty and Wingfield 1986b).

What happens if exogenous testosterone is given to individuals? Do they rise in status, gain a territory, or enlarge an existing one? If a testosterone implant was given to an identified subordinate of a regularly matched pair of Japanese quail, he became more aggressive and fought more persistently with other males. Nevertheless, these subordinates did not win a sufficient number of fights to be considered dominant (Ramenofsky 1982). This suggests that testosterone is not sufficient in itself to heighten aggressive displays to the point of overthrowing previously established relationships. Similar results have been obtained in dominance hierarchies of California quail (*Lophortyx californica*), free-living Harris's sparrows (*Zonotrichia querula*), and sharp-tailed grouse (*Tympanuchus phasianellus*) (Emlen and Lorenz 1942; Trobec and Oring 1972; Rohwer and Rohwer 1978).

Another laboratory experiment sheds more light on the challenge hypothesis. Castrated male white-crowned sparrows (*Zonotrichia leucophrys gambelii*) were given implants of testosterone that maintained circulating levels very similar to those observed during the spring (see Wingfield and Farner 1978a, 1978b). Castrated controls were given empty implants. Songs and threat displays often seen during the establishment of territories in the field increased in both groups after treatment, but there was no significant difference in the frequency of these actions between the two groups despite the wide difference in testosterone level (Wingfield 1985a).

This apparent contradiction of the challenge hypothesis can perhaps be attributed to the fact that the birds had been housed together for over six months. It was thus likely that social relationships among individuals had been established for some time. When a new male was introduced in an adjacent cage, there was an immediate increase in aggression in both groups, and the males with higher levels of testosterone showed significantly more aggressive displays than did the controls. By the next day, the frequency of aggression had dropped dramatically, illustrating how quickly social relationships can be established and emphasizing the ephemeral nature of the correlation between testosterone and aggressive behavior.

It is of little surprise that some investigations have identified hormone-behavior relationships and some have not, particularly since social contexts vary across the studies. Experiential factors such as the development of dominance relationships among individuals can exert a strong influence on the degree to which the circulating levels of testosterone affect frequency and intensity of aggressive behavior. Nevertheless, there is little doubt that testosterone is requisite for increased frequency of aggressive behavior when an individual is challenged in a territorial or other reproductive context.

Environmental cues and testosterone

What controls the timing and amplitude of changes in plasma levels of testosterone so that they occur at appropriate stages in the reproductive period? Clearly, environmental cues play a major role, and one obvious candidate is the annual change in the length of day. It is well known that the vernal increase in length of day promotes secretion of luteinizing hormone and steroid hormones such as testosterone (e.g., Farner and Follett 1979; Wingfield and Farner 1980). Experiments with male white-crowned and song sparrows demonstrated that spermatogenesis is completed, secondary sex characteristics are developed, and the full repertoire of reproductive behaviors (both territorial and sexual) are expressed when birds are transferred from short to long days (see Wingfield and Moore 1986). However, the seasonal changes in testosterone in free-living males are dramatically different from those generated solely by exposing captive males to long days in the laboratory, and the absolute levels can reach an order of magnitude higher than those of males maintained in captivity. Since it has also been shown that high circulating levels of testosterone are not required for the expression of sexual behavior (Moore and Kranz 1983), it is possible that elevated levels in free-living males are involved solely in the regulation of aggression.

What other environmental cues influence the secretion of testosterone and aggressive behavior? Two possibilities spring to mind: stimuli from the territory itself or signals emanating from a challenging male. To evaluate these possibilities, male song sparrows were captured

and removed from their territories, thus creating a vacant spot within the local population. Usually another male claimed the spot within 12 to 72 hours. The result was an increase in conflict between the replacement male and the neighbors, who reestablished territorial boundaries with the newcomer. During this period of social instability, blood samples were collected from replacement males and neighbors. Samples were also

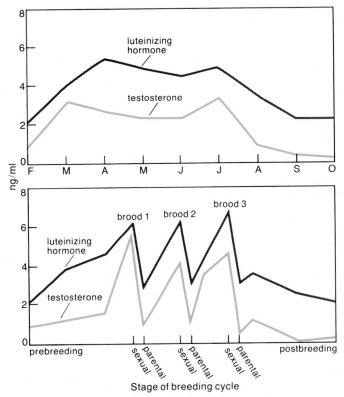

Figure 3. It is important when analyzing seasonal changes in luteinizing hormone and testosterone to distinguish between organization of data by calendar date and by stage in the breeding cycle. If plasma levels of a number of individuals are organized by calendar date (*top*), the various stages in the breeding cycle average out, and the result for both luteinizing hormone and testosterone is a curve with two peaks. If on the other hand the data are organized by stages in the breeding cycle (*bottom*), a much more complicated pattern of hormone variation appears. The data displayed here are from free-living male house sparrows (*Passer domesticus*). (After Hegner and Wingfield 1986.)

collected from control males in a separate area in which boundaries had been stable for some time.

The results were quite clear: plasma levels of testosterone were higher in the replacement males and in their otherwise untreated neighbors than in the controls. Both the neighbors of the replacement and the controls had territories, yet the latter had much lower levels of testosterone. These two groups differed only in that the neighbors were reestablishing territorial boundaries whereas the controls were not. This suggests that the stimulus for increased secretion of testosterone may be not the territory per se (although the data do not disprove a possible effect) but the challenging behavior of the replacement male as he attempts to establish new territorial boundaries (Wingfield 1985b).

To test this further, intrusions were simulated with a decoy male song sparrow in a cage placed in the center

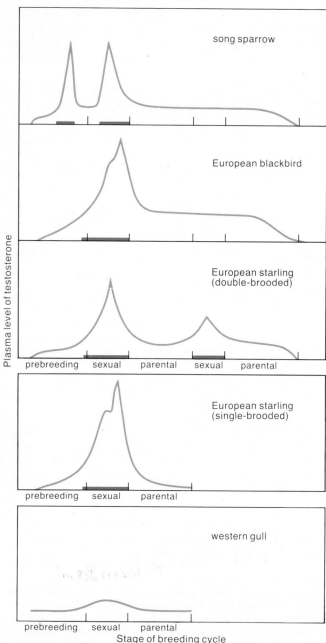

Figure 4. The plasma levels of testosterone in males of four other monogamous species are quite different from those of the house sparrow shown in Figure 3, although in all cases a relationship between testosterone and aggressive behavior is discernible. (Periods when confrontations between males are most frequent are indicated by bars.) There are two peaks for the song sparrow (*Melospiza melodia*), the first associated with the establishment of territory and the second with the egg-laying period of the first brood. The European blackbird (*Turdus merula*) has a single peak during the first brood. Neither the song sparrow nor the European blackbird has a peak during subsequent broods, because these species nest in open cups, for which there are unlimited sites and thus little competition. In these species, competition between males is most intense early in the season. The European starling (*Sturnus vulgaris*), on the other hand, nests in holes, for which competition is keen, and so starlings with double broods have a second peak during their second brood; those with single broods have the expected single peak. The western gull (*Larus occidentalis wymani*) has a distinctively different pattern because of the relative lack of conflict in its breeding colonies. (After Schwabl et al. 1980; Wingfield et al. 1982; Dawson 1983; Wingfield 1984a; Ball and Wingfield 1986.)

of a territory (see Fig. 1). Tape-recorded songs also were broadcast through a speaker placed alongside the decoy. The territorial male almost invariably attacked and attempted to drive the simulated intruder away. He was captured after skirmishing with the intruder for 5 to 60 minutes, and a blood sample was drawn. Controls were captured at the same time of day as the simulated intrusions. Males exposed to a challenge from a simulated intruder showed an increase in testosterone compared with controls. Essentially the same result was obtained in early April and in May through June, indicating that this effect could occur at any time during the breeding period.

It is important to note that the response required about ten minutes before the increase in testosterone was significant. We know that males tend to trespass on other territories regularly and are quickly chased out when seen by the owner (Wingfield 1984b). The confrontations usually last only a few seconds, and thus an increase in testosterone level is unnecessary. However, if an intruder persists and attempts to take over the territory, prolonged fights lasting several hours or even days may result. In such cases an increase in plasma testosterone is appropriate.

There is a third line of evidence suggesting that encounters between males can result in an increase in plasma levels of testosterone. Implants of testosterone in free-living male song sparrows resulted in heightened aggression for longer periods than in control males. In turn, plasma levels of testosterone were elevated in neighbors of testosterone-implanted males compared with neighbors of controls. This effect was most apparent early and late in the season. At other times no effect was noted, because factors such as the presence of young possibly overrode the effect of the aggressive male neighbor.

It was also found that males who had a territory at least one removed from a testosterone-implanted male did not have elevated levels, even though they could hear and see encounters between their immediate neighbors and the testosterone-implanted males (Wingfield 1984b). It appears that an individual male must be involved directly in an agonistic encounter for a hormonal response to be initiated. This blocking of a ripple effect may be adaptive; otherwise, a wave of responses would pass indiscriminately through the local population, affecting males that were not involved in the original skirmish. Moreover, functionally irrelevant surges of testosterone could interfere severely with other reproductive activities such as the feeding of young.

The environmental stimuli generated in the course of an agonistic encounter could enter the central nervous system by several routes: visual, auditory (songs and other vocalizations), tactile (fights), or chemical (pheromones). We can rule out tactile stimuli, because several of the experiments outlined above show that testosterone levels increase in response to a caged male with whom contact is precluded. Also we can probably rule out pheromonal cues, since these are largely regarded as being absent in birds (although it is important to note that this point has not been rigorously investigated). Thus we are left with visual and auditory information influencing secretion of testosterone.

Are both components required for the response?

Recent field experiments showed that if male song sparrows are exposed to a playback of tape-recorded songs (auditory but no visual stimulus), a devocalized male (visual but no auditory stimulus), or a playback plus a devocalized male (visual and auditory stimuli), only those males exposed to both visual and auditory cues have elevated levels of testosterone. Auditory or visual cues alone do not result in a significant increase in testosterone. It was also found that the response is specific: captive male song sparrows showed an increase in plasma levels of testosterone following a challenge from another song sparrow but not following a challenge from a house sparrow (a heterospecific).

Now that the external receptors for environmental cues have been identified and the endocrine response and the specificity of that response determined, we can investigate the neural pathways by which environmental information controls reproductive function.

What is testosterone doing?

This may appear to be an odd question, since it is well established that testosterone has direct effects on aggressive territorial behavior. There is no doubt that testosterone has organizational effects insofar as it influences the formation of song control nuclei in the brain during development and seasonal breeding (e.g., Nottebohm 1981). It is also clear that high levels of testosterone during establishment of a territory are playing some activational role, at least early in the breeding season. However, many of our observations do not fit neatly into these categories (see also Arnold and Breedlove 1985).

Responses to challenges outside the normal seasonal pattern suggest another role for testosterone in the arousal of aggressive behavior. The initial response to a challenging male is to attack vigorously even though the circulating level of testosterone may be much lower than in early spring. Only *after* the attack does testosterone increase, and this appears to take at least ten minutes. Clearly testosterone cannot be playing an activational role in the literal sense of the word, since it increases after the fact. Is it possible that testosterone is playing a facilitative role for the neurons involved during extended periods of intense aggression?

This role would require a very rapid action of a steroid hormone on a target cell. The classical mode of action is through the genome, a process that can take many hours (typically 16 to 30). But recently compiled evidence from mammals suggests that steroid hormones can also have very rapid effects. For example, steroids have been shown to influence rates of gene transcription in rats within 15 minutes, and estradiol can have morphological effects on neuronal cell nuclei within two hours (Jones et al. 1985; McEwen and Pfaff 1985). Even more striking is the demonstration in vitro that estradiol injected directly onto the membrane of an excitable cell induces action potentials within one minute (Dufy et al. 1979). Furthermore, Towle and Sze (1983) found that several steroid hormones, including testosterone, bind to synaptic membranes in the rat brain with high specificity and affinity. Thus the potential exists for very rapid actions of testosterone on the central nervous system through membrane receptors, although more research is required to confirm this in avian systems.

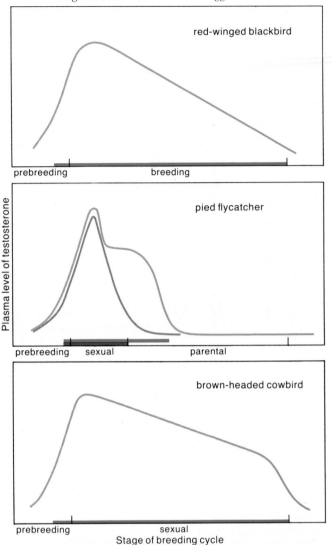

Figure 5. Males of polygynous and promiscuous species have their own characteristic patterns of circulating testosterone: levels remain high for longer periods than they do in monogamous species (see Fig. 4). These correlate with the greater amount of time that such males must spend defending territories and females. (Periods of frequent conflicts between males are again indicated by bars.) In red-winged blackbirds (*Agelaius phoeniceus*), the breeding period includes both sexual and parental stages, since each male may have several females on his territory, some of which may be in the sexual or parental stage at any one time. The pied flycatcher (*Ficedula hypoleuca*) includes both monogamous (*gray line and bar*) and polygynous males (*colored line and bar*). The brown-headed cowbird (*Moluthrus ater*) is a brood parasite that has no territory and performs no parental duties but rather spends the breeding season guarding females from competing males. (After Silverin and Wingfield 1982; Dufty and Wingfield 1986a. Additional data supplied by W. A. Searcy.)

Such a concept is speculative, but the possibility arises that in addition to the two classical modes of genomic action of steroid hormones, involving organizational and activational effects, a third mode of action—supporting or facilitative—could arise during periods of heightened agonistic encounters. Mediated either through rapid-acting membrane receptors or genomically, the third mode would influence the function of brain nuclei involved in the control of aggression. Whether this may ultimately prove to be simply a form of activational effects of testosterone, or indeed a separate mode of action, remains to be seen.

References

Arnold, A. P., and S. M. Breedlove. 1985. Organizational and activational effects of sex steroids on brain and behavior: A reanalysis. *Horm. Beh.* 19:469–98.

Ball, G. F., and J. C. Wingfield. 1986. Changes in plasma levels of sex steroids in relation to multiple broodedness and nest site density in male starlings. *Physiol. Zool.* 60:191–99.

Balthazart, J. 1983. Hormonal correlates of behavior. In *Avian Biology*, ed. D. S. Farner, J. R. King, and K. C. Parkes, vol. 7, pp. 221–366. Academic Press.

Balthazart, J., R. Massa, and P. Negri-Cesi. 1979. Photoperiodic control of testosterone metabolism, plasma gonadotropins, cloacal gland growth, and reproductive behavior in the Japanese quail. *Gen. Comp. Endocrinol.* 39:222–35.

Bernstein, I. S., T. P. Gordon, and R. M. Rose. 1983. The interaction of hormones, behavior, and social context in non-human primates. In *Hormones and Aggressive Behavior*, ed. B. Svare, pp. 535–62. Plenum.

Brain, P. F. 1983. Pituitary-gonadal influences on social aggression. In *Hormones and Aggressive Behavior*, ed. B. Svare, pp. 3–26. Plenum.

Dawson, A. 1983. Plasma gonadal steroid levels in wild starlings (*Sturnus vulgaris*) during the annual cycle and in relation to the stages of breeding. *Gen. Comp. Endocrinol.* 49:286–94.

Dixson, A. F. 1980. Androgens and aggressive behavior in primates: A review. *Aggressive Beh.* 6:37–67.

Dufty, A. M., and J. C. Wingfield. 1986a. Temporal patterns of circulating LH and steroid hormones in a brood parasite, the brown-headed cowbird, *Molothrus ater*. I. Males. *J. Zool. London (A)* 208:191–203.

———. 1986b. Endocrine changes in breeding brown-headed cowbirds and their implications for the evolution of brood parasitism. In *Behavioural Rhythms*, ed. Y. Quéinnec and N. Delvolvé, pp. 93–108. Toulouse: Université Paul Sabatier.

Dufy, B., et al. 1979. Membrane effects of thyrotropin-releasing hormone and estrogen shown by intracellular recording from pituitary cells. *Science* 204:509–11.

Eaton, G. G., and J. A. Resko. 1974. Plasma testosterone and male dominance in Japanese macaque troops with repeated measures of testosterone in laboratory males. *Horm. Beh.* 5:251–59.

Emlen, J. T., and F. W. Lorenz. 1942. Pairing responses of free-living valley quail to sex-hormone pellets. *Auk* 59:369–78.

Farner, D. S., and B. K. Follett. 1979. Reproductive periodicity in birds. In *Hormones and Evolution*, ed. E. J. W. Barrington, pp. 829–72. Academic Press.

Harding, C. F. 1981. Social modulation of circulating hormone levels in the male. *Am. Zool.* 21:223–32.

Hegner, R. E., and J. C. Wingfield. 1986. Behavioral and endocrine correlates of multiple brooding in the semi-colonial house sparrow *Passer domesticus*. I. Males. *Horm. Beh.* 20:294–312.

Hunt, G. L., Jr., J. C. Wingfield, A. Newman, and D. S. Farner. 1980. Sex ratio of western gulls on Santa Barbara Island, California. *Auk* 97:473–79.

Jones, K. J., D. W. Pfaff, and B. S. McEwen. 1985. Early estrogen-induced nuclear changes in rat hypothalamic ventromedial neurons: An ultrastructural and morphometric analysis. *J. Comp. Neurol.* 239:255–66.

McEwen, B. S., and D. W. Pfaff. 1985. Hormone effects on hypothalamic neurons: Analysing gene expression and neuromodulator action. *Trends Neurosci.*, March, pp. 105–10.

Moore, M. C., and R. Kranz. 1983. Evidence for androgen independence of male mounting behavior in white-crowned sparrows (*Zonotrichia leucophrys gambelii*). *Horm. Beh.* 17:414–23.

Nottebohm, F. 1981. A brain for all seasons: Cyclical anatomical changes in song control nuclei of the canary brain. *Science* 214:1368–70.

Phoenix, C. H. 1980. Copulation, dominance, and plasma androgen levels in adult rhesus males born and reared in the laboratory. *Archives Sexual Beh.* 9:149–68.

Ramenofsky, M. 1982. Endogenous plasma hormones and agonistic behavior in male Japanese quail, *Coturnix coturnix*. Ph.D. diss., Univ. of Washington.

———. 1984. Agonistic behavior and endogenous plasma hormones in male Japanese quail. *Animal Beh.* 32:698–708.

Rohwer, S., and F. C. Rohwer. 1978. Status signalling in Harris' sparrows: Experimental deceptions achieved. *Animal Beh.* 26:1012–22.

Sachser, N., and E. Pröve. 1984. Short-term effects of residence on the testosterone responses to fighting in alpha male guinea pigs. *Aggressive Beh.* 10:285–92.

Sapolsky, R. M. 1984. The endocrine stress-response and social status in the wild baboon. *Horm. Beh.* 16:279–92.

Schuurman, T. 1980. Hormonal correlates of agonistic behavior in adult male rats. *Prog. Brain Res.* 53:415–520.

Schwabl, H., J. C. Wingfield, and D. S. Farner. 1980. Seasonal variation in plasma levels of luteinizing hormone and steroid hormones in the European blackbird, *Turdus merula*. *Vogelwarte* 30:283–94.

Silverin, B., and J. C. Wingfield. 1982. Patterns of breeding behaviour and plasma levels of hormones in a free-living population of pied flycatchers, *Ficedula hypoleuca*. *J. Zool. London (A)* 198:117–29.

Towle, A. C., and P. Y. Sze. 1983. Steroid binding to synaptic plasma membrane: Differential binding of glucocorticoids and gonadal steroids. *J. Steroid Biochem.* 18:135–43.

Trobec, R. J., and L. W. Oring. 1972. Effects of testosterone propionate implantation on lek behavior of sharp-tailed grouse. *Am. Midland Nat.* 87:531–36.

Tsutsui, K., and S. Ishii. 1981. Effects of sex steroids on aggressive behavior of adult male Japanese quail. *Gen. Comp. Endocrinol.* 44:480–86.

Wingfield, J. C. 1984a. Environmental and endocrine control of reproduction in the song sparrow, *Melospiza melodia*. I. Temporal organization of the breeding cycle. *Gen. Comp. Endocrinol.* 56:406–16.

———. 1984b. Environmental and endocrine control of reproduction in the song sparrow, *Melospiza melodia*. II. Agonistic interactions as environmental information stimulating secretion of testosterone. *Gen. Comp. Endocrinol.* 56:417–24.

———. 1985a. Environmental and endocrine control of territorial behavior in birds. In *Hormones and the Environment*, ed. B. K. Follett, S. Ishii, and A. Chandola, pp. 265–77. Springer-Verlag.

———. 1985b. Short-term changes in plasma levels of hormones during establishment and defense of a breeding territory in male song sparrows, *Melospiza melodia*. *Horm. Beh.* 19:174–87.

Wingfield, J. C., and D. S. Farner. 1976. Avian endocrinology—field investigations and methods. *Condor* 78:570–73.

———. 1978a. The endocrinology of a naturally breeding population of the white-crowned sparrow (*Zonotrichia leucophrys pugetensis*). *Physiol. Zool.* 51:188–205.

———. 1978b. The annual cycle in plasma irLH and steroid hormones in feral populations of the white-crowned sparrow, *Zonotrichia leucophrys gambelii*. *Biol. Reprod.* 19:1046–56.

———. 1980. Environmental and endocrine control of seasonal reproduction in temperate-zone birds. *Prog. Reprod. Biol.* 5:62–101.

Wingfield, J. C., and M. C. Moore. 1986. Hormonal, social, and environmental factors in the reproductive biology of free-living male birds. In *Psychobiology of Reproductive Behavior: An Evolutionary Perspective*, ed. D. Crews, pp. 149–75. Prentice-Hall.

Wingfield, J. C., A. Newman, G. L. Hunt, and D. S. Farner. 1982. Endocrine aspects of female-female pairing in the western gull (*Larus occidentalis wymani*). *Animal Beh.* 30:9–22.

Wingfield, J. C., and M. Ramenofsky. 1985. Testosterone and aggressive behavior during the reproductive cycle of male birds. In *Neurobiology*, ed. R. Gilles and J. Balthazart, pp. 92–104. Springer-Verlag.

Witschi, E. 1961. Sex and secondary sexual characters. In *Biology and Comparative Physiology of Birds*, ed. A. J. Marshall, pp. 115–68. Academic Press.

Shaping Brain Sexuality

Male plainfin midshipman fish exercise alternative reproductive tactics.
The developmental trade-offs involved shape two brain phenotypes

Andrew H. Bass

Viewed from the perspective of an evolutionary biologist, life is a game whose object is to maximize the number of individuals carrying your genes into subsequent generations. Reproductive strategies vary with the species, but they always represent some kind of trade-off. Some animals try to beat the odds by producing as many offspring as possible, leaving little time or metabolic energy to care for them. Other animals try to maximize the survival rates of those they do produce by having fewer young, but tending to them until they go off on their own. Particular strategies adopted by animals to win this game have evolved as a response to the specific selection pressures on particular species.

Biologists have come to appreciate that selection pressures may act on individuals within a species, so that different individuals of the same sex may employ very different reproductive tactics. This view comes in part from studies of teleost fishes, where males may engage in one of several alternative reproductive tactics. The sequentially hermaphroditic reef fishes, such as wrasses, sea bass, gobies, parrotfish and anemonefish include individuals that can permanently change their sex. Behavioral sex change begins within minutes of a social cue, and complete

sex change can be achieved within days. Other teleosts, such as sunfish, swordtails, platyfish, salmon and the plainfin midshipman—the fish I study in my laboratory—cannot change sex. Rather, in these species, individual males develop into one of two types, or morphs.

For the midshipman, type I males are the larger of the two, and the only morph capable of attracting females.

Type I males build the nests in which females deposit their eggs and attract the females with their almost indefatigable humming, which has earned them the nicknames "California singing fish" and "canary bird fish." To acquire these abilities, type I males take longer to reach sexual maturity than do the second male morph, the type II males.

Type II males may become sexually mature earlier than type I's, but they

Figure 1. Underneath rocks along the intertidal and subtidal zones of the western coast of North America from Canada down to northern California, type I male plainfin midshipman fish build their nests. These fish are teleosts, many species of which manifest unusual reproductive strategies. Some are hermaphrodites, able to alter their sex in a matter of minutes. Others, such as the midshipman, have two forms of males—the nest-building type I, and the "sneaker males," or type II. Type I males also attract the females to the nest, coax them to lay their eggs and guard the nests, as they can be seen doing above and on the following page. Type II males, on the other hand, do not build nests or attract females on their own; they merely sneak into the type I's nests and deposit their sperm. The differing behavior of these two distinct reproductive morphs provides neuroscientists with a rare opportunity to study whether and how behavioral differences translate into differences in the brains of these fish. (All photographs courtesy of Margaret Ann Marchaterre.)

Andrew H. Bass is a professor in and chair of the Department of Neurobiology and Behavior at Cornell University and a research associate at the Bodega Marine Laboratory of the University of California at Davis. He is interested in the evolution of vertebrate brain and behavior with a focus on mechanisms of acoustic communication and reproductive plasticity in teleost fishes. He received his Ph. D. in zoology from the University of Michigan in 1979. Address: Section of Neurobiology and Behavior, Cornell University, Mudd Hall, Ithaca, NY 14853. Internet: ahb3@cornell.edu.

lose something in the bargain. They are smaller and have never been found to build nests or attract females. Their reproductive strategy is to sneak into the Type I's nest or lie perched outside the nest's entrance and deposit their sperm there, earning them the nickname of "sneaker males" or "satellite males."

Having two distinct male forms—each exhibiting distinct behaviors—presents neurobiologists such as myself with a unique opportunity to study a brain-behavior relationship. At the root of this issue is whether behavioral differences translate into differences in the structure and function of

the nervous system. I have found that in fact they do.

The behavioral trade-offs exhibited by the two male morphs essentially reflect the sexual phenotype of the nervous system, which in turn directs the expression of an adult individual's sexual behavior. Armed with this knowledge, my colleagues and I have been exploring the factors that shape the sexual differentiation of the type I and type II brains throughout the animal's sexual development. We have focused our efforts on understanding how the structure and function of the brain might be shaped by early events in de-

velopment that involve trade-offs between individual characters, such as growth rate and age or size at sexual maturity. An understanding of how and why early developmental events lead to alternative phenotypes for individuals within a species provides fertile ground for examining the linkages between neurobiology and behavioral ecology within a modern evolutionary framework.

Nests and Songfests
Along the western coast of North America, from southern Canada into northern California, from late spring

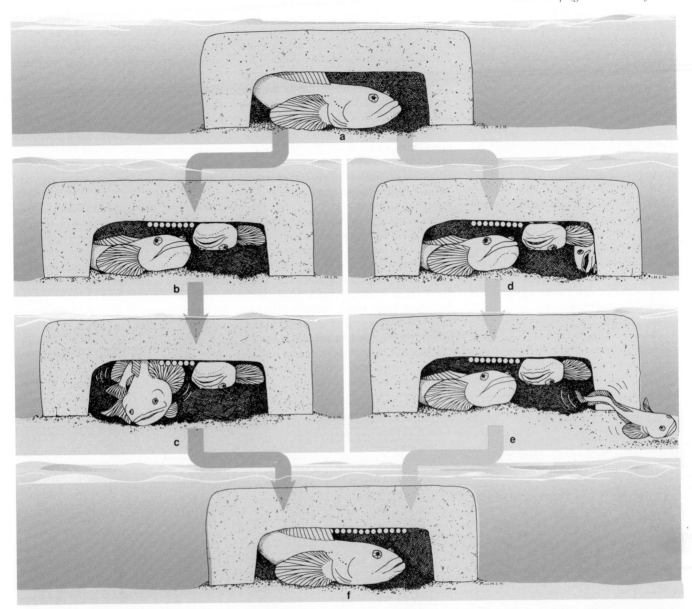

Figure 2. Spawning behavior of type I male plainfin midshipman differs dramatically from that of type II. A type I male generates advertisement calls in the form of low-frequency hums from inside his nest after nightfall (*a*). The male stops his humming after a gravid female enters the nest and they spawn. The female deposits eggs on the roof of the nest (*b*). The type I male rolls and quivers and releases sperm near the eggs (*c*). A variation on this sequence takes place when a type II male enters the picture. If he can get inside the nest, the type II male (*far right*) sneak spawns (*d*). Otherwise, the type II male releases sperm while fanning water towards the nests's opening (*e*). After egg laying is completed, the female leaves the nest, and the type I male remains to guard the eggs (*f*). He hums again the next evening to attract another female, and the sequence begins all over again. This sequence is based on a series of photographs of captive, reproductively active specimens that have taken up residence in artificial nests in aquaria. (Illustration adapted from Brantley and Bass 1994; courtesy of Margaret Nelson.)

through summer, the "song" of the male plainfin midshipman can be heard after nightfall. This song, really a low-frequency hum, may drone on to human ears, but it is highly attractive to female midshipman fish, who seek out the singers and mate with them. The males await the arrival of the females in nests that they build under rocks in the intertidal and subtidal zones. Females deposit their eggs and leave the male to guard the nest soon after spawning.

Not all male midshipman are vocal lotharios or nest builders. As mentioned before, these activities are the sole province of type I males. Richard Brantley, a former graduate student in my lab who is now at Vanderbilt University discovered that type II males exploit the type I's reproductive tactics. As mentioned earlier, they lie perched outside of or sneak into a type I male's nest and shed sperm in a competition with the type I male for the eggs. It is unknown whether type II males remain affiliated with one or several type I nests.

In addition to behavioral differences, each reproductive morph also has a characteristic suite of morphological traits. On average, type I males are about two times longer and eight times heavier than are type II males at the time of sexual maturity. Surprisingly, although they are smaller overall, type II males have the advantage in gonad size. The average ratio of gonad to body weight in type II males is nine times greater than in type I males. Type II males may therefore invest up to 15 percent of their weight in testes, compared with only one percent in type I males.

Gravid females resemble type II males in having a large gonad-to-body weight ratio. Both gravid females and type II males have a distended and firm belly, reflecting the large size of their gonads. In fact, it is easy for the untrained observer to mistake a type II male for a small, gravid female. Adding to the confusion, type II males and females are similarly colored. Although the backs of all three reproductive morphs are olive-gray, the bellies of the fish differ during breeding season. Type I males are typically light to dark gray on the underside, whereas type II males are mottled yellow. Gravid females have a bronze or golden color on their bellies, and spent females are more like type II males in their appearance.

The females, for their part, apparently select only one type I male to mate with each season. The number of eggs per female increases with body size and may approach 200. Each female leaves her entire clutch in the nest of the cho-

sen male. Midshipman have large eggs, approximately 5 millimeters in diameter, which are attached by an adhesive disk to the roof of the nest. Embryos develop upside-down while attached to the yolk sac. After the fry hatch, they remain in the nest. But during their early pre-reproductive months, juveniles aged 5–12 months are found only in eelgrass beds, where adult morphs are also occasionally found. Whereas females apparently mate with only one male per season, type I males, and presumably type II's, mate with several different females. Each nest typically contains several thousand eggs—obviously originating with many different females.

Nesting type I males generate two major classes of vocalizations. They make short grunts, which, along with their large body mass, form an effective threat to any potential intruder males. But the sounds that have been of particular interest to members of my laboratory are the mating sounds, the monotonic hums that can last from minutes to over an hour at a time. Those of us who

study this phenomenon know that these are mating calls by observing the fish in experimental situations. For example, Jessica McKibben, a graduate student currently in my lab, played computer-synthesized acoustic signals that approximate these hums to females, who are then attracted to underwater speakers in outdoor aquariums. Simulated grunts, on the other hand, seem to do nothing for the females. The hum may help females select the best male to mate with, or it may just serve as a beacon for females looking for nest sites. It may also attract the type II males to these sites. Type II males, like females, do not produce hums at all, and only produce the occasional grunt in non-spawning situations.

Just as the differences between the two male types in their acoustic repertoire are quite pronounced, so too are the differences in the level of development of the organs that produce those sounds. The vocal organ of a midshipman consists of a pair of sonic muscles attached to the walls of the swimblad-

Figure 3. Seen from above, the three reproductive morphs—type I males (large fish, *lower right*), type II males (four smaller fish, *left and center*) and females *(topmost fish)*—appear the same olive-gray. The undersides of the fish differ during breeding season. Type I males are typically light to dark gray on the underside, whereas type II males are mottled yellow. Gravid females have a bronze or golden color on their bellies, and spent females are more like type II males in their appearance.

sexually polymorphic traits	type I male	type II male	female
nest building	yes	no	no
egg-guarding	yes	no	no
body size	large	small	intermediate
gonad-size/body-size ratio	small	large	large(gravid), small(spent)
ventral coloration	olive-gray	mottled yellow	bronze (gravid), mottled (spent)
circulating steroids	testosterone, 11-ketotestosterone	testosterone	testosterone, estradiol
vocal behavior	hums, grunt trains	isolated grunts	isolated grunts
vocal muscle	large	small	small
vocal neurons	large	small	small
vocal discharge frequency	high	low	low

Figure 4. Traits of type I and type II males differ markedly. In many respects, type II males more closely resemble females than they do type I males.

der. Contracting the muscles causes the swimbladder to act like a drum—which produces the type I male's low-frequency hums and grunts. One would expect the huge disparity in the vocal capabilities of the two male morphs to be reflected in differences in the sonic musculature. Indeed this is the case.

The ratio of vocal muscle to body weight is six times greater in type I males than in type II males or in females. Furthermore, type I males have four times as many muscle fibers and these are five times larger in diameter than those of type II males and females. Margaret Marchaterre, an electron microscopist in my lab, and I discovered that the disparities are evident even at the subcellular level.

For example, the Z-lines of muscles, the points at which the actin filaments of muscle overlap, which show up under the electron microscope as a dark band, are much wider in the sonic muscles of type I than in type II males or females. The reservoir called the sarcoplasmic reticulum, which contains calcium ions required to mediate muscle-cell activities, is more highly branched in type I males. In addition, the muscles of these males contain vastly higher numbers of the subcellular energy-producing mitochondria than do the muscles of type II males and females. All of this—both the gross appearance of the muscle and the subcellular features—suggests that the sonic muscles of type I males are in every way better equipped than are the muscles of type II males and females

to sustain the continuous singing. (Imagine trying to sing without stopping for over an hour.)

The relative differences in muscle development in the two male morphs are impressive, but there is nothing especially surprising about it given the vocal prowess of the type I males. My interest as a neuroscientist is in determining the neuronal input into this differential behavior. To do that, my colleagues and I looked at the brains of these animals to see whether the difference in muscle size and use in the two male morphs is in any discernible way reflected in their control by nerve cells in the brain.

Brainwork

Before my colleagues and I could start to parse differences in the brains of type I and type II males, we had first to identify the neuronal pathway controlling vocalizations. In this task we were aided by the recent discovery in my lab that a tracer compound called biocytin, which consists of the amino acid lysine and biotin, a naturally occurring protein in neurons, turns out to be perfect for delineating entire circuits in the midshipman's brain. We learned that we could apply biocytin crystals to the cut ends of the motoneurons, the cells that innervate and stimulate the sonic muscles, and the biocytin would be carried backwards from the nerve ending along an axon to its parent cell body. Biocytin also completely filled the arbor of dendrites that extends from the cell body and receives inputs from other neurons. Furthermore, the biocytin did not stop in the first cell it encoun-

tered. Rather, it crossed into the synaptic space between that cell and the end-terminal of the one before it in the circuit. The biocytin travels backwards this way, all the way up to the first cells in the brain that we have so far identified as the sites that initiate vocalizations. Since the biocytin stains the cells brown, my colleagues and I could actually see the entire vocal neuronal circuit required for stimulating the sonic muscle. Biocytin staining confirmed what we had previously found using standard neurophysiological techniques and helped us find new components.

The cell bodies of the sonic motoneurons that stimulate the sonic muscle lie in two sausage-shaped clusters on both sides of the midline of the midshipman's brain, close to the junction of the brain and spinal cord. In adults, about 2,000 cells are found in each cluster, or nucleus. Axons exiting from the sonic motor nucleus bundle together and leave the brain to form the sonic nerve, which stimulates the activity of the sonic muscle.

Robert Baker, at the New York University Medical Center, and I demonstrated that the sonic motoneurons in midshipman receive direct input from a set of pacemaker cells that lie just adjacent to the motoneurons. Each pacemaker neuron connects to motoneurons on both sides of the brain and fire in a constant rhythm, setting the pace at which the sonic motor cells fire. The rhythm set by pacemaker cells corresponds exactly with the rhythm at which sonic motoneurons stimulate the sonic muscle. This in turn determines the frequency at which the muscles contract, which ultimately determines the pitch of the sound.

Our mapping studies with biocytin allowed us to discover another set of neurons in this circuit that had previously been unknown. We found a cluster of cells just in front of the sonic motor nucleus, which we called the ventral medullary neurons. These neurons form the major route connecting the two sides of the pacemaker–motoneuron circuit and so likely make a major contribution to coordinating the activities of both sides of the brain. This eventually leads to the simultaneous contraction of both sonic muscles.

We are just beginning to investigate the sensory stimuli that might activate the vocal motor system. One obvious candidate would be activation of this system by a neighboring midshipman's vocal signal. The inner ear of all vertebrates has a number of divisions. In the midshipman, the largest one is known as the sacculus, which is considered to be the main organ of hearing. It includes a palette of sensory cells linked by the eighth nerve to neurons in the hindbrain, which are the first in a chain of neurons forming a central auditory pathway that extends through all levels of the brain. Deana Bodnar, a research associate in my lab, has recently identified neurons in a midbrain auditory nucleus that encode information that could be used to recognize differences between the hums of neighboring males. This discovery, together with other anatomical data collected from our biocytin studies, suggest that midbrain auditory neurons along with neurons of the paraventricular and tegmental nuclei may form a vocal-acoustic network, which provides a circuit for vocalizations to be elicited by the sounds of neighboring midshipman.

Having worked out the entire circuit, members of my lab group were in a good position to make comparisons of the brains of the different morphs to determine whether there were any obvious differences between them. The first thing we discovered was that male morphs and females possess identical circuitry. We also found that they have the same ratio of nerve cells to body weight, so any differences in behavior could not be due to the number of cells.

We did find, however, that the pacemaker-motoneuron circuit in type I males fires at a frequency that is about 15 to 20 percent higher compared with type II males and females. This parallels sex differences in the frequency of natural vocalizations. We also found that the cell bodies, dendrites and axons are one to three times larger in type I males than in females and type II males. The junction between the nerve and muscle is also larger in type I males. Therefore differences within and between sexes in the organization of the vocal motor system depend upon a divergence in the morphological and physiological properties of individual nerve cells. It seems likely that the larger cells of the type I male are specifically adapted to fire more frequently and without attenuation for a longer period to support the activity of their much enlarged sonic muscle during prespawning periods of singing.

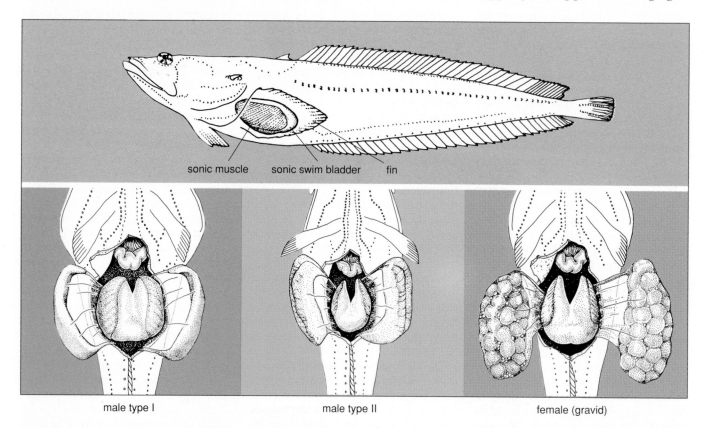

sonic muscle   sonic swim bladder   fin

male type I   male type II   female (gravid)

Figure 5. To attract females to their nests, type I males perform low-frequency hums throughout the night during breeding season. The vocal organ of the midshipman fish is a pair of sonic muscles attached to the walls of the swimbladder (top). Contracting the sonic muscles makes the swimbladder act as a kind of drum. Sonic muscles of type I males (bottom left) are extremely well developed in comparison with muscles from type II males (bottom center) or females (bottom right). The ratio of sonic muscle to body weight is six times greater in type I males than in the other reproductive morphs. In contrast, the gonad to body weight ratio is nine times larger in type II males and 20 times larger in gravid females than in type I males and juveniles. (Illustration courtesy of Margaret Nelson.)

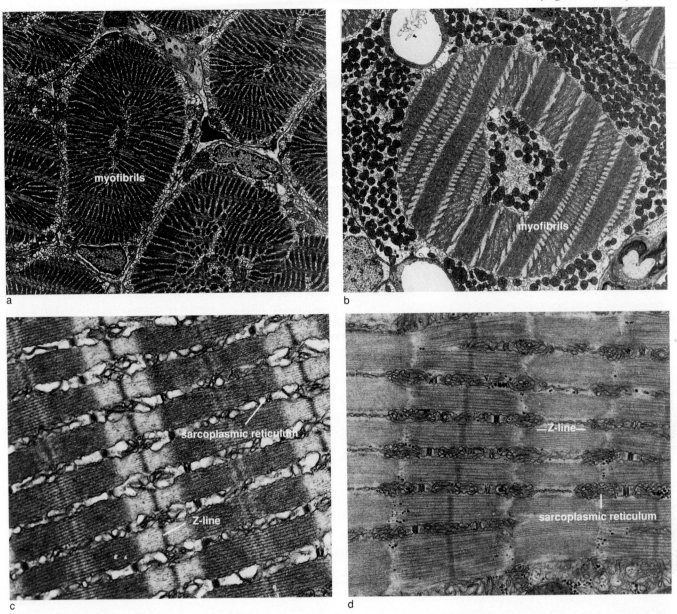

Figure 6. Morph-related differences in sonic muscles can be seen clearly in the electron microscope. In cross section, sonic muscles from type II males (*a*) and type I males (*b*) appear tubular. The myofibrils of juveniles, females and adult type II males are densely packed, surrounded by a thin rind of sarcoplasm, which contains the muscle-cell nuclei and a few mitochondria, which mediate muscle activity. Type I males have fibers with an inner doughnut-shaped core of myofibrils bordered by a large volume of sarcoplasm densely filled with mitochondria. When viewed in longitudinal section, additional differences between muscle fibers of type II (*c*) and type I males (*d*) become apparent. The width of Z-lines, the point at which the actin filaments of muscle overlap, and the degree of branching of sarcoplasmic reticulum are greater in magnitude in type I males. Females and juveniles resemble type II males. The bar scale is 6 micrometers in panels a and b, 0.8 micrometers in panels c and d. (Electron micrographs courtesy of the author; adapted from Bass and Marchaterre 1989.)

Getting Big

The studies performed by my colleagues and I thus suggest that alternative mating tactics among sexually mature males are paralleled by alternative phenotypes for the neurons in the relevant circuit. My colleagues and I wanted to learn the origins of these differences. We could envision at least two distinct scenarios.

In the first, juvenile males may transform into either type I or type II males. That is, both types of males fol-low mutually exclusive, nonoverlapping growth patterns. The other possibility is that the smaller type II males eventually change into type I males. To determine which of these was happening, my coworkers and I followed the development of these fish in the early juvenile stages just prior to and on into sexual maturity.

We exploited the properties of biocytin to map and trace the development of the neuronal vocal circuit in juvenile midshipman. These studies revealed that for type I males sexual maturation is preceded by growth of the mate-calling circuit and the sonic muscle. The size of motoneurons and the volume of the entire sonic motor nucleus, which likely reflects the growth of motoneuron dendrites as well, increase twofold just prior to the type I's sexual maturation. At the same time, the number of sonic muscle fibers increases by four.

With the onset of sexual maturity, the type I male experiences an additional, albeit more modest, growth in the size of motoneurons. This, however, is coupled with a large increase in the size of pacemaker neurons although the size of this increase does not compare to that of motoneurons. At this time, the sonic muscle undergoes its major expansion—a remarkable fivefold increase in the size of the muscle fibers, which accounts mostly for the large increase in sonic-muscle weight in type I males. It was surpising to see that the largest increase in muscle growth followed, rather than coincided with, the largest growth in the neural circuit. The neurons in the ventral medulla show similar growth increments during both stages.

In contrast, the transformation from juvenile to type II male or to adult female is not accompanied by the dramatic changes seen for type I males in the sizes of vocal neurons and muscle cells. In fact, these cells change little or not at all as type II males and females mature. The sum of these findings suggests that type I males and type II males and females grow along alternative growth trajectories, at least as it con-

cerns the neurons and muscles that determine morph-specific vocal behaviors.

As a next step in our research, my colleagues and I were interested in learning the rate at which each reproductive morph achieved sexual maturity. To do this, we had to be able to determine the precise age of the individual fish we studied, which turned out to be surprisingly easy.

The sacculus division of the inner ear of teleost fishes contains a structure called an otolith, which is mainly composed of calcium carbonate. As a fish grows, new layers of calcium carbonate are added to the otolith. These layers of calcium carbonate can be read, like the rings of a tree, to determine the individual's age. By reading the growth increments in midshipman otoliths, Ed Brothers of EFS Consultants in Ithaca, New York, and I have shown that type I and type II males not only overlap in age, but type II's become sexually mature earlier than do type I males. This finding, we believe, adds support for our hypothesis that alternative male

morphs in midshipman fish adopt non-sequential, mutually exclusive growth patterns during their first year of life.

An Early Start
At this point in our research, we were reasonably convinced that the path leading to one or another male morph was set very early in the animal's development. We know that for fish, as for all vertebrates, hormones are necessary for the development of secondary sexual characteristics. We were interested to see whether differences between type I and type II males relating to their reproductive strategies might be reflected in hormonal differences acting during maturation.

A number of studies have indicated that the hormonal cascade leading to sexual maturation is initiated in a part of the forebrain known as the preoptic area. Neurons in this region release a neurochemical called gonadotropin-releasing hormone (GnRH). GnRH is a 10 amino-acid-long peptide that has been identified in a wide range of vertebrate

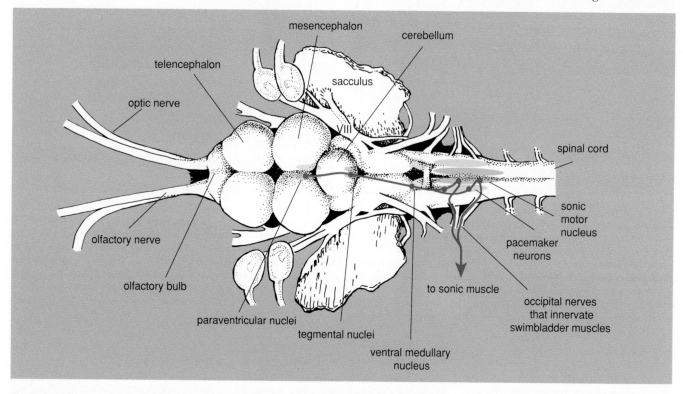

Figure 7. Differences in the vocal-motor circuit in the brain controlling vocalization parallel behavioral differences between the morphs. Here a schematic illustrates the entire circuit beginning in the brain. The initial stimulation comes from sites in the mesencephalon and hindbrain called paraventricular and tegmental nuclei. These stimulate ventral medullary neurons, which in turn stimulate pacemaker neurons. Pacemaker neurons fire at a set frequency equal to the one at which the sonic motoneurons fire and the sonic muscle contracts. In effect, the pacemaker neurons determine the pitch of sounds made by midshipman fish. The axons of the sonic motoneurons bundle together and eventually form the sonic nerve, after exiting the brain via paired occipital nerve roots. The sonic nerve stimulates the sonic muscle on that side of the body. The activity of the two sonic nerves emerging from each side of the brain are coordinated by ventral medullary and pacemaker neurons. In this way, both sonic muscles contract synchronously. The circuit in the three reproductive morphs is the same as is the ratio of neurons relative to body weight. The neurons in the circuit of type I males have larger cell bodies, dendrites and axons than do those in the other two morphs. These features likely help the neurons of type I males fire frequently without fatigue or attenuation. (Adapted from an illustration by Margaret Nelson.)

species. Working with Dean and Tami Myers, Cornell colleagues now at the University of Oklahoma, our lab showed that the gene sequence encoding the GnRH decapeptide in midshipman is remarkably similar to the GnRH coding sequence in other fishes, amphibians, birds and mammals, including people. This finding argues for a highly conserved function for the GnRH peptide among all vertebrates.

GnRH stimulates another set of cells in the pituitary gland, a structure found at the base of the brain. Once stimulated, the pituitary releases a family of hormones called gonadotropins. Gonadotropins act directly on the developing gonads, be they female or male. The gonads, in turn, release steroid hormones—androgens, such as testosterones and estrogens—that stimulate the development of secondary sexual characteristics. In midshipman, one of the effects of the steroid hormones is to mediate the development of the vocal motor system.

One of our first projects was to compare the steroid hormones produced by the three reproductive morphs. Working with Brantley and John Wingfield from the University of Washington, we found that the different morphs did indeed produce different levels of the various hormones.

Testosterone is detectable in all three morphs, although at progressively lower levels, with type II males producing the highest levels followed by females. Type I males produce the lowest amounts. Estrogen (in the form of 17β-estradiol) is detectable only in females, but at much lower concentrations than testosterone.

In addition to these common steroids, teleosts produce a unique form of testosterone known as 11-ketotestosterone. On average, type I males have five times as much 11-ketotestosterone as they do testosterone; 11-ketotestosterone is undetectable in type II males and females. This hormonal distribution is similar in all teleosts with two distinct male morphs. It seems likely, then, that 11-ketotestosterone is more potent than testosterone in supporting courtship behaviors, such as humming.

We were also interested in knowing whether hormonal differences could explain why type II males reach sexual maturity earlier than type I males. Martin Schreibman and his colleagues at Brooklyn College first showed in platyfish, which also have two male morphs that differ in age at sexual maturity, that the fish start to become sexually mature when the GnRH cascade is initiated. Our studies of midshipman sexual development have yielded similar results. Matthew Grober, a former postdoctoral associate in my lab and now at Arizona State University, led a study that found that the number and size of neurons releasing GnRH increase as the animal is making the transition from juvenile to adult. At the point of sexual maturity, this region is equally as developed in all three morphs. That is, it is not the case that the more sexually precocious morphs have better-developed hormonal circuits. The difference, as indicated by otolith-aging studies, is that the cascade is turned on at least three to four months earlier in type II males and females than in type I males.

The question that remains for us to answer is whether these differences are genetic or environmental. When I began these studies, I believed firmly that the difference between type I and type II males was programmed into their genes. But recent studies with Christy Foran, a current graduate student in my laboratory, are casting some doubt on that assertion. We have some preliminary experimental data to suggest that the number of type II males produced is a function of population density. Our work suggests that under sparsely populated conditions, more type I males are produced. As the population density increases, so too does the percentage of type II males.

Alternative Male Morphs: Trade-Offs

Our work suggests that type II males reach sexual maturity earlier than type I males, but they remain physically and behaviorally immature with regard to their ability to vocalize. On the other

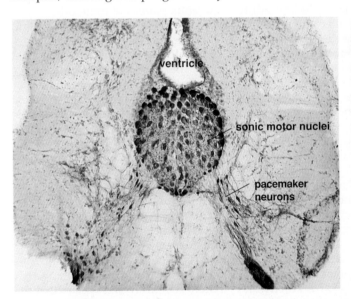

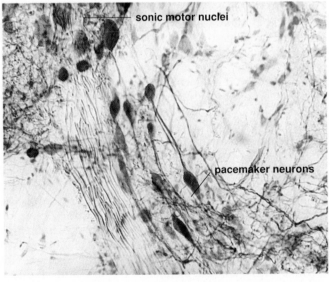

Figure 8. Neuronal circuitry of the vocal motor system can be traced using biocytin, a compound made of the amino acid lysine and the protein biotin. Biocytin crystals are applied to the cut ends of the axons innervating the sonic muscle. The compound is carried backwards to the cell bodies of these neurons, and further backwards still into the nerve cell innervating that one and on back to the first nerve cells in the circuit. In a low-power photomicrograph of the circuit in a type I male *(left)*, sonic motor nuclei, which contain the cell bodies and dendrites of the sonic motoneurons, appear brown when stained with biocytin. Also shown are the pacemaker neurons, which lie along the sonic motor nuclei. Biocytin transport results in extensive filling of the cell bodies, dendrites and axons of vocal pacemaker neurons, as shown in the right panel. (Photomicrographs courtesy of the author.)

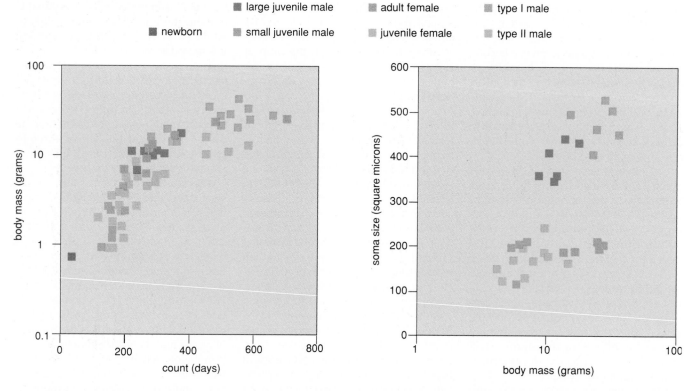

Figure 9. Alternative male morphs represent nonsequential, mutually exclusive life-history tactics, according to studies of their developmental trajectories. The age of fish can be determined by counting the growth increments on otoliths, calcium carbonate structures in the ear. The left graph includes data from 72 individuals representative of all juvenile and adult morphs. It is clear that type II males and females can reach sexual maturity earlier than do type I males and that one type does not change into the other. The right graph is a scatter plot of average values for the size of cell bodies of sonic motoneurons of juvenile and adult morphs. The data indicate alternate growth trajectories for the neurons that determine morph-specific vocal behaviors. This further supports the notion that type I and II males adopt mutually exclusive patterns of growth during their first year of life. (Adapted from Bass *et al.*, in press.)

hand, type I males delay maturation, but have a fully developed sexual behavioral repertoire.

In his book *Ontogeny and Phylogeny*, Stephen Jay Gould reviewed the extensive literature suggesting that the dissociation in time between sexual and physical maturity, referred to as heterochrony, is often characteristic of speciation events. But from our work, we see that heterochrony may also lead to behavioral innovation within a species. In a 1986 review article in the *Proceedings of the National Academy of Sciences*, Mary Jane West-Eberhard proposed a switch mechanism giving rise to alternative phenotypes within a species.

The evolutionary significance of a switch is that it determines which of an array of potential phenotypes will be expressed and, therefore, exposed to selection in a particular timespan and context. Insofar as one set of characters is independently expressed relative to another, it is independently molded by selection. Therefore, different covariant character sets evolve semi-independently, taking on different forms in accord with their different functions. Like juvenile and adult forms, different alternative phenotypes of the same species may show dramatic differences in morphology, behavior and ecological niche. This is possible because once a switch mechanism is established, contrasting phenotypes can evolve simultaneously within the same genome—without reproductive isolation between forms.

It seems likely that the type II male morph developed under conditions of intense sexual selection, namely competition between males for access to females and nest sites. The switch mechanism is associated with a trade-off among midshipman males in the age and size at sexual maturity, as well as a multidimensional suite of secondary sexual characteristics. Thus whereas type I and II males share gonadal sex, they are highly divergent in behavioral, cellular, hormonal and vocal-motor traits. The convergence or monomorphism in behavioral and physical traits between type II males and females reflects a common developmental pattern of trade-offs.

This implies that the type I male morph represents the ancestral behavioral state for male plainfin midshipman. This is supported by the available comparative data that show that other species of midshipman and their closest phylogenetic relative, the toadfish, have a single male reproductive phenotype resembling type I males. Therefore the most parsimonious conclusion is that the type II male morph represents a derived character state for this group of teleosts.

The close temporal onset of sexual maturity in type II male and female midshipman resembles many other teleosts. Graham Bell at McGill University shows in a review of approximately 100 freshwater species that the age at sexual maturity in males is close to or less than that of females. The exceptions are in species where males, like the type I midshipman morph, guard a territory or suitable substrate for a female's eggs. By adopting early maturity and thus an

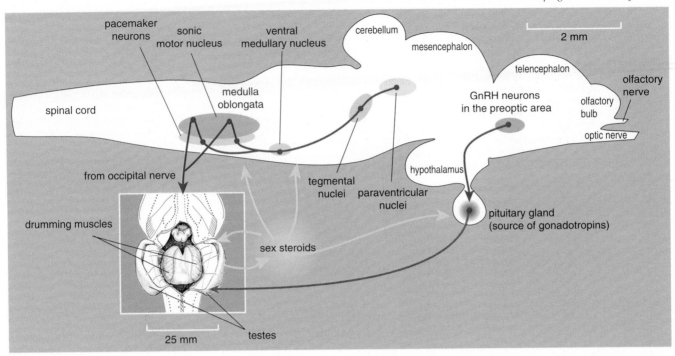

Figure 10. Proposed cascade of hormonal events leading to the expression of type I male, type II male and female reproductive morphs is shown in this schematic. Gonadotropin-releasing hormone (GnRH) is released from a region in the forebrain. GnRH stimulates the pituitary gland, at the underside of the midshipman brain, to release gonadotropins. Gonadotropins stimulate the gonads to release steroid hormones, such as testosterone. These steroids stimulate the differential development of secondary sexual characteristics, such as the ability to vocalize and hence the neuronal circuitry of the vocal motor system. Type I males have a type of testosterone, called 11-ketotestosterone, not found in the other two morphs. This hormone might be especially potent in the development of the sonic musculature and the neural circuitry. The hormonal cascade leading to sexual maturity is initiated three to four months earlier in sexually precocious type II males and females than it is in type I males. The contributions of ecological, behavioral and genetic factors to the activation of a GnRH-gate remain to be defined.

increased chance of surviving to become reproductively active, type II male midshipman essentially forgo direct competition with type I males to establish nest sites. Their investment in gonad development and in an expanded hormonal system is likely to be among the principal costs associated with early maturation.

Type I male midshipman derive at least two major growth-related benefits from delayed maturity. The first is increased body size, which would provide an advantage in combat and nest competition. The second is hypertrophy of a vocal motor system that generates vocalizations important, at the very least, for female attraction to nest sites and probably useful for intimidating other males. The principal costs for type I males in delaying maturity are the considerable structural and metabolic investment in physical and behavioral development and an increased chance of *not* surviving to sexual maturity.

Michael Taborsky at the Konrad Lorenz Institute in Vienna, Austria, points to external fertilization as a trait favoring the evolution of alternative reproductive tactics and morphs among teleosts because it "makes it difficult for male fishes to monopolize access to fertilizable eggs." The latter leads to males trying to "control preferential access to fertilizable eggs," as exemplified by the nest-building and vocal courtship tactic of type I males who alone guard substrate-attached eggs. (Egg-guarding would also incur additional costs to a type I male tactic.) Intense competition among males for nest sites, and thus access to a female's eggs, would then be considered a major selective force favoring the adoption of alternative type II-like male reproductive tactics.

Usually, when theorists consider the effects of evolution on an animal's physical traits, they consider body size or color, which in fact do differ between the reproductive morphs of the midshipman. My colleagues and I have considered how evolution also affects the neural substrates of display behaviors, which are also part of an animal's life history. Our studies of alternative phenotypes in midshipman fish show that sexuality for each reproductive morph or for that matter, each individual, can be defined by developmental, sexual maturity-dependent trade-offs between suites of species-typical traits. As we have seen, these trade-offs shape the brain and its behavioral sex, and provide an important link between neurobiology and behavioral ecology.

Acknowledgments

The author is indebted to the collaboration of many colleagues at Cornell University and other institutions, as well as the generous financial support of Cornell University, the University of California's Bodega Marine Laboratory and the National Science Foundation, all of whom helped make this research possible. Special thanks to Peter Klimley for encouragement to write this essay, to Margaret Ann Marchaterre for the field photography and to Deana Bodnar, Matthew Grober, Rosemary Knapp and Margaret Ann Marchaterre for help with the manuscript.

Bibliography

Bass, A. H. 1990. Sounds from the intertidal zone: Vocalizing fish. *Bioscience* 40:249–258.

Bass, A. H. 1992. Dimorphic male brains and alternative reproductive tactics in a vocalizing fish. *Trends in Neurosciences* 15:139–145.

Bass, A. H. 1995. Alternative life history strategies and dimorphic males in an acoustic communication system. *Fifth International Symposium on Reproductive Physiology in Fish*, pp. 258–260.

Bass, A. H., B. J. Horvath and E. B. Brothers. In press. Non-sequential developmental trajectories lead to dimorphic vocal circuitry for males with alternative reproductive tactics. *Journal of Neurobiology.*

Bass, A. H., and M. A. Marchaterre. 1989. Sound-generating (sonic) motor system in a teleost fish (*Porichthys notatus*): Sexual polymorphism in the ultrastructure of myofibrils. *Journal of Comparative Neurology* 286:141–153.

Bass, A. H., M. A. Marchaterre and R. Baker. 1994. Vocal-acoustic pathways in a teleost fish. *Journal of Neuroscience* 14:4025–4039.

Bell, G. 1980. The costs of reproduction and their consequences. *American Naturalist* 116:45–76.

Bodnar, D. and A. H. Bass. 1996. The coding of concurrent signals (beats) within the auditory midbrain of a sound producing fish, the plainfin midshipman. *Association for Research in Otolaryngology.*

Brantley, R. K., and A. H. Bass. 1994. Alternative male spawning tactics and acoustic signalling in the plainfin midshipman fish, *Porichthys notatus*. *Ethology* 96:213–232.

Brantley, R. K., J. Tseng and A. H. Bass. 1993. The ontogeny of inter- and intrasexual vocal muscle dimorphisms in a sound-producing fish. *Brain, Behavior and Evolution* 42:336–349.

Brantley, R. K., M. A. Marchaterre and A. H. Bass. 1993. Androgen effects on vocal muscle structure in a teleost fish with inter and intrasexual dimorphisms. *Journal of Morphology* 216:305–318.

Brantley, R. K., J. Wingfield and A. H. Bass. 1993. Hormonal bases for male teleost dimorphisms: Sex steroid levels in *Porichthys notatus*, a fish with alternative reproductive tactics. *Hormones and Behavior* 27:332–347.

Caro, T. M. and P. Bateson. 1986. Organization and ontogeny of alternative tactics. *Animal Behaviour* 34:1483–1499.

Crews, D. 1987. Animal sexuality. *Scientific American* (January) 106–114.

DeMartini, E. E. 1990. Annual variations in fecundity, egg size and condition of the plainfin midshipman (*Porichthys notatus*). *Copeia* 3:850–855.

Emlen, S. T., and L. W. Oring. 1977. Ecology, sexual selection and the evoluton of mating systems. *Science* 197:215–223.

Fine, M. L., H. Winn and B. L. Olla 1977. Communication in fishes. In *How Animals Communicate*, ed. T. Seboek. Bloomington, Indiana: Indiana University Press, pp. 472–518.

Gould, S. J. 1977. *Ontogeny and Phylogeny.* Cambridge, Mass.: Belknap Press.

Grober, M. S., S. Fox, C. Laughlin and A. H. Bass. 1994. GnRh cell size and number in a teleost fish with two male reproductive morphs: Sexual maturation, final sexual status and body size allometry. *Brain, Behavior and Evolution* 43:61–78.

Grober, M. S., T. R. Myers, M. A. Marchaterre, A. H. Bass and D. A. Myers. 1995. Structure, localization and molecular phylogeny of a GnRH cDNA from a paracanthopterygian fish, the plainfin midshipman (*Porchthys no-*

tatus). *General and Comparative Endocrinology* 99: 85–99.

Gross, M. R. 1996. Alternative reproductive strategies and tactics: Diversity within sexes. *Trends in Ecology and Evolution* 11:92–98.

Gross, M. R., and R. C. Sargent. 1985. The evolution of male and female parental care in fishes. *American Zoologist* 25:807–822.

Halpern–Sebold, L. R., M. P. Schreibman and H. Margolis-Nunno. 1986. Differences between early- and late-maturing genotpyes of the platyfish (*Xiphophorus maculatus*) in the morphometry of their immunoreactive luteinizing hormone releasing hormone-containing cells. A developmental study. *Journal of Experimental Zoology* 240:245–257.

Ibara, R. M., L. T. Penny, A. W. Ebeling, G. van Dykhuizen and G. Cailliet. 1983. The mating call of the plainfin midshipman fish, *Porichthys notatus*. In *Predators and Prey in Fishes*, eds. D. L. G. Noakes et al. The Hague, The Netherlands: Dr. W. Junk Publishers, pp. 205–212.

Kelly, D. B. 1988. Sexually dimorphic behaviors. *Annual Review of Neuroscience* 11:225–251.

McKibben J., D. Bodnar and A. H. Bass. 1995. Everybody's humming but is anybody listening: Acoustic communication in a marine teleost fish. *Fourth International Congress of Neuroethology*, p. 351.

Moore, M. C. 1991. Application of organization-activation theory to alternative male reproductive strategies: a review. *Hormones and Behavior* 25:154–179.

Shapiro, D. Y. 1992. Plasticity of gonadal development and protandry in fishes. *Journal of Experimental Zoology* 261:194–203.

Thresher, R. E. 1984. *Reproduction in Reef Fishes.* Neptune, New Jersey: T. F. H. Publications, Ltd.

Stearns, S. C. 1992. *The Evolution of Life Histories.* New York:Oxford University Press.

Taborsky, M. 1994. Sneakers, satellites, and helpers: Parasitic and cooperative behavior in fish reproduction. In *Advances in the Study of Behavior*, volume 23, ed. P. J. B. Slater, J. S. Rosenblatt, C. T. Snowdon and M. Milinski. New York: Academic Press, pp. 1–100.

Walsh, P. J., T. P. Mommsen and A. H. Bass. 1995. Biochemical and molecular aspects of singing in batrachoidid fishes. In *Biochemistry and Molecular Biology of Fishes*, volume 4, ed. P. W. Hochachka and T. P. Mommsen. Amsterdam: Elsevier, pp. 279–289.

Warner, R. R. 1984. Mating behavior and hermaphroditism in coral reef fishes. *American Scientist* 72:128–136.

West-Eberhard, M. J. 1986. Alternative adaptations, speciation, and phylogeny. *Proceedings of the National Academy of Sciences* 83:1388–1392.

Aerial Defense Tactics of Flying Insects

Preyed upon by echolocating bats, some night-flying insects have developed acrobatic countermeasures to evade capture

Mike May

Walking home late one summer night, I glimpsed a small mass slip through the air, past the halo of a street lamp, and into the darkness. Although I was fatigued by a long day, my curiosity was piqued; I crouched down in the darkness and waited for another sign of movement. Within minutes the elusive flyer returned, swerving momentarily in the light, and then shooting back into the night. It was a bat—apparently foraging for its nightly meal of insects. Soon there were others, darting and weaving by the lamp as they attempted to scoop up the insects attracted to the light. As I watched the aerial display, I was impressed by the remarkable speed at which a bat could change its flight path. I tossed a few pebbles into the air and watched as the bats easily pursued the decoys, but turned away when the deception became apparent. Surely, I thought, there was little hope for an insect once a bat had homed in on it.

I gave the matter no more thought until several years later, when I began my doctoral research—perhaps not coincidentally concerned with the flying abilities of insects. As a graduate stu-

Mike May is a free-lance science writer. He acquired a taste for the breadth of biology as an undergraduate at Earlham College in Richmond, Indiana. While completing an M.S. in biological engineering at the University of Connecticut at Storrs, he discovered some electronic answers to biological questions. After pursuing bicycle mechanics for a year, he returned to biology and earned a Ph.D. as a biomechanic at Cornell University. Address: P.O. Box 141, Etna, New York 13062.

dent I learned there is a considerable history to the study of the aerial encounters between bat and insect. It proves to be a story with a number of surprising turns, and it begins almost 200 years ago with the discovery that bats use their ears, and not their eyes, to navigate.

Lazaro Spallanzani, an 18th-century pioneer of experimental biology, showed that blinded bats are not only able to avoid obstacles in their flight path—such as fine silk threads—but are also able to snag insects in mid-flight. After hearing of Spallanzani's research, Charles Jurine, a surgeon and entomologist, demonstrated that when the bats' ears are plugged, the animals collide with even relatively large objects in their path, and they are incapable of catching insects.

For over a century the observations of Spallanzani and Jurine were not widely accepted, primarily because no one could imagine how it was that a bat could hear the precise location of such small, essentially silent objects. No advance was made in understanding "Spallanzani's bat problem" until 1920, when the English physiologist H. Hartridge suggested that bats might somehow use sounds of very high frequency to detect the objects. Perhaps the frequencies might even extend beyond the upper limit of human hearing—about 20 kilohertz—to the part of the spectrum called ultrasound.

The mystery of bat navigation was ultimately solved by a Harvard undergraduate, Donald Griffin, in collaboration with the Harvard physicist G. W. Pierce—who invented a device that could detect ultrasound—and the Harvard physiologist Robert Galambos. In

1938 Pierce and Griffin pointed a "sonic detector" at bats flying in a room and found that the animals were, in fact, emitting signals at ultrasonic frequencies. Griffin and Galambos later showed that bats emit ultrasonic cries from their mouths and use their ears to detect the echoes of the sounds reflected from objects in their flight paths. Griffin called this process of navigation *echolocation*.

Echolocation turns out to be an extremely precise and effective method by which bats navigate and identify objects in the dark. In the early 1980s Hans-Ulrich Schnitzler and his colleagues at the Institute for Biology in Tübingen, and Nobuo Suga of Washington University, found that bats are able to analyze the ultrasonic echoes reflected from the bodies and wings of flying insects in such a way as to determine not only the location but also the speed and, perhaps, the type of insect that produces the echoes. All the evidence suggests that the echolocating bat is a very sophisticated hunter; not only is it an adept flyer, but it is equipped with a sensitive auditory system designed to locate and identify potential targets.

However, the bat's ability to find and capture a flying insect is just one side of the story. Some flying insects are able to detect the ultrasonic cries of a bat and take evasive action. Flying insects pursued by a bat do not follow simple ballistic trajectories; they are not such easy targets. To the contrary, the encounter between a bat and an insect is one that might rival the tactics of modern air-to-air combat, involving an efficient early-warning system, some clever aerodynamic engineering and

Figure 1. Aerial encounter between an insect-eating bat and a green lacewing reveals one of the evasive maneuvers—a passive dive—that an insect will use to escape a bat. Hunting bats locate their prey by emitting high-frequency (ultrasonic) cries and detecting the echoes of the sounds reflected from the insect's wings and body—a system of navigation called echolocation. Some insects are able to hear the bat's high-frequency sounds and can respond with rapid changes in their flight trajectories. In this stop-motion photograph, a stroboscopic flash reveals the relative positions of the insect and the bat—at intervals of less than one-tenth of a second—as they move from left to right in the scene. (Photograph courtesy of Lee Miller, Odense University, Denmark.)

the simple economics of making do with what is available.

Dodging a Speeding Bat
Almost 70 years before Griffin and Galambos demonstrated that bats can locate objects with ultrasound, F. Buchanan White of Perth, Scotland, proposed that moths can detect bats through the sense of hearing. Although White had no evidence for this conjecture, his idea was ultimately confirmed by behavioral studies in the 1950s, and especially by the work of Kenneth Roeder of Tufts University in the early 1960s. Roeder made hundreds of long-exposure photographs of

free-flying moths and recorded their aerial maneuvers in response to a stationary source of artificial ultrasound. He found that if the moths were more than 10 feet from the source of the ultrasound, they simply turned away. But if they were closer to the sound, the moths performed a variety of acrobatic maneuvers, including rapid turns, power dives, looping dives and spirals. For a more natural touch, Roeder photographed wild bats attacking the flying moths; clearly visible in the photographs is the track of the bat zipping across the scene and the evasive path of the moth as it escapes the attack—sometimes.

Since Roeder's studies, a number of nocturnal flying insects have been found to perform evasive aerial maneuvers in response to the ultrasonic cries of bats. In 1979 Lee Miller and Jens Olesen of Odense University found that hunting bats or artificial ultrasonic pulses induce erratic flight in free-flying green lacewings. More recently, David Yager of Cornell University, Brock Fenton of York University in Toronto, and I have shown that some species of praying mantis will perform different types of escape procedures depending on the loudness of the ultrasound emission—not unlike Roeder's moths. In these experiments we

Figure 2. Failure to escape results in death by devourment for a green lacewing performing a diving arc into the embracing wings and tail membrane of an approaching bat. The wings of the captured insect can be seen in the mouth of the bat as it descends to the right. (Photograph courtesy of Lee Miller, Odense University, Denmark.)

used an artificial source of ultrasound—sportively called a "batgun"—with which we "shot" free-flying mantises. At distances greater than 10 meters, most of the mantises did not respond to the ultrasound. Within seven to nine meters of the batgun, however, the mantises would make a slight turn or a shallow dive. At still closer range—within five meters—the mantises would perform steep dives ranging from 45 degrees to nearly a vertical drop, occasionally even in a spiral. Just as Roeder had found in his studies of the moths, we found that the praying mantis will make its most drastic evasive maneuvers when the ultrasound is loudest.

Artificial ultrasound has also been shown to induce changes in the flying patterns of other insects. Daniel Robert, now at Cornell University, found that flying locusts respond to ultrasonic

pulses by steering away from the source of the sound and by increasing the rate at which they beat their wings. Similarly, Hayward Spangler of the Carl Hayden Bee Research Center in Tucson showed that, immediately after hearing artificial ultrasonic pulses, tiger beetles fly toward the ground and land. Frederic Libersat, now at the Hebrew University in Jerusalem, and Ronald Hoy of Cornell University discovered that a tethered, flying katydid will stop flying immediately after hearing an ultrasonic stimulus, suggesting that it would perform a dive. Although no one has reported interactions between these insects and bats, it seems likely that such bat-avoidance responses will be found in many night-flying insects that hear ultrasound.

Certainly the value of a rapid escape mechanism for the survival of a flying insect is no longer in doubt. Roeder's

studies showed that moths that dive in response to an ultrasonic stimulus were 40 percent less likely to be captured by a bat. The green lacewings, studied by Miller and Olesen, were even more successful at escaping bats—being captured only 30 percent of the time. On the other hand, deafened green lacewings were captured about 90 percent of the time.

More recently, Yager and members of Fenton's research group performed a series of field experiments in which they exposed two species of praying mantis to wild, hunting bats. One species, *Parasphendale agrionina*, makes rapid changes in its flight path in response to artificial pulses of ultrasound or in response to hunting bats. The other species, *Miomantis paykullii*, is an excellent flyer, but does not change its flying pattern in response to artificial ultrasound or in response to hunting bats. Of five attacks on *P. agrionina* in which the mantises performed evasive maneuvers the insects successfully escaped the hunting bats in every case. In contrast, during three attacks on *M. paykullii* and three attacks on *P. agrionina* in which neither species performed evasive maneuvers the insects were captured in five of the six cases. These experiments provide strong evidence that ultrasonic hearing and rapid changes in trajectory can help a flying insect evade an attack by a predatory bat.

Mating, Death and Phonotaxis

It seems clear that some insects are able to detect the ultrasonic pulses emitted by bats, and then use this information as an early warning system—much the same way a combat pilot in a fighter plane might detect the radar of an enemy plane. And like the combat pilot, the targeted insect must perform evasive maneuvers or suffer the consequences of being captured. In the case of the fighter pilot, however, we know the physics and the engineering behind the detection of radar, and the aerodynamics of flight maneuvers is the stuff of textbooks in flight school. But how does an insect do it? How does it detect the ultrasound, convert this signal into a message that says "take evasive action," and then perform its spectacular acrobatic maneuvers? We don't as yet know all the answers, but bits and pieces of the story are coming to light.

Part of the answer lies in the behavior known as phonotaxis, the movement of an animal in a direction determined by the location of a sound source. Phonotactic behavior of insects has been especially well studied in certain species of crickets. Phonotaxis takes two forms in these animals, based on the direction the cricket moves with respect to the sound source. When a female cricket moves toward the source of the calling song of a courting male cricket—the familiar chirp we hear on summer nights—the locomotory behavior of the female is described as positive phonotaxis. On the other hand, the same female will respond to another sound, of a higher frequency, by flying away from the source—a display of negative phonotaxis. Positive and negative phonotaxis in these instances suggest that the cricket is able to discern at least two distinct aspects of the sound: its fre-

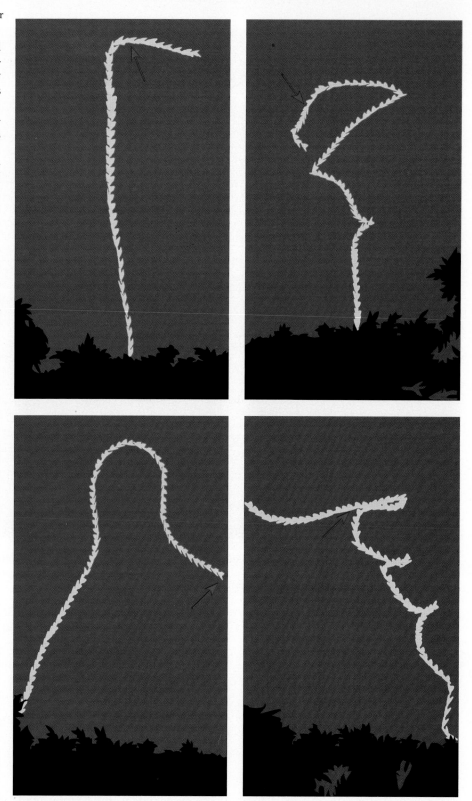

Figure 3. Flight paths of various insects in response to the onset of artificial pulses of ultrasound *(red arrows)* demonstrate some of the tactics used to escape a hunting bat. A single species of insect will often have several different evasive maneuvers in its behavioral repertoire—preventing bats from anticipating any single response. A passive dive *(upper left)*, resulting from the absence of any wing motion, is the simplest type of response. Erratic flight movements *(upper right)*, consisting of a looping turn and ending with a passive dive, is one of the evasive maneuvers performed by a small geometer moth. A powered dive (assisted by wingbeats) may be preceded by a rapid ascent *(lower left)*. A series of tight turns may also make the insect's descent to the ground somewhat more gradual *(lower right)*. (Adapted from Roeder 1962.)

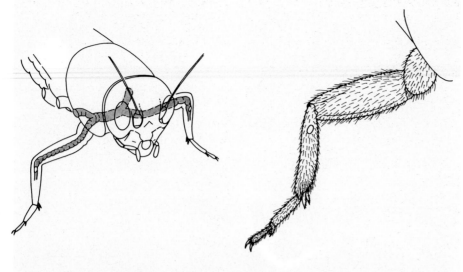

Figure 4. Crickets detect sounds—such as the cries of hunting bats—with a pair of ears located on the forelegs, just below the "knees." The ears are connected via air-filled tubes that meet at the insect's midline. Each of the tubes also has a branch leading to an opening, called a spiracle, behind each of the forelegs. The connections between the ears and the spiracles suggest that sound may reach an ear through separate channels. The presence of these different sound paths is thought to produce a differential response in the left and right ears that varies according to the location of the sound source. (Left illustration adapted from Hill and Boyan 1976.)

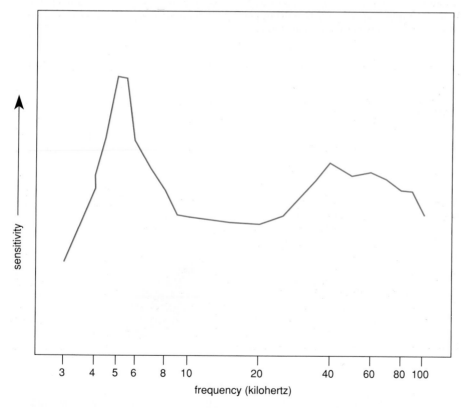

Figure 5. Frequency-sensitivity curve of a cricket's hearing exhibits two distinct peaks. A female cricket is most sensitive to sounds of about 5 kilohertz, which corresponds to the calling song of a courting male cricket. A second, broadly tuned peak—between 20 kilohertz and 100 kilohertz—lies within the frequency range of the ultrasonic cries emitted by hunting bats. The curve represents the average sensitivity of several animals as determined by observations of steering behavior toward or away from the sound. (Adapted from Moiseff, Pollack and Hoy 1978.)

quency and the location of its source.

An elegant demonstration of the frequency dependence of phonotactic behavior was devised in 1978 by Andrew Moiseff and his colleagues at Cornell. They investigated the behavior of flying female crickets (of the species *Teleogryllus oceanicus*) in response to electronically synthesized sounds ranging in frequency from 3 kilohertz to 100 kilohertz. The crickets were attached to a tether so that they were able to fly in place while suspended in an airstream—the aerial equivalent of a treadmill. Moiseff and his colleagues noticed that the sound stimulation caused the crickets to move their abdomens to one side or the other, an indication that they were attempting to steer in a particular direction. The female crickets seemed particularly responsive to two frequency ranges. They were sensitive to sounds with a frequency of about 5 kilohertz, but they also had a second sensitivity band that ranged from about 30 kilohertz to 90 kilohertz. The females steered toward the source of the 5-kilohertz sound—which corresponds to the frequency of the natural calling song of the male cricket—but steered away from the high-frequency sounds—which lie within the frequency range of the ultrasonic cries of echolocating bats. The phonotactic responses of the female cricket suggest that the auditory system of these animals is specialized not only for communication with other members of the same species but also for the detection and avoidance of the predatory bat.

The structure of the cricket's auditory system may help us to understand how it is that the cricket is able to determine the location of a sound source. The cricket's ears—consisting essentially of membranous eardrums—are not located on either side of its head but just below the "knee" on its foremost pair of legs. The ears on the left and right legs are connected via air-filled tubes that meet at the animal's midline. Each of the tubes also has a branch leading to an external opening, called a spiracle, on the cricket's body behind each of the forelegs.

Because the cricket's ears are connected to each other and to the spiracles, sound may reach the ear through any of three channels. First, of course, is the direct path, in which sound pressure waves strike the outside of the eardrum. But there are also two in-

happened one day while I was toying with the ultrasound stimulus and watching the responses of a tethered cricket. I was looking at the cricket's wings—which appear to be a transparent blur, beating about 30 times a second—when I noticed that the hindwing farther from the source of the ultrasound didn't go down as far. With every burst of ultrasound, the blurred wing seemed to stop short. Was this another mechanism by which the cricket was trying to turn? If the wing farther from the ultrasound did not make a full stroke, it should produce less thrust—inducing a turn toward that side.

On closer inspection, more than the hindwing was changing. For every pulse of ultrasound, one of the cricket's hindlegs would swing up, and appear to collide with the hindwing. Was the hindleg impeding the movement of the hindwing? Jeffrey Camhi of the Hebrew University in Jerusalem had noticed a similar phenomenon in tethered locusts, but had not shown that there were any aerodynamic consequences.

To establish that there was an interaction between the hindwing and the hindleg I needed photographic evidence. I posed the crickets in three situations: an intact cricket flying without ultrasound, an intact cricket flying with ultrasound, and a cricket that had a hindleg removed and that was flying with ultrasound. The photographs confirmed my suspicions. When there is no ultrasonic stimulus, the hindleg does not interfere with the hindwing, but with the ultrasound the hindleg farther from the sound does impede the hindwing's downstroke. In a cricket without the hindleg, the hindwing completes a full stroke even during an ultrasound stimulus.

Is there an aerodynamic effect to sticking a leg into a beating wing? To test this possibility I tethered some crickets in the device I had used to measure yaw motions. The ultrasonic stimulus was given from the left and the right sides of the cricket, both before and after removing the cricket's left hindleg. Removing the left hindleg means that all of the ultrasonic pulses from the left serve as controls (the hindleg only hits the hindwing when the ultrasound comes from the *opposite* side). And indeed, the yaw induced by an ultrasound stimulus from the left looked the same before and after removing the left hindleg. Removing the

Figure 14. Hypothetical flight path shows how a free-flying cricket might respond to the ultrasonic cries of an echolocating bat. Normal flight *(top)* is interrupted after the cricket detects the ultrasound coming from the insect's left. Within 40 milliseconds *(middle)* the cricket swings its right hindleg into its right wing, thereby reducing the thrust on that side. At about the same time, the cricket tilts its forewings, increases its speed, and pitches downward while rolling and yawing to its right. All factors should contribute to a powered dive away from the hunting bat *(bottom)*.

hindleg that was on the same side as the ultrasound stimulus did not change the flight behavior. But when the ultrasound came from the right, there was a large change in the cricket's responses. When the cricket had both hindlegs, ultrasound from the right would cause the cricket to yaw after 100 milliseconds; after removing the left hindleg, the yaw did not begin until after 140 milliseconds. The latency of the yaw response increased by 40 percent. The magnitude of the yaw also changed: it was 17 degrees before removing the leg and only 9 degrees after removing the leg. The hindleg does affect steering—it makes the yaw bigger and faster.

The hindleg might affect steering in two different ways. It might work alone to cause some drag, or its interaction with the hindwing could be the crucial factor. I repeated the yaw experiments with just one change—I removed the hindwings. This time there was no change in the yaw; it looked the same whether the ultrasound came from the left or the right, or whether the hindleg was present or absent. The result was clear: The hindleg steers the cricket by impeding the hindwing's downstroke.

Although the mechanism may appear to be crude—somewhat like stopping a bicycle by sticking a tire pump into the spokes of the rear wheel—it is functional and economical engineering design. To appreciate this, consider another example from baseball. During a fast-ball pitch, the pitcher's arm is pulled through a high-speed arc to throw the ball toward the plate. Attempting to stop this motion halfway through the pitch is humanly impossible. A cricket winging its way from an echolocating bat faces the same difficulty. Its wing muscles are not designed to halt the downstroke in the brief time needed to steer out of the bat's way. But the hindleg is able to block the downstroke of the hindwing in such a way as to produce a nearly instantaneous change in the length of the stroke—which causes the cricket to veer away. It is a clever yet simple solution.

All told, however, the swing of the hindleg is merely one small part of the cricket's acrobatic maneuvers. We now know that the cascade of responses consists of swinging legs, tilting forewings, twisting thoraxes and rapidly beating wings. It all adds up—in a fraction of a second—to an elegant ballet that whisks the cricket beyond the grasp of a hungry bat.

Bibliography

Brodfuehrer, P. D. and R. Hoy. 1989. Integration of ultrasound and flight inputs on descending neurons in the cricket brain. *Journal of Experimental Biology* 146:157-171.

Brodfuehrer, P. D. and R. Hoy. 1990. Ultrasound-sensitive neurons in the cricket brain. *Journal of Comparative Physiology, A* 166:651-662.

Camhi, J. M. 1970. Yaw-correcting postural changes in locusts. *Journal of Experimental Biology* 52:519-531.

Cooter, R. J. 1979. Visually induced yaw movements in the flying locust, *Schistocerca gregaria* (Forsk). *Journal of Comparative Physiology* 99:1-66.

Cranbrook, T. E. O., and H. G. Barrett. 1965. Observations of nocturnal bats (*Nyctalus noctula*) captured while feeding. *Proceedings of the Zoological Society of London* 144:1-24.

Easteria, D. A. and J. O. Whitaker, Jr. 1972. Food habits of some bats from Big Bend National Park, Texas. *Journal of Mammalogy* 53(4):887-890.

Griffin, D. R. and R. Galambos. 1941. The sensory basis of obstacle avoidance by flying bats. *Journal of Experimental Zoology* 86:481-506.

Griffin, Donald R. 1984. *Listening in the Dark.* Dover Publications.

Hill, K. G. and G. S. Boyan. 1976. Directional hearing in crickets. *Nature* 262:390-391.

May, M. L., Brodfuehrer, P. D., and R. R. Hoy. 1988. Kinematic and aerodynamic aspects of ultrasound-induced negative phonotaxis in flying Australian field crickets (*Teleogryllus oceanicus*). *Journal of Comparative Physiology* 164:243-249.

May, M. L. and R. R. Hoy. 1990a. Ultrasound-induced yaw movements in the flying Australian field cricket (*Teleogryllus oceanicus*). *Journal of Experimental Biology* 149:177-189.

May, M. L. and R. R. Hoy. 1990b. Leg-induced steering in flying crickets. *Journal of Experimental Biology* 151:485-488.

Miller, L. A. and J. Olesen. 1979. Avoidance behavior in green lacewings. I. Behavior of free flying green lacewings to hunting bats and ultrasound. *Journal of Comparative Physiology* 131:113-120.

Moiseff, A., Pollack, G. S. and R. R. Hoy. 1978. Steering responses of flying crickets to sound and ultrasound: mate attraction and predator avoidance. *Proceedings of the National Academy of Sciences of the U.S.A. Biological Sciences.* 75:4052-4056.

Moiseff, A. and R. R. Hoy. 1983. Sensitivity to ultrasound in an identified auditory interneuron in the cricket: a possible neural link to phonotactic behavior. *Journal of Comparative Physiology* 152:155-167.

Nachtigall, W. and D. M. Wilson. 1967. Neuromuscular control of dipteran flight. *Journal of Experimental Biology* 47:77-97.

Nolen, T. G. and R. Hoy. 1984. Initiation of behavior by single neurons: The role of behavioral context. *Science* 226:992-994.

Pollack, G. S. and N. Plourde. 1982. Phonotaxis in flying crickets: Neural correlates. *Journal of Insect Physiology* 146:207-215.

Popov, A. V. and V. F. Shuvalov. 1977. Phonotactic behavior of crickets. *Journal of Comparative Physiology* 119:111-126.

Robert, D. 1989. The auditory behavior of flying locusts. *Journal of Experimental Biology* 147:279-301.

Roeder, K. D. 1962. The behavior of free flying moths in the presence of artificial ultrasonic pulses. *Animal Behavior* 10:300-304.

Roeder, K. D. 1967. Turning tendency of moths exposed to ultrasound while in stationary flight. *Journal of Insect Physiology* 13:873-888.

Roeder, K. D. and A. E. Treat. 1961. The detection and evasion of bats by moths. *American Scientist* 49:135-148.

Rüppell, G. 1989. Kinematic analysis of symmetrical flight manoeuvres of Odonata. *Journal of Experimental Biology* 144:13-42.

Schnitzler, H.-U., D. Menne, R. Kober and K. Heblich. 1983. The acoustical image of fluttering insects in echolocating bats. pp. 235-250, in *Neuroethology and Behavioral Physiology*, F. Huber and H. Markl (eds.). Springer: Heidelberg.

Spangler, H. G. 1988. Hearing in tiger beetles (Cicindelidae). *Physiological Entomology* 13:447-452.

Suga, N. 1984. Neural mechanisms of complex-sound processing for echolocation. *Trends in Neurosciences* 7:20-27.

Whitaker, J. O., Jr. and H. Black. 1976. Food habits of cave bats from Zambia, Africa. *Journal of Mammalogy* 57(1):199-205.

Yager, D. D., May, M. L., and B. M. Fenton. 1990. Ultrasound-triggered, flight-gated evasive maneuvers in the flying praying mantis, *Parasphendale agrionina*. I: Free flight. *Journal of Experimental Biology* 152:17-39.

Yager, D. D. and M. L. May. 1990. Ultrasound-triggered, flight-gated evasive maneuvers in the flying praying mantis, *Parasphendale agrionina*. II: Tethered flight. *Journal of Experimental Biology* 152:41-58.

M. Brock Fenton
James H. Fullard

Moth Hearing and the Feeding Strategies of Bats

Variations in the hunting and echolocation behavior of bats may reflect a response to hearing-based defenses evolved by their insect prey

In the eighteenth century several natural historians suggested that the seemingly chaotic flight of moths was somehow related to attacks by hunting bats (see Kirby and Spence 1826), but the details of this relationship were clarified only after we understood something about moth hearing and about bat echolocation, or biosonar—the bat's use of echoes of the sounds it produces to detect objects from which these echoes rebound. Moths in several families have ears sensitive to ultrasonic sound located variously on their thoraxes, abdomens, or mouth parts (see Michelsen 1979), and these moths respond neurologically and behaviorally to the orientation cries of insectivorous bats. Moths that can hear such cries are 40% less likely to be caught than those that cannot (Roeder 1967), and there are similar data for some lacewings (Miller and Olesen 1979). The evasive response of moths is thus an example of a predator-specific defense (Edmunds 1974).

Two main lines of study have contributed to our understanding of the interaction between moths and bats: a detailed consideration of what moths hear and a close examination of the echolocation, prey selection, and feeding behavior of bats (see Fig. 1). This research suggests that moths may, by their ability to detect marauding bats, exert a selective pressure on echolocation strategies. Here we review the data on moth hearing and bat echolocation, analyze the implications of these data for detectability of bats by moths, and examine available information on the diets of insectivorous bats to find out if our predictions are correct. We also consider some of the intriguing questions raised by this research. If all bats are not equally detectable by all moths, do bats that are acoustically inconspicuous take advantage of this characteristic by specializing in moths? What is the function of the clicks produced by some tiger moths (Arctiidae) immediately before they are attacked by a bat?

Moth hearing

Roeder (1967) used the behavioral responses of flying moths—both individuals mounted in stationary flight on pins in the laboratory and free-flying moths in the field—to show that the response of a flying moth to an approaching bat varies according to the situation. Impulses, or spikes, from a pair of auditory neurons inform the moth of the direction and proximity of an echolocating bat. Weak stimulation, encoded as a few nerve spikes, implies the presence of a distant bat and results in negative phonotaxis: the moth turns and flies away from the bat. Strong stimulation, perceived as many nerve spikes from more than one neuron, causes the moth to fold its wings and dive to the ground.

Roeder and his coworkers have dem-

onstrated this pattern of response through experiments in which they monitored activity in the auditory nerves, usually by attaching a stainless steel extracellular hook electrode, and have described the anatomy and neurology of various moth ears. The ear of a typical moth, such as an arctiid or a noctuid, includes three neurons: two A cells that transmit data about sound, and a B cell that may function mainly as a stretch receptor for monitoring changes in pressure on the ear. Some moths—notodontids, for example—have simpler wiring, with only one auditory neuron, while geometrids, at the other extreme, have four auditory neurons.

The relatively simple wiring of moth ears and the fact that they respond to any sound of appropriate intensity and frequency suggest that they react to ultrasonic stimuli rather than to specific bat sounds. However, echolocating bats are the major source of ultrasonic sound in the nocturnal environment, and they are moreover important predators of nocturnal insects. Significantly, the ears of moths are not equally sensitive to sounds at all frequencies, but seem to have their greatest sensitivity in the range of frequencies used by most of the bats that share their habitat (Roeder 1967; Fullard 1979a).

In a recent study, we further explored the ability of moth ears to detect sounds at various frequencies by monitoring activity in the auditory nerves of moths while presenting signals of known intensity and frequency (Fenton and Fullard 1979). The threshold for response was defined as the first detectable activity occurring in the auditory nerve cells in conjunction with presentation of the stimuli; some of the resulting data

M. Brock Fenton has been a member of the Department of Biology at Carleton University since 1969. Trained at the Royal Ontario Museum and at the University of Toronto, where he received his Ph.D. in 1969, he is currently studying the ecology and behavior of bats. James H. Fullard received his Ph.D. from Carleton University in 1979 for work on the role of audition in moth defensive behavior. He is now an Assistant Professor of Zoology at Erindale College, University of Toronto, and is continuing his studies of moth defensive behavior and auditory systems. Address: M. Brock Fenton, Department of Biology, Carleton University, Ottawa, Canada K1S 5B6.

Figure 1. The little brown bat (*Myotis lucifugus*), the subject of early research on bat echolocation and its use in locating insect prey, continues to offer insights into the workings of the predator-prey relationship between bat and moth. The high-intensity echolocation calls produced by this bat have been found to be readily detectable by moths sharing its habitat at distances of up to 40 m. However, the little brown bat appears to be capable of detecting the moths only at a range of 2 m or less. (Photo by C. Hill.)

are shown here as audiograms, or tuning curves (Fig. 2).

Moths from two different zoogeographic regions—North America and Africa—showed significant differences in their audiograms. Species from Ontario were generally similar in their responses, although there was some variation among individuals and species. By contrast, moths from Ivory Coast showed greater interspecific variation in their hearing ability, and differed from the Ontario species in other ways as well: their range of best frequencies—frequencies at which they were most sensitive—was broader, and they were more sensitive over the entire range of frequencies tested (5 to 110 kHz), particularly to sounds above 65 kHz (Fullard 1979a).

The neural audiograms appear to reflect the behavioral thresholds for response to sounds. Dogbane tiger moths (*Cycnia tenera*) show three gradations in their response to sound: the threshold for continuous neural response requires the weakest auditory signal, a stronger signal leads to the production of characteristic clicking sounds, and a signal 15–25 dB more intense results in cessation of flight (Fullard 1979b). Thus there is a continuum of behavioral responses from negative phonotaxis to cessation of flight and, in the case of some arctiids, a burst of sound just before the moth stops flying and the bat completes its attack.

The data presented in the audiograms can be used to predict the distances over which a moth could detect sound at given frequencies. The maximum distance of detection (Fig. 3) is calculated by taking into account the neural threshold of the moth's

ear, the initial intensity of the sound, and the rate of attenuation of sound at different frequencies and air humidities (Griffin 1971). This distance is probably a crucial element in bat-moth interaction, since it represents the time available to predator and prey for appropriate subsequent action. It will be useful at this point to review some aspects of bat echolocation and to consider in detail the options available to an echolocating bat, given the maximum distances of detection for different moths and the effect of sound intensity and frequency on these values.

Bat echolocation

In the late 1700s Lazarro Spallanzani conducted a series of experiments in an effort to determine how bats orient themselves in total darkness. From his tests, which eliminated the use of smell, vision, and touch, he concluded that bats had a "sixth sense." In the 1930s Griffin repeated some of Spallanzani's experiments while monitoring bats with microphones sensitive to ultrasonic sound. He demonstrated that the little brown bat (*Myotis lucifugus*) produces pulses of ultrasonic sound and that it uses the echoes of these sounds to orient itself with respect to obstacles (see Griffin 1958). He also showed (Griffin et al. 1960) that bats use calls and their

echoes to locate insect prey. Griffin provided a basic understanding of echolocation and some appreciation of the diversity involved in this form of orientation. Recent work (e.g. Busnel and Fish 1980) has embellished the picture considerably by providing data on bats hunting in the field and on how information is processed in their brains.

Several aspects of bat echolocation calls, including intensity, the pattern of frequency change over time, the presence of harmonics, and the rate at which calls are produced, are relevant to the interaction between moths and bats.

Differences in the intensities of bat echolocation calls have been recognized for some time, because not all bats proved equally easy to detect with microphones sensitive to ultrasonic sound (Griffin 1958). Intensities measured 10 cm in front of the bats' mouths or nostrils range from very high (≥110 dB) to low (65 dB) and include a variety of intermediate readings (75 to 100 dB). For comparison, consider that a typical smoke detector produces a signal of 113 dB at 10 cm. It is difficult to measure the intensities of bat calls accurately in the field or in the laboratory—among other problems, the calls tend to be short in duration—and many recent

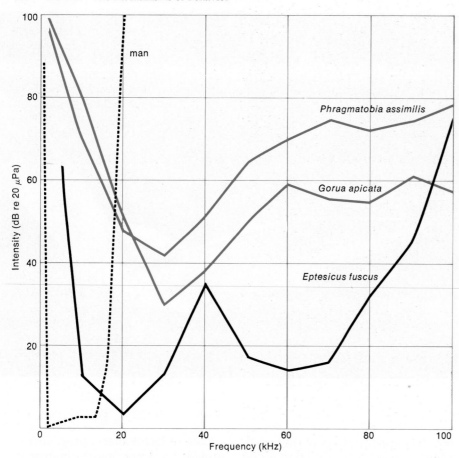

Figure 2. Audiograms for a human ear, an insectivorous bat (*Eptesicus fuscus*), and moths from North America (*Phragmatobia assimilis*) and Africa (*Gorua apicata*) reveal varying patterns of sensitivity to sound. Whereas the maximum sensitivity of human hearing is well below 10 kHz, the other animals have good hearing in the ultrasonic range. The bat has more sensitive ears than either of the moths, and the African moth is generally more sensitive than the North American one, particularly above 65 kHz. In reading an audiogram it is important to remember that the lower the dB value on the vertical axis, the more sensitive the ear is to sound at a given frequency. For example, the threshold for response from the A cell of the arctiid *P. assimilis* is a signal of 50 dB at 40 kHz; at 80 kHz the strength of the signal had to be increased 10 times—to 70 dB—before the moth's ear responded. (Data from Dalland 1965; Fullard and Barclay 1980; and Fenton and Fullard 1979.)

up to, and in some cases more than, 10 m when the bat is facing the microphone, while the faint "whispering bats" (65 dB) are usually detectable only at 0.5 m or less. Bats using calls of intermediate intensity (75 to 100 dB) are detectable at distances of 1.5 to 2.0 m. This arbitrary classification is convenient but must be treated with caution, since the intensity of these emissions for one species varies with the situation, making the continuum from very high to low intensity even more complex.

At one time it seemed appropriate to classify bats by the frequency-time patterns of their echolocation calls, distinguishing frequency-modulated (FM) from constant-frequency (CF) species or from bats using some combination of the two components, for example CF-FM (Simmons et al. 1975). However, examination of the calls emitted by different species under different conditions has shown that although calls or their components can be classified in this way, the designations are not necessarily correlated with the bats' use of the calls (Simmons et al. 1978; Gustafson and Schnitzler 1979).

studies that deal in depth with echolocation (e.g. Gustafson and Schnitzler 1979) avoid the topic completely.

However, by using a broadband bat detector that is sensitive to sounds from 10 to over 200 kHz (Simmons et al. 1979a) it is possible to classify bats in a general way by the intensity of their orientation calls (Fig. 4). Species characterized by calls of high intensity (≥110 dB) are readily detectable by these microphones at distances of

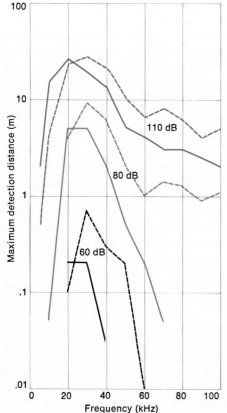

Figure 3. The maximum distance at which a moth can detect a sound of known intensity and frequency is an important strategic factor in the moth's confrontation with its predator. Using data provided by the audiograms shown in Fig. 2, maximum detection distances are calculated for *P. assimilis* (*solid line*) and *G. apicata* (*dashed line*) at three levels of sound corresponding to low, intermediate, and high-intensity echolocation calls. The dB values correspond to intensities measured 10 cm from the bats' mouths or nostrils.

Several basic patterns of frequency change over time provide the bat with various kinds of information about targets (Fig. 5). Steep FM calls rapidly sweep from higher to lower frequencies, while shallow FM calls are longer and cover a smaller range of frequencies (Simmons and Stein 1980). Steep FM components have been detected in some segments of

Figure 4. Bats exhibit a wide range of echolocation strategies, varying such crucial elements as the intensity of the call. From top to bottom are shown five species whose calls are representative of the variety of intensities used. Geoffroy's horseshoe bat (*Rhinolopus clivosus*) emits high-intensity pulses through its nostrils, whereas the Western big-eared bat (*Plecotus townsendii*) is commonly considered to produce a low-intensity call. The short-eared trident bat (*Cloeotis percivali*) uses an intermediate-intensity call, also emitted through the nostrils, while the Mexican long-eared bat (*Myotis auriculus*) emits pulses of intermediate intensity through its open mouth. The Mexican free-tailed bat (*Tadarida brasiliensis*) produces high-intensity calls, adjusting them according to the conditions under which it is hunting. (Photos by M. B. Fenton.)

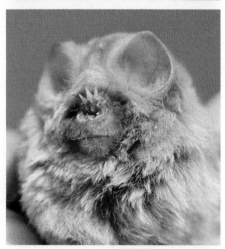

the hunting sequences of all bats examined to date, although they may only occur immediately before the bat attacks its target. The difference in the information available to the bat is clear: the steeper FM components provide more precise data about both the location of the target and its surface details.

CF components are less common in the echolocation calls of bats, but provide an excellent means of detection in some species. An integral part of the Doppler-shift compensating system found in some bats, they serve as carrier or marker frequencies, permitting the monitoring of small changes in frequency associated with insect wing beats. This Doppler-shift compensation is accomplished by a zone of maximal frequency sensitivity in the inner ear known as an acoustic fovea (Schuller and Pollak 1979), and species capable of it appear to be well equipped to detect flying insects (Griffin and Simmons 1974; Goldman and Henson 1977; Schnitzler 1978; Simmons et al. 1979b).

Echolocating bats may alter the harmonic structure of their orientation calls. Some species use multiple harmonics of calls, apparently to increase precision of target resolution (Fig. 5). Others, for example the Mexican free-tailed bat (*Tadarida brasiliensis*), add harmonics when hunting in cluttered surroundings but omit them in more open terrain (Simmons et al. 1978).

The sequence of calls produced by an

echolocating bat as it searches for, locates, and then closes with its prey (Fig. 6) demonstrates two important points: (1) the design of the call changes as this process takes place, and (2) so does the rate of pulse production. Changes in the design of the call include differences in duration and frequency that produce information required by the attacking bat. The characteristic increase in the rate of pulse production appears to apprise the bat of last-millisecond changes in the position of the target. Mexican long-eared *Myotis* (*Myotis auriculus*) did not increase their rate of pulse repetition when taking stationary moths from a wall, but did exhibit an increase when hunting flying prey (Fenton and Bell 1979). By using their wing or tail membranes to grab insect prey, bats can compensate for imprecise information about target location, perhaps the result of last-ditch defensive maneuvers by prey.

Although most biologists associate echolocation in bats with ultrasonic sound, many species use lower-frequency components in their orientation calls. In some instances the components of bat echolocation calls audible to human observers are the "Ticklauts" associated with the production of each pulse—the low-frequency transients that helped the Dutch biologist Dijkgraaf unravel the role of sound in bat orientation. In other cases the human ear hears only the lowest-frequency components in a call whose energy is mainly in the ultrasonic range. However, some bats in temperate and tropical areas—for example the spotted bat (*Euderma maculatum*) of western North America or Martienssen's free-tailed bat (*Otomops martiensseni*) of East Africa—use echolocation calls that are entirely within the range of human hearing (Fenton and Bell, in press).

Since most bats appear regularly to include an FM component in their echolocation calls, it is not possible to associate a particular frequency with a given species of bat. The frequency range covered by species studied to date is 8 to 215 kHz. It is possible, however, to use pulse repetition rate and the pattern of frequency change over time to identify bats in the field (Fenton and Bell, in press); the most convenient means of "seeing" these characteristics is through the oscil-

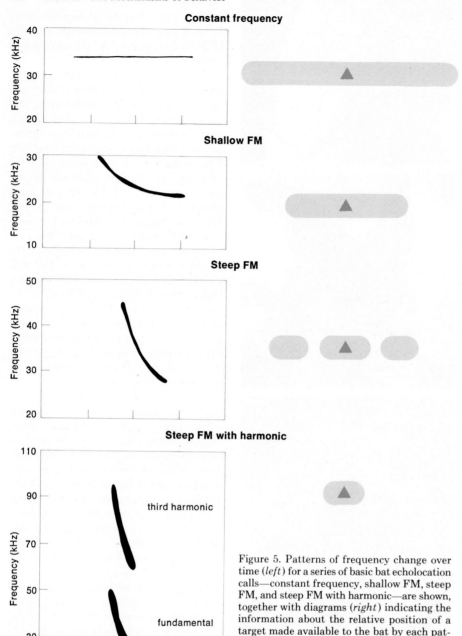

Figure 5. Patterns of frequency change over time (*left*) for a series of basic bat echolocation calls—constant frequency, shallow FM, steep FM, and steep FM with harmonic—are shown, together with diagrams (*right*) indicating the information about the relative position of a target made available to the bat by each pattern. The area shown in color indicates the target area as perceived by the bat; the triangular shape shows the position of the target itself. The addition of harmonics to a steep FM sweep results in a greatly heightened resolution of the target.

bats coincide with predictions based on the hearing sensitivity of the typical North American moth (Fig. 3), but the data on maximum distances of detection and echolocation suggest that not all bats would be equally conspicuous to all moths.

African insectivorous bats show greater variety in the design of their calls than do their North American counterparts (Fenton and Bell, in press). We therefore tested the ability of the African notodontid moth *Desmeocraera graminosa* to detect bats using echolocation calls of different intensities and frequencies by monitoring the activity in the moth's auditory nerve as various known species of bats flew in the laboratory (Fenton and Fullard 1979).

The effect of the design of the echolocation calls on the ability of the moth to detect a given bat was easily demonstrated. The banana bat (*Pipistrellus nanus*), whose high-intensity FM echolocation calls sweep from 120 to 75 kHz, was detectable anywhere in the small laboratory (up to a distance of 3 m from the moth), whereas Noack's African leaf-nosed bat (*Hipposideros ruber*), whose high-frequency (139 kHz in the CF portion of the call) calls are of intermediate intensity (75–80 dB), was detected only when it flew within 1.5 m of the moth. Large-eared slit-faced bats (*Nycteris macrotis*), which use echolocation calls of low intensity (60 dB) with frequency components from 120 to 50 kHz, were detected only when they flew within 0.2 m of the moth. In the field the ears of *D. graminosa* apparently detected echolocation calls in the 15 to 25 kHz range at distances exceeding 25 m, but there the identity of the bats was unknown, and precise figures on range of detection are lacking.

The results of such experiments generally support predictions of maximum distance of detection (Fig. 3). They suggest, moreover, that bats using sounds over 65 kHz would be less conspicuous to moths than those using sounds from 20 to 50 kHz, and that the differences in the hearing sensitivities of North American and African moths coincide with differences in the range of echolocation calls to which the two groups are exposed (Fenton and Fullard 1979). In North America most echolocating bats include FM components of be-

loscope display of the output of a zero-crossing period meter (Simmons et al. 1979a).

Echolocating bats, then, vary the intensity, rate of pulse repetition, frequency-time structure, and harmonic components of their echolocation calls according to the situation encountered. It is appropriate to consider the implications of this circumstance for the hearing-based defense of moths.

Detection of bats by moths

Roeder has demonstrated (1967) that several species of moths detect approaching bats at distances of up to 40 m. Indeed, evidence suggests that moths' ears are much more sensitive to bat calls than the most sensitive bat-detecting microphones now available (Griffin and Roeder, pers. comm.). The distances at which Roeder's moths detected approaching tween 25 and 60 kHz in their calls, whereas in Africa many species use components above or below this range (Fenton and Bell, in press).

The possible costs associated with the use of higher-frequency calls are greater atmospheric attenuation and heightened directionality of the sound. Greater directionality narrows the area from which the bat gathers information, increasing its chances of losing contact with a moving target. Either limitation could reduce the effective range of the bat's echolocation and affect its flight speed. Lower intensity would also reduce the effective range of echolocation; Griffin (1971) has estimated that the fawn horseshoe bat (*Hipposideros galeritus*), whose calls are of intermediate intensity, cannot detect targets beyond 2 m.

The distances at which echolocating bats respond to targets vary from about 1 m to at least 10 m (Griffin et al. 1960; Novick 1977; Fenton and Bell 1979; Fenton 1980). However, using behavioral responses to determine the range of echolocation is questionable, since the feature being measured is the reaction, not the detection distance. Our knowledge of the maximum distance at which a bat can detect a target is poor, but at present there is no evidence for detection of targets beyond 20 m, or even beyond 10 m for most bats (Fenton 1980).

Distance of detection is the essential point in the predator-prey confrontation between bats and moths. If a bat detects a moth at the same time or before the bat itself is detected by the moth, then the bat has an opportunity to lock onto its target and pursue it through subsequent evasive maneuvers. The maximum distance at which some moths detect bats greatly exceeds the range of echolocation in many bats, especially those using high-intensity calls in the area of maximum acoustical sensitivity of the moths.

In determining the optimal intensities and frequencies that a bat might use to locate prey, several factors must be taken into account. The first consideration is the incidence of prey equipped with bat detectors. If the existence of functional ears in some species implies a broader occurrence of this feature within families or su-

Big brown bat   50 msec

Frequency (kHz)

Mauritian tomb bat   5 msec

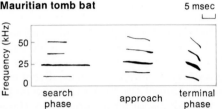

Frequency (kHz)

search phase   approach   terminal phase

Figure 6. The sequence of echolocation calls produced by a big brown bat (*Eptesicus fuscus*) as it searches for, approaches, and attacks a June bug is compared with calls used by a Mauritian tomb bat (*Taphozous mauritianus*) as it closes with a handful of sand thrown up into the air. Notice how both bats alter the design of their calls as the attack unfolds, progressively shortening the calls, decreasing the CF or shallow FM components, and adjusting the presence of harmonics. Calls in the search phase tend to be longer and either shallow FM (*above*) or CF (*below*) in nature. In approaching its target the big brown bat relied more heavily on steep FM calls than did the tomb bat, but in the terminal phase both species used short calls with steep FM components, the big brown bat exclusively, the tomb bat more intermittently. The modulation of the design of the call from phase to phase is accompanied in both cases by an increased rate of pulse production.

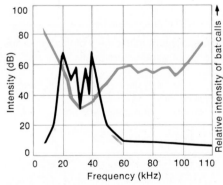

Figure 7. The frequencies used by echolocating bats at a site in Ivory Coast are those to which *G. apicata*, a noctuid moth that shares this habitat, is most sensitive. The bat calls (*black line*) show increasing power progressing up the vertical axis, while moth hearing sensitivity (*colored line*) increases down the vertical axis. Characteristically, the calls of this community of African bats extend both above and below the frequency range of 25 to 60 kHz used by comparable bat communities in Ontario.

perfamilies, then most moths have ears that are sensitive in some degree to the calls of echolocating bats. In Ontario over 95% of the species of macrolepidopteran moths possess functional ears, while in southern Africa we estimate that 85% of the species in families for which there are any data have bat detectors, although either of these figures could be influenced by variables such as infestation by ear mites (Treat 1975).

Outside the Lepidoptera many insects have ears (Michelsen 1979). There are ultrasonic components in the calls of many Orthoptera (Sales and Pye 1974), suggesting that these insects could hear bats, and there is some behavioral evidence that crickets avoid high-frequency sound (Moiseff et al. 1978). Some lacewings have bat detectors, but there are few data on the many other insects that are potential bat food. A number of beetles, notably some scarabs, possess sound-producing structures, although data on their ears are lacking. While we do not yet have enough information to evaluate the incidence of bat detectors among nocturnal insects precisely, it appears that a bat feeding heavily on moths equipped with bat detectors should have some means of foiling the insects' warning system.

Second, it is important to determine which bats make the greatest contribution to sounds in a given habitat from the insect's point of view. In monitoring bat activity with a broadband microphone we found (1979) that with one exception FM calls of molossids and vespertilionids with frequency components from 15 to 60 kHz accounted for most of the ultrasonic sounds in a variety of habitats. The exception was Gallery Forest in Ivory Coast, where bats using high-intensity calls with FM components were absent and species using intermediate-intensity calls with CF and FM components dominated. In both North America and West Africa the frequencies most commonly used by the bats were those to which the moths were most sensitive (Fig. 7).

The tuning of moth ears to the frequencies most commonly used by sympatric bats is of potential strategic importance because it leaves the way open for other bats to exploit this tuning curve by using calls outside the region of maximum sensitivity. In some tropical and subtropical areas, especially in the Ethiopian, Oriental, and Australasian tropics, about a third of the insectivorous species use calls that by virtue of their intensity and frequency might be outside the moths' early warning systems. These species could be more effective predators of moths because of their relative acoustic inconspicuousness.

A third important point to be considered is variation in the hearing ability of insects. The moths we studied showed considerable interspecific and somewhat less intraspecific variability in hearing sensitivity. Roeder (1975) has commented on the "evitability" of moth behavioral and neurological responses in general, pointing out that there was more variability in the former than in the latter in the moths he studied.

The predatory behavior of bats would ensure continuous selection for variability in the defensive response of prey. Echolocation clearly permits a bat to recognize a moth by the echo it creates (Simmons et al. 1979a), and as soon as a moth's response to an approaching bat can be predicted by the bat, the defense would be less effective. Miller (pers. comm.) found that Pipistrelles (*Pipistrellus pipistrellus*) feeding on green lacewings learned to exploit the defensive response of the prey and thus increased their rate of capture. Bats also learn to exploit predictable food sources such as the ultraviolet lights of resident entomologists. Unpredictability of re-

sponse on the part of the individual prey could therefore make it more difficult for a hunting bat to circumvent or exploit a defensive response.

Diets of insectivorous bats

The information about moth hearing ability and bat echolocation presented here might lead one to suspect that some bats would specialize in moths, whereas others would not. But studies of the diets of insectivorous bats do not unequivocally support this suspicion. Such diets have been analyzed by examination of the contents of stomachs or feces, direct observation of the hunting behavior of individuals, or a combination of these techniques.

With each of these methods it is important to know where the bats have fed and the incidence of different insects there. Although we know of no study that has obtained this array of data for insectivorous bats, some broad outlines are emerging. Analyses of prey in stomach contents or feces have demonstrated a combination of variability (several prey taxa per sample) and selectivity (few prey taxa per sample), usually within a relatively narrow size range (see Buchler 1976; Anthony and Kunz 1977). However, in the case of the large slit-faced bat (*Nycteris grandis*) such analysis showed variability in the type, weight, and size of prey consumed, with prey ranging from 10-g sunbirds to 15-mm caddisflies (Fen-

ton et al., in press). By contrast, evidence of specialization comes from a study that tabulated the presence of specific insect parts (beetle pieces or moth scales), rather than analyzing complete samples (Black 1974).

The diet of the little brown bat, a species for which our data are relatively complete (Belwood and Fenton 1976; Buchler 1976; Anthony and Kunz 1977), shows a combination of selectivity and variation. The species occasionally feeds selectively on particular types of insects, usually aquatic species, but at other times it consumes a range of prey. Variability may reflect differences in sex, reproductive condition, or age; lactating females include more moths in their diets than do males and nonparous females, and subadults show great variation in the prey they capture.

Such variability in diet also appears to be related to prey availability, and is moreover compatible with what we know of bats' energy budgets (Anthony and Kunz 1977; Kunz 1980). Insectivorous bats tend to be small and to employ an energy-intensive means of locomotion, and so incur a high cost in foraging. Given this situation, they minimize the intervals between captures by feeding in swarms of insects (Fenton and Morris 1976; Bell 1980) and exhibit masticatory specializations that permit rapid feeding (Kallen and Gans 1972).

An overall assessment of a wide range of feeding studies (Table 1) suggests four important points: (1) bats with high-intensity FM calls tend to show greater variation in their diets than species mainly using CF-FM calls; (2) some bats with low-intensity calls (e.g. *Nycteris* spp. and *Micronycteris* spp.) have extremely variable diets; (3) many species show a mosaic of extreme selectivity and great variation in prey taxa consumed per feeding period; and (4) there is a tendency among some rhinolophids and hipposiderids to feed heavily on moths.

The incomplete data we now have on the diets of some rhinolophids and hipposiderids support predictions, based on the intensities and frequencies of their echolocation calls, that these bats are well suited to thwart the auditory defenses of some moths. Dent's horseshoe bat (*Rhinolophus denti*), a rhinolophid, uses

Table 1. Results of studies of the incidence of Lepidoptera in the diets of insectivorous bats, arranged according to intensity and frequency of echolocation calls

Echolocation strategy	Number of species studied	Number of studies	Incidence of Lepidoptera		
			Mean % by volume ± SD[a]	N$_{100\%}$[b]	N$_{0\%}$[c]
High-intensity FM	35	45	32.4 ± 29	1	8
High-intensity CF	8	15	77.2 ± 21	4	1
Intermediate-intensity CF	3[d]	3	80.2 ± 15	1	0
Low intensity	3	11	25.8 ± 28	0	2

SOURCE: see Fenton and Fullard 1979
[a] SD = standard deviation
[b] Number of studies where Lepidoptera constituted 100% of diet
[c] Number of studies where Lepidoptera constituted 0% of diet
[d] *Hipposideros ruber, Cloeotis percivali*, and *Rhinolophus denti*

high-frequency calls (~ 100 kHz in the CF portion) of intermediate intensity, as does Noack's African leaf-nosed bat (CF at 139 kHz), a hipposiderid, and both bats feed heavily on moths.

Another hipposiderid, the small short-eared trident bat (*Cloeotis percivali*), also feeds heavily on moths (Whitaker and Black 1976), and from its size and diet we predicted that it too would use calls of high frequency and intermediate intensity—a prediction confirmed by recent field observations in Zimbabwe, which showed that the CF portion of its call was at 210 kHz (Fenton and Bell, in press). By contrast, the larger Hildebrandt's horseshoe bat (*R. hildebrandti*), a rhinolophid using more intense calls of lower frequency (CF at 50 kHz), commonly takes a variety of beetles, moths, and other insects.

It may be significant that rhinolophids and hipposiderids are probably capable of Doppler-shift compensation, which would permit them to exploit information about the wing beats of targets by means of an acoustic fovea (Schuller and Pollak 1979). The acoustic fovea of the greater horseshoe bat (*R. ferrumequinum*) may allow this species to distinguish between different species of insects by their wing-beat signatures. However, fluctuations in insect populations from night to night in many habitats may deny bats the opportunity to learn to recognize prey by wing-beat signatures even though they have the neural capacity for such identification. Present data on the diets of rhinolophids and hipposiderids, while indicating that they feed on moths, do not include any information about where or when these bats feed, what the insect populations of their habitats are, or what strategy they use to capture prey. Until there are field observations of the feeding behavior of the bats and the associated responses of the prey, definite conclusions are impossible.

Jamming bats

Some arctiid moths produce trains of ultrasonic clicks when presented with high-intensity acoustic stimulation typical of a bat in the terminal phase of its attack. Not all arctiids click, however, and many silent species have functional ears. In southeastern

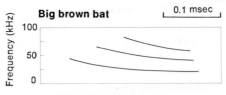

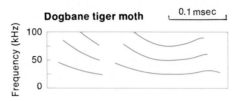

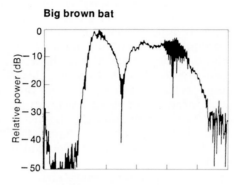

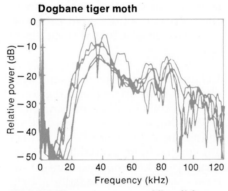

Figure 8. The power spectra of five clicks produced by the tymbal of the dogbane tiger moth (*Cycnia tenera*), an arctiid, appear to mimic closely the power spectra of the echolocation call of a big brown bat as it closes with a target (*above*). Moreover, both the intensity and the frequency-time structure of the clicks resemble comparable features of the echolocation call (*below*).

Ontario, clicking arctiids are more common in summer than in spring, and in genera such as *Phragmatobia* the spring species have vestigial and the summer forms functional clickers (Fullard 1977). Clicking is typical of arctiids emerging during times of highest bat-predation pressure (Fullard and Barclay 1980).

Dunning and Roeder (1965) have demonstrated that little brown bats in the terminal stage of approach to their prey veer sharply away from

their targets when presented with arctiid clicks. A variety of functions have been ascribed to these clicks, ranging from jamming the echolocation system of the bat, to startling the bat, to conveying a warning that arctiids are distasteful.

In the past, jamming has been rejected as an explanation for the response of the bat on the grounds that the moths' clicks are too low in intensity and the bats' echolocation systems too sophisticated. However, a recent reconsideration of the situation (Fullard et al. 1979) has led to renewed support for the jamming theory. The power spectra of the clicks of the dogbane tiger moth (*Cycnia tenera*) are surprisingly similar to those of calls produced by some bats during their feeding buzz (Fig. 8). Furthermore, the clicks are similar in intensity to echoes the bats would be receiving from a target at close range, and the frequency-time structures of the clicks resemble those of the bat calls (Fig. 8).

Taken together, these data suggest that the moth click could penetrate the information-processing system of the bat by passing through filters designed to remove extraneous information. The moth clicks could then disrupt the bat's processing of information and elicit the startle response reported by Dunning and Roeder (1965).

Roeder (pers. comm.) has suggested that moths could not mimic bats. He compares the situation to a cocktail party. Although it would be difficult for someone to convince you that you said something you did not say at such an event, chemical stimuli notwithstanding, if in the middle of a conversation you heard someone else talking about you, this would interfere with your concentration and your ability to process information. In this way moth clicks could interfere with bats' processing of information and effectively jam them.

Moth vs. bat

The interaction between moth and bat is an example of an eye-opening discovery about animal behavior. It involves much more than a simple response to predator-specific signals, and the theory has complications from either the moth's or the bat's point of view.

Nocturnal activity allows moths to avoid a range of diurnal predators, and possession of ears sensitive to the frequencies most commonly used by echolocating bats further reduces the odds against them. The average moth in North America or Africa detects the average bat long before the bat is aware of the moth, and negative phonotaxis may keep the alerted moth out of harm's way. Closer, more immediate threats, perceived by the moth as more intense stimuli, lead to complicated and unpredictable flight patterns, effective at close range against attacking bats.

There is variability in the responses of flying moths at both the individual and the faunal level, in part reflecting variation in hearing sensitivity. This variability enhances the defensive effect of bat detection by making it impossible for an attacking bat to predict the evasive response of its target.

Yet another dimension is added to the contest between moth and bat by the clicks of some tiger moths. These clicks are well suited to jam the echolocation of some bats by interfering with echo processing.

On the other side, bats appear to have several ways of countering the hearing-based defense of insects. Since the ears of moths are most effective at detecting the average bat, echolocation calls differing from the average may make an attacking bat less conspicuous acoustically. The use of calls either above or below the range of frequencies to which moths are most sensitive appreciably reduces the distance at which most moths can detect bats, as does the use of calls of lower intensity. Either alternative, however, imposes costs on the bat.

Another way to circumvent insect defenses is to stop echolocating and rely on other cues, a strategy adopted by some bats (Vaughan 1976; Fiedler 1979; Barclay et al., in press). Fringe-lipped bats (*Trachops cirrhosus*) may use the sounds produced by their prey in conjunction with echolocation to locate their targets, while Indian false vampire bats (*Megaderma lyra*) often use the sounds of their prey alone. We still do not know the extent to which animal-eating bats exploit other cues from prey, or how much they rely on their vision in finding food.

Observations of Mexican long-eared bats feeding on moths in Arizona (Fenton and Bell 1979) suggested that they might circumvent hearing-based defenses by taking moths resting on surfaces. However, more recent studies (Werner 1981) have shown that moths sitting or walking on surfaces also respond behaviorally to ultrasonic sound.

Although we know that bats use echolocation calls to detect prey and that some prey avoid attack at least some of the time by listening for bat calls, we lack several items of information needed to complete the picture. It appears that some bats—for example, rhinolophids and hipposiderids—may exploit the hearing sensitivities of moths to permit closer undetected approach, but more precise information about bat hunting, selection of prey, and the defensive behavior of the moths is required to confirm this possibility. It would also be interesting to find out whether the clicks of arctiids are more effective against some bats than against others; the properties of different echolocation calls suggest that this should be true.

Echolocation may have been a vital element in the origin and radiation of insectivorous bats, permitting them or their immediate ancestors to hunt flying insects regardless of the amount of light available. The appearance of hearing-based insect defenses may in turn have reduced bats' effectiveness against some prey. We are now making progress in exploring the third twist in this story: the possibility that the variety we find in bat hunting and echolocation behavior may be in part a response to these insect defenses.

References

Anthony, E. L. P., and T. H. Kunz. 1977. Feeding strategies of the little brown bat, *Myotis lucifugus*, in southern New Hampshire. *Ecology* 58:775–86.

Barclay, R. M. R., M. B. Fenton, M. D. Tuttle, and M. J. Ryan. In press. Echolocation calls produced by *Trachops cirrhosus* (Chiroptera: Phyllostomatidae) while hunting for frogs. *Can. J. Zool.*

Bell, G. P. 1980. Habitat use and response to patches of prey by desert insectivorous bats. *Can. J. Zool.* 58:1876–83.

Belwood, J. J., and M. B. Fenton. 1976. Variation in the diet of *Myotis lucifugus* (Chir-

optera: Vespertilionidae). *Can. J. Zool.* 54: 1674–78.

Black, H. L. 1974. A north temperate bat community: Structure and prey populations. *J. Mamm.* 55:138–57.

Buchler, E. R. 1976. Prey selection by *Myotis lucifugus* (Chiroptera: Vespertilionidae). *Amer. Nat.* 110:619–28.

Busnel, R. -G., and J. F. Fish, eds. 1980. *Animal Sonar Systems*. NATO Advanced Study Institute Series A, vol. 28. Plenum Press.

Dalland, J. I. 1965. Hearing sensitivity in bats. *Science* 150:1185–86.

Dunning, D. C., and K. D. Roeder. 1965. Moth sounds and the insect-catching behavior of bats. *Science* 147:173–74.

Edmunds, M. 1974. *Defence in Animals*. Longman.

Fenton, M. B. 1980. Adaptiveness and ecology of echolocation in terrestrial (aerial) systems. In *Animal Sonar Systems*, ed. R. -G. Busnel and J. F. Fish, pp. 427–46. NATO Advanced Study Institute Series A, vol. 28. Plenum Press.

Fenton, M. B., and G. P. Bell. 1979. Echolocation and feeding behaviour in four species of *Myotis* (Chiroptera). *Can. J. Zool.* 57: 1271–77.

———. In press. Recognition of species of insectivorous bats by their echolocation calls. *J. Mamm.*

Fenton, M. B., and J. H. Fullard. 1979. The influence of moth hearing on bat echolocation strategies. *J. Comp. Physiol.* 132:77–86.

Fenton, M. B., and G. K. Morris. 1976. Opportunistic feeding by desert bats (*Myotis* spp.). *Can. J. Zool.* 54:526–30.

Fenton, M. B., G. P. Bell, and D. W. Thomas. 1980. Echolocation and feeding behaviour of *Taphozous mauritianus* (Chiroptera: Emballonuridae). *Can. J. Zool.* 58:1774–77.

Fenton, M. B., D. W. Thomas, and R. Sasseen. In press. *Nycteris grandis* (Nycteridae): An African carnivorous bat. *J. Zool. London.*

Fiedler, J. 1979. Prey catching with and without echolocation in the Indian false vampire (*Megaderma lyra*). *Behav. Ecol. Sociobiol.* 6:155–60.

Fullard, J. H. 1977. Phenology of sound-producing arctiid moths and the activity of insectivorous bats. *Nature* 267:42–43.

———. 1979a. Auditory components in the defensive behaviour of certain tympanate moths. Ph.D. Thesis, Carleton University.

———. 1979b. Behavioral analyses of auditory sensitivity in *Cycnia tenera* Hübner (Lepidoptera: Arctiidae). *J. Comp. Physiol.* 129:79–83.

Fullard, J. H., and R. M. R. Barclay. 1980. Audition in spring species of arctiid moths as a possible response to differential levels of insectivorous bat predation. *Can. J. Zool.* 58:1745–50.

Fullard, J. H., M. B. Fenton, and J. A. Simmons. 1979. Jamming bat echolocation: The clicks of arctiid moths. *Can. J. Zool.* 57: 647–49.

Goldman, L. J., and O. W. Henson, Jr. 1977. Prey recognition and selection by the constant frequency bat, *Pteronotus parnellii parnellii*. *Behav. Ecol. Sociobiol.* 2:411–20.

Griffin, D. R. 1958. *Listening in the Dark*. Yale Univ. Press.

———. 1971. The importance of atmospheric

attenuation for the echolocation of bats (Chiroptera). *Anim. Behav.* 19:55–61.

Griffin, D. R., and J. A. Simmons. 1974. Echolocation of insects by horseshoe bats. *Nature* 250:731–32.

Griffin, D. R., F. A. Webster, and C. R. Michael. 1960. The echolocation of flying insects by bats. *Anim. Behav.* 8:141–54.

Gustafson, Y., and H. -U. Schnitzler. 1979. Echolocation and obstacle avoidance in the hipposiderid bat *Asellia tridens. J. Comp. Physiol.* 131:161–67.

Kallen, F. C., and C. Gans. 1972. Mastication in the little brown bat, *Myotis lucifugus. J. Morph.* 136:385–420.

Kirby, W., and W. Spence. 1826. *An Introduction to Entomology.* London.

Kunz, T. H. 1980. *Daily Energy Budgets of Free-Living Bats.* Proc. Fifth Int. Bat Res. Conf. Lubbock: Texas Tech Press.

Michelsen, A. 1979. Insect ears as mechanical systems. *Amer. Sci.* 67:696–706.

Miller, L. A., and J. Olesen. 1979. Avoidance behavior in green lacewings. I. Behavior of free flying green lacewings to hunting bats and to ultrasound. *J. Comp. Physiol.* 131:113–20.

Moiseff, A., G. S. Pollack, and D. R. Hoy. 1978. Steering responses of flying crickets to sound and ultrasound: Mate attraction and predator avoidance. *Proc. NAS:* 75:4052–56.

Novick, A. 1977. Acoustic orientation. In *Biology of Bats*, ed. W. A. Wimsatt, vol. 3, pp. 74–289. Academic Press.

Roeder, K. D. 1967. *Nerve Cells and Insect Behavior.* Rev. ed. Harvard Univ. Press.

———. 1975. Neural factors and evitability in insect behavior. *J. Exp. Zool.* 194:75–88.

Roeder, K. D., and M. B. Fenton. 1973. Acoustic responsiveness of *Scoliopteryx libatrix* (Lepidoptera: Noctuidae), a moth that shares hibernacula with some insectivorous bats. *Can. J. Zool.* 51:291–99.

Sales, G., and J. D. Pye. 1974. Ultrasonic communication by animals. London: Chapman and Hall.

Schnitzler, H. -U. 1978. Die detektion von Bewegungen durch Echoortung bei Fledermäusen. *Verh. Deut. Zool. Ges.* 1978:16–33.

Schuller, G., and G. Pollack. 1979. Disproportionate frequency representation in the inferior colliculus of Doppler-compensating greater horseshoe bats: Evidence for an acoustic fovea. *J. Comp. Physiol.* 132:47–57.

Simmons, J. A., and R. A. Stein. 1980. Acoustic imaging in bat sonar: Echolocation signals and the evolution of echolocation. *J. Comp. Physiol.* 135:61–84.

Simmons, J. A., M. B. Fenton, W. R. Ferguson, M. Jutting, and J. Palin. 1979a. *Apparatus for Research on Animal Ultrasonic Signals.* Life Sci. Misc. Pub. Royal Ontario Museum.

Simmons, J. A., M. B. Fenton, and M. J. O'Farrell. 1979b. Echolocation and pursuit of prey by bats. *Science* 203:16–21.

Simmons, J. A., D. J. Howell, and N. Suga. 1975. Information content of bat sonar echoes. *Amer. Sci.* 63:204–215.

Simmons, J. A., W. A. Lavender, B. A. Lavender, J. E. Childs, K. Hulebak, M. R. Rigden, J. Sherman, B. Woolman, and M. J. O'Farrell. 1978. Echolocation by free-tailed bats (*Tadarida*). *J. Comp. Physiol.* 125:291–99.

Treat, A. E. 1975. *Mites of Moths and Butterflies.* Cornell Univ. Press.

Vaughan, T. A. 1976. Nocturnal behavior of the African false vampire bat (*Cardioderma cor*). *J. Mamm.* 57:227–48.

Werner, T. K. 1981. Responses of non-flying moths to ultrasound: The threat of gleaning bats. *Can. J. Zool.* 59:525–29.

Whitaker, J. O., Jr., and H. L. Black. 1976. Food habits of bats from Zambia, Africa. *J. Mamm.* 57:199–204.

PART III
The Evolutionary History of Behavior

Behavioral mechanisms and the behaviors they control have a history, which can sometimes be reconstructed by evolutionary biologists. The five articles in Part III introduce readers to the methods used to trace the historical pathways leading to some particularly odd and fascinating behaviors. Note that the reconstruction of historical pathways complements studies on how internal mechanisms control animal behavior (Part II) as well as studies on the adaptive value of behavior (Parts IV and V). These different components of behavioral research constitute the different levels of analysis outlined by Holekamp and Sherman in Part II.

We begin with an article by Stephen Jay Gould, who claims that evolution is like a bush, rather than a ladder—contrary to the common view that evolutionary history is characterized by steady progress upward toward higher and higher forms. According to Gould, each living species belongs to a lineage with a long and convoluted history whose course was affected by many chance events and quirks of fate. To look at any one living species and see a linear progression with one ancient species giving rise to another, which in turn produces yet another in an inexorable march toward the current living species of interest (often ourselves), is to ignore all the side branches that went nowhere, and thus, to invent an orderly pattern where none existed.

But despite the unique and convoluted history behind each species and trait, Gould believes that one can reconstruct that history using methods as rigorous and scientific as those used by any other kind of biologist. Although many persons think that the "scientific method" depends entirely on carefully controlled experiments, Gould argues that Darwin had it right when he developed the comparative method, a nonexperimental approach to evolutionary hypothesis testing. This method takes advantage of the thousands of "natural experiments" that have occurred over evolutionary time: the formation of new species and the changes leading to their special characteristics. How does this approach fit into the framework outlined by Woodward and Goodstein in Part I? How is Popperian falsification employed when the comparative method is used by evolutionary biologists?

According to Gould, it is possible to establish the evolutionary relationships among species by identifying shared traits with distinctive similarities that were probably inherited from the common ancestor of those species. For example, one can see oddities and apparent imperfections in the flower parts of many beautiful and bizarre orchids, indicating that these parts must have been derived from ordinary petals and sepals possessed by less spectacular plants. Thus, we may be able to infer the history of a weirdly wonderful orchid flower by comparing it with other less elaborate plant species alive today.

An interesting problem, however, is how to check our inferences about history when fossil evidence is absent, as we often must when dealing with behavioral evolution, since behaviors rarely leave fossil evidence. Gould provides a prescription for dealing with this problem based on "grounds of mathematical probability alone." What do you think of his remedy in view of the bushlike history of lineages?

Bert Hölldobler and Edward Wilson use the comparative method to get at the history behind the amazing behaviors of modern weaver ants, several of which build elaborate silken nests in trees, rather like the constructions of the more familiar tent caterpillars. The nest builders belong to a group of four living ant genera whose members weave with silk to varying degrees. Hölldobler and Wilson examine the shared behavioral characteristics of these genera as they try to figure out what series of changes led to the complex nest-weaving behavior of certain species. Detailed comparisons of adult and larval behaviors enabled the authors to array the genera according to "phylogenetic grade," from least to most complex. For example, in the "advanced" genus *Oecophylla*, middle-aged larvae produce silk, and workers use them as living shuttles, carrying them to and fro to bind leaves together into a nest. By contrast, in the "primitive" genus *Dendromyrmex*, larvae produce silk only when they are about to pupate, and workers simply plop them down and let them create small patches of silk as they thrash about.

Hölldobler and Wilson suggest that the primitive *Dendromyrmex* weaver ants may be on an evolutionary trajectory that will carry them through stages exemplified by the genera of intermediate complexity until they can attain the advanced nest-weaving accomplishments of *Oecophylla*. What are the merits of arraying species according to phylogenetic grade? If Gould is correct that evolution is more like a bush than a ladder, are Hölldobler and Wilson mistaken to infer "progress" toward more complex forms of behavior? Does natural selection tend to promote complexity, or is this view a result of our admiration for the technological advances that have occurred during human cultural evolution? Are "advanced" species with complex behaviors or social structures really better adapted? If so, why are there so many "intermediate" and "primitive" species that appear to possess "imperfect" adaptations? Finally, why do you suppose only a handful of the estimated 20,000 species of ants have taken to building silken nests in the tree-tops? Is it because most ants have never had the capacity to take the first steps toward silk weaving, or is tree-top nesting actually disadvantageous for almost all ant species? If so, why?

William Shear does not use the concept of phylogenetic grade in his article on the evolution of another amazing user of silk, the spiders that build orb webs. However, Shear does make extensive, disciplined use of comparisons among living species of spiders to evaluate competing hypotheses on the origin and modification of these marvelous devices. Shear takes advantage of knowledge about the relationships among spiders, based on shared characteristics other than web design, to construct a phylogenetic tree—a diagram that outlines the probable evolutionary relationships among the main groups of spiders. He assumes that the living species in the first lineage to split off from the common ancestor of all spiders exhibit the silk-using behavior of that ancient common ancestor. He also assumes that more recently evolved lineages share the silk-using trait of their more recent common ancestor. By assembling the evidence on the time of evolution of different lineages and their current web-building characteristics, Shear can give tentative answers to questions about what the first silk-using spider used its silk for, and whether different families of orb-weaving spiders evolved their special devices independently or as a result of having a shared common ancestor that built an orb web.

Shear is quick to acknowledge that his answers are not certain, because shared characteristics in two species or families of spiders might reflect shared ancestry, simultaneous parallel changes in related lineages, or they might have arisen when two very different lineages were subjected to similar selection pressures, causing them to evolve similar traits totally independently. Likewise, differences in the characteristics of any two taxonomic groups can be difficult to interpret. Are certain differences between them great enough to indicate that these organisms had very different ancestors and thus should be placed in two different families, or are they slight differences that do not indicate a long period of separate ancestry? What do you think of the idea of using the behavior of living species to infer what their evolutionary ancestors did, in light of Gould's "evolution is a bush" argument? Would your answer to this question depend on whether the ancestral environment was markedly different from, or much the same as, the current environment occupied by the living descendants of extinct species?

In any event, Shear's approach differs considerably from that of Hölldobler and Wilson, which you can demonstrate for yourself by categorizing spiders in terms of the complexity of their webs and then ordering the types into a series from least complex to most complex. How would such an analysis differ from the one

that Shear presents? Which analysis would be more palatable to Gould, and why? Which one do you feel is more likely to be correct, and why?

Bowerbirds are the avian equivalent of orb-weaving spiders in that some species build highly complex structures using behavior patterns whose origin and subsequent modification are far from obvious, thus providing an intriguing historical puzzle for evolutionary biologists. Gerald Borgia has a go at solving this puzzle in his article on why bowerbirds build bowers. Although a phylogenetic tree of bowerbird relationships does not accompany the article, you should be able to draw one based on the information that Borgia provides. You can then check your diagram against one that Borgia and his colleagues constructed based on molecular data (see *Proceedings of the Royal Society of London, Series B*, 264:307–314, 1997). Your diagram should reveal that among the living bowerbirds, there are species whose ancestors probably engaged in complex behaviors that have been lost in these modern species. What significance does this fact have for persons who assume that living species with "simple" behavior patterns have retained the simple traits of their ancestors?

Borgia is interested in more than just reconstructing the pattern of changes in bower-building. He also wants to address another level of analysis, namely, the adaptive value of the characteristics that have evolved in one lineage or another. What do you think of Borgia's explanation that male bowerbirds build bowers to protect females against forced copulation? What does this hypothesis imply about the importance of female choice of mates in bower-building species? How could males evolve a trait that made it hard for them to reproduce compared with their male ancestors that sometimes engaged in forced matings? How could a trait that reduced a male's chances of siring offspring have spread through an ancestral population? And what about other bird species? Doesn't the "protection against forced copulation" hypothesis predict that (1) forced copulation will be a feature of the reproductive lives of other birds, at least those species closely related to bowerbirds, and that if so, (2) building of bowers, or their analogs, will have evolved in other species whose ancestors engaged in occasional forced copulations? Can you acquire the information needed to test these predictions?

The evolutionary history of human behavior, a topic of great interest to our own species, is explored by Kim Hill and Magdalena Hurtado. Their article describes foraging behavior and food distribution patterns in a tribe of hunter–gatherers, the Ache, with whom they lived for many years in the rainforests of Paraguay. Many persons have argued that studies of modern hunter–gatherers offer a window on our evolutionary past, since it is almost certain that ancestral *Homo sapiens* were hunters and gatherers. Thus, by examining modern hunting and gathering societies, we might gain insight into the conditions our ancestors confronted and the behavior patterns they used to survive and reproduce.

One way in which ancestral humans might have improved their chances of surviving to reproduce would have been to collect food as efficiently as possible, since food consumption and reproductive success may well be correlated, in humans and in many other animals. Hill and Hurtado followed foraging groups of male hunters and female gatherers to find out whether the Ache foraged in an optimal manner. They found that both men and women were highly selective, seeking animals and plants that yielded the highest caloric return per unit of time devoted to the chase or harvest, while ignoring many species that are edible but hard to obtain. However, the sexes differed in what they did with the foods they collected. Women kept some of what they gathered for themselves and their offspring, whereas when successful hunters returned to the group's encampment, the men butchered their kill, divided it up, and then *gave it all away!* If the point of food gathering is to maximize reproductive success, why wouldn't hunters reserve at least some of their hard-earned protein for themselves and their families? And what prevents a man from taking it easy and allowing other tribesmen to provide for his family's needs? Hill and Hurtado offer one possible explanation, but there are others. Can you come up with one of your own?

Although maximizing the collection of food resources probably plays some role in Ache foraging, the relationship between this goal and reproductive success among the Ache is not immediately obvious. Furthermore, according to Hill and Hurtado, other groups of modern hunter–gatherers do not collect and share food in the same way as the tribe they studied. In fact, there is no one typical hunter–gatherer way of life, which makes it difficult to infer how our ancestors hunted and gathered. However, a sexual division of labor in foraging does appear to be universal, with women gathering vegetable foods and men doing the hunting. Does this fact imply something about the genetic basis of male and female foraging tactics, or about when these traits originated? If there are few other universals among living hunter–gatherers, can we say much more about the history of human behavior? What do the many differences among modern hunter–gatherers reveal about the ability of our ancestors to cope with variation in ecological conditions? How could you test whether humans have universally evolved the capacity to alter their behavior in different environments in ways that advance individual reproductive success?

Although we leave it as an open question whether a comparative analysis of the behavior of living species—including human beings—can completely reveal the history behind given behaviors, the right kinds of comparisons among different species are clearly instructive. Any approach that offers a way to figure out how such odd and wonderful behavior patterns as spider orb-weaving and bowerbird bower-building came into being is an approach well worth knowing about.

Evolution and the Triumph of Homology, or Why History Matters

Stephen Jay Gould

In 1912, when the nation both needed and still had a good five cent cigar, Sigma Xi spent three dollars to rent a hall for its annual banquet. Receipts for 1912 totaled $646.42 against expenses of $160.22 (including that three bucks), leaving a balance of $486.20, a fine improvement from the 1911 surplus of $295.67. Our society then included 8,200 members, 2,176 listed as active. In that year, Sigma Xi also decided, for the first time, to publish a journal, the *Sigma Xi Quarterly* (renamed *American Scientist* in 1942). In his "Salutatory" to the very first issue, president (and paleontologist) S. W. Williston wrote on page 1, volume 1, number 1: "Since its beginning Sigma Xi has stood for the encouragement of investigation, of research, rather than for the mere acquisition of knowledge."

In 1886 the founders of Sigma Xi had chosen for their motto "Companions in Zealous Research"—a phrase that we have happily retained despite its archaic ring. The original zealots were an uncompromising lot. Some roamed public places with hidden daggers to strike down supporters of Rome; others committed mass suicide at Masada. They were, above all, men of action—the *doers* of their generation. Our founders chose their words well. Science is doing, not just clever thinking. As Williston noted, our society stands for action expressed as research.

I have been assigned the impossible task of encapsulating the intellectual impact of evolution, both on other sciences and upon society in general, during the past 100 years. I have chosen this fundamental definition of Sigma Xi as prologue because I want to argue that Darwin's most enduring impact has generally been underestimated (or underesteemed). I will hold that his theory is, first and foremost, a guide to action in research—the first *workable* program ever presented for evolution. Darwin was, above all, a *historical methodologist*. His theory taught us the importance of history, expressed in doing as the triumph of homology over other causes of order. History is science of a different kind—pursued, when done well, with all the power and rigor ascribed to more traditional styles of science.

Stephen Jay Gould is Alexander Agassiz Professor of Zoology at Harvard University.

How Darwin's "long argument" has changed the path of scientific thought during the past 100 years

Darwin taught us why history matters and established the methodology for an entire second style of science.

Darwin as a historical methodologist

Introducing the final chapter of the *Origin of Species* (1859), Darwin writes: "this whole volume is one long argument" (p. 459). Since Darwin was not a conscious or explicit philosopher, and since he crammed the *Origin* so full of particulars collected during twenty long years of preparation, readers often miss the unity of intellectual design. Indeed, Huxley commented that readers often misinterpret the *Origin* as a "sort of intellectual pemmican—a mass of facts crushed and pounded into shape, rather than held together by the ordinary medium of an obvious logical bond" (1893, p. 25).

What, then, is Darwin's "long argument," so deftly hidden amidst his particulars? It is not merely the specific defense of "natural selection," for most of the *Origin* is a basic argument for descent, not for any particular mechanism governing the process. But neither is it the most general marshaling of support for evolution—for transmutation was among the commonest of nineteenth-century heresies, and Darwin had something more special and personal to say.

The "long argument," as I read it, is the claim that *history* stands as the coordinating reason for relationships among organisms. Darwin's argument possesses a simple and beautiful elegance. Before the *Origin*, scientists had sought intrinsic purpose and meaning in taxonomic order. Darwin replied that the ordering reflects a historical pathway, pure and simple. (As just one example, numerological systems of taxonomy flourished in the decades before Darwin. These attempts to arrange all creatures in groups neatly ordered and numbered according to simple mathematical formulae—see Oken 1809–11 or Swainson 1835, for example—make sense if a rational intelligence created all organisms in an ordered scheme, but devolve to absurdity if taxonomy must classify the results of a complex and contingent history.) The eminent historian Edward H. Carr writes: "The real importance of the Darwinian revolution was that Darwin, completing what Lyell had already begun in geology, brought history into science" (1961, p. 71).

So much so good. But the simple statement that Darwin made history matter contains a dilemma, especially if we wish to defend the cardinal premise of Sigma Xi: that science is productive doing, not just clever thinking. History is the domain of narrative—unique, unrepeatable, unobservable, large-scale, singular events. One of the oldest saws of freshman philosophy classes asks: Can history be science? Many professors answer "no" because science seeks immanence by experiment and prediction, while the narrative quality of history seems to preclude just these defining features.

How then can we marry these two apparently contradictory statements—the claim that Darwin's "long argument" made history matter and the usual impression that Darwin was a great scientist? The problem vanishes when we locate Darwin's singular greatness in his extended campaign to establish a scientific methodology for history—to make history doable for the zealous researchers of science. Darwin was, more than anything else, a historical methodologist.

Michael Ghiselin's landmark book (1969) was the first to analyze all of Darwin's writing (not just the central trilogy of the *Origin*, the *Descent of Man*, and the *Expression of the Emotions*). He was also the first to suggest with proper documentation that Darwin's greatness as a scientist lay in the middle ground between his most basic elucidation of evolution as a fact, and his most general development of the radical implications (randomness, materialism, nonprogressionism) that so upset Western culture (but produced less immediate impact upon the day-to-day practice of science). This middle ground embodies Darwin's arguments for a methodology of research, for the actual *doing* itself. For Ghiselin, Darwin succeeded because he consistently used the hypothetico-deductive method so celebrated by recent philosophers of science—even though, as a loyal Victorian, he usually misportrayed himself (primarily in his *Autobiography*) as a patient and rigorous Baconian inductivist. (Ruse 1979 supports Ghiselin in the major work on Darwin's methodology written since.)

While I applaud Ghiselin's insight that Darwin must be viewed primarily as a methodologist—as someone who taught scientists how to proceed—I disagree that the central theme and sustaining power of Darwin's methodology lies in its hypothetico-deductive format. Philosopher of science Philip Kitcher has recently written that "if Darwin was a scientist practicing by the canons favored by Ghiselin and Ruse, then he was a poor practitioner"; in Kitcher's account, Darwin "answers to rather different methodological ideals" (1985, footnote 11). Kitcher views Darwin's theories as sets of "problem-solving patterns, aimed at answering families of questions about organisms, by describing the histories of these organisms" (p. 135)—the very aspect of life that had no relevance in the pre-Darwinian world of created permanence. But how can a naturalist do history in a scientific way, especially given the poor reputation of history as a ground for testable hypotheses? How can history be incorporated into science?

Darwin's long argument is not a simple brief for evolution; it is, above all, a claim for *knowability*—a set of methods that subjected evolution, for the first time, to zealous research. Previous briefs for evolution—and many had appeared to much comment (Lamarck 1809; Chambers 1844)—had presented speculative systems suggesting little in the way of doable research. Lyell's strong distaste for Lamarck (an opinion shared by Darwin) centered upon the methodological vacuousness of his system: "There were no examples to be found. . . . When Lamarck talks of 'the efforts of internal sentiment,' 'the influence of subtle fluids,' and 'acts of organization,' as causes . . . he substitutes names for things; and, with a disregard to the strict rules of induction, resorts to fictions, as ideal as the 'plastic virtue,' and other phantoms, of the geologists of the middle ages" (1842, pp. 10–11).

Darwin's claim for knowability centers upon two themes: first, the uniformitarian argument that one should work by extrapolating from small-scale phenomena that can be seen and investigated; second, the establishment of a graded set of methods for inferring history when only large-scale results are available for study.

The uniformitarian argument

In a famous letter for once not overly immodest, Darwin stated that half his work came out of Lyell's brain. Lyell's insistent argument through three volumes of the *Principles of Geology*—that a historical scientist must work with observable, gradual, small-scale changes and extrapolate their effects through immense time to encompass the grand phenomena of history—won Darwin's allegiance, with a central commitment for its transfer, in toto, to biological realms. But, as a fateful event in the history of nineteenth-century science (Gould 1965; Rudwick 1972; Hooykaas 1963), Lyell advanced uniformity as more than a methodological postulate—work with small-scale events when you can, because they are all you have. It became a strong substantive claim as well—the world really works that way, all the time.

In Lamarck's system, small-scale adaptations (the giraffe's neck, the long legs and webbed feet of wading birds) are tangential—literally orthogonal—to a different, virtually unobservable, and more essential process that moves organisms up the ladder of life toward ever-greater complexity. Savor the paradox for a scientist committed by definition to doing: what you can know and manipulate is unimportant or irrelevant; what is essential cannot be directly observed.

Darwin broke through this disabling paradox by proclaiming that the tangible small-scale evidences of change—artificial selection as practiced by breeders and farmers, tiny differences in geographic variation among races of a species, for example— are, by smooth extrapolation, *the* stuff of all evolution. Darwin, for the first time, made evolution a workable research program, not just an absorbing subject for speculation. This methodological breakthrough was his finest achievement.

But Darwin, like Lyell, then ventured beyond the methodological issue, thereby setting the pathways of evolutionary debate ever since (including all the hubbub of the last decade). He argued that all change, at whatever apparent level, really did arise as the extrapolated result of accumulated selection within populations. The distinctive features of strict Darwinism—particularly its location of causality in struggles among organisms (denial of hierarchy), and its argument for continuity in rate, style, and effect from the smallest observable to the largest inferred events of change (see Gingerich 1983 for a modern defense; Gould 1984 for a rebuttal)—emerge from this substantive extension of the uniformitarian argument.

The questioning of this extension unifies the apparently diverse critiques prominently discussed during the past decade. Critics are denying the reductionistic causal premise (struggles among organisms) by outlining a hierarchical theory of selection, independent at several levels of genes, organisms, demes, and species, but with complex interactions across all levels (Gould 1982a; Vrba and Eldredge 1984; for a philosopher's defense of hierarchy as logically sound see Sober 1984). They are also denying causal continuity from competition in a crowded world (Darwin's "wedging") to relays and replacements of faunas in mass extinctions (Raup and Sepkoski 1984), thereby defending more randomness and discontinuity in *change* (rather than merely for raw material) than Darwin's vision allowed. (Punctuated equilibrium does not challenge the continuationist claim per se, since paleontological punctuations proceed tolerably slowly in ecological time, but rather the reductionistic argument about the primacy of selection on organisms, since trends mediated by differential speciation offer such scope for true species selection—see Eldredge and Gould 1972; Gould 1982b. Thus punctuated equilibrium leads to hierarchy, not saltationism.)

In short, Darwin made evolution doable for the first time, but by holding so strongly to the substantive side of uniformitarianism, ultimately offered a restrictive version that hierarchies of causal levels and tiers of time (Gould 1985a) must extend. This discussion has raged for ten years and continues unabated, but I shall pursue it no farther here because I want to concentrate on the character and meaning of Darwin's second great contribution to a science of history.

Inferring history from its results

The uniformitarian argument constructs history from an observable, small-scale present. But how can scientists proceed when they have only results before them? Past processes are, in principle, unobservable, yet science traffics in process. How, then, can we make history doable if our data feature only its results?

I have come to view Darwin's sequence of books as proceeding at two levels—an explicit and conscious treatment of diverse subjects (from coral reefs, to orchids, to insectivorous plants, to climbing plants, to worms, to evolution); and a covert, perhaps unconscious extended treatise on historical methodology, with each book featuring a different principle of historical reconstruction. We may arrange these principles—three in number—in terms of decreasing information for making inferences.

First, the large-scale results may lie before us, and we can also measure the rate and effect of the process that presumably produces them. In such cases of maximal information, we can use the uniformitarian method in its purest form: make rigorous measurements of the

modern process and extrapolate into available time to render the full result. *The Formation of Vegetable Mould, Through the Action of Worms* (1881) is Darwin's finest example of this method. This book, Darwin's last, may also be his most misunderstood. Often seen as a pleasant trifle of old age, Darwin's worm book is a consciously chosen exemplar of historical reasoning at its most complete. What better choice of object than the humble, insignificant worm, working unnoticed literally beneath our feet? Could something so small really be responsible both for England's characteristic topography and for the upper layer of its soil?

Punch's commentary on Darwin's last work on worms (*Punch* 22 October 1881, p. 190)

Darwin's argument is pure uniformitarianism, carefully extended in stages. He counts worms to see if the soil contains enough for the work needed. He collects castings to measure the rate of churning (about 1/10 inch per year). He then extends the time scale to decades, via natural experiments on layers, once at the surface, that now lie coherently below material brought up by worms (burned coals, rubble from demolition, flints on his own ploughed field)—and then even farther by measuring the rate of foundering for historical objects (the "Druidical" rocks of Stonehenge, for example).

Second, we may have insufficient data about modern rates and processes simply to extrapolate their effects, but we can document several kinds or categories of results and seek relationships among them. Here we face a problem of taxonomy. Darwin argues that we may still proceed in the absence of direct data for uniformitarian extrapolation. We must formulate a historical hypothesis and then arrange the observed results as stages of its operation. The historical process, in other words, becomes the thread that ties all results together causally. This method succeeds because the process works on so many sequences simultaneously, but beginning at different times and proceeding at different rates in its various manifestations; therefore, all stages exist somewhere in the world at any one time (just as we may infer the course of a star's life by finding different stars in various stages of a general process, even though we trace no actual history of any individual star).

Darwin's first theoretical book, *The Structure and Distribution of Coral Reefs* (1842), illustrates this powerful guide to history. Its argument rests upon a classification of reefs into three basic categories of fringing, barrier, and atoll. Darwin proposed a common theme—subsidence of islands—to portray all three as sequential stages of a single historical process. Since corals build up and out from the edges of oceanic platforms, reefs begin by fringing their islands, become barriers as their islands subside, and finally atolls when their platforms submerge completely. But the taxonomy itself guarantees no history. During the nineteenth century, Darwin's opponents developed a series of counterproposals that may be called, collectively, "antecedent platform" theories. They argued that since corals build at the edges of platforms, these fringing reefs, barriers, and atolls only record the extent of previous planation, not a historical sequence in coral growth—platforms eroded to a small notch develop fringing reefs; those planed flat by waves become substrates for atolls. (Lyell, before Darwin convinced him otherwise, had advanced an even simpler ahistorical theory: that atolls develop on the circular rims of volcanoes.) The two theories can be distinguished by a crucial test not available in Darwin's time: no correlation between vertical extent of the reef and its form for antecedent platforms, progressive thickening from fringing reef to atoll in Darwin's subsidence. Twentieth-century drilling into Pacific atolls has affirmed Darwin's view.

Third, we must sometimes infer history from single objects; we have neither data for extrapolation from modern processes, nor even a series of stages to arrange in historical sequence. But how can a scientist infer history from single objects? This most common of historical dilemmas has a somewhat paradoxical solution. Darwin answers that we must look for imperfections and oddities, because any perfection in organic design or ecology obliterates the paths of history and might have been created as we find it. This principle of imperfection became Darwin's most common guide (if only because the fragmentary evidence of history often fails to provide better data in the preceding categories). I like to call it the "panda principle" in honor of my favorite example—the highly inefficient, but serviceable, false thumb of the panda, fashioned from the wrist's radial sesamoid bone because the true anatomical first digit had irrevocably evolved, in carnivorous ancestors of the herbivorous panda, to limited motility in running and clawing. (The herbivorous panda uses its sesamoid "thumb" for stripping leaves off bamboo shoots.) I titled one of my books *The Panda's Thumb* (1980) to honor this principle of historical reasoning.

The panda principle is a basic method of all histori-

cal science, linguistics, and history itself, for example, not just a principle for evolutionary reasoning. In *The Various Contrivances by which Orchids Are Fertilized by Insects* (1862), the book that followed the *Origin of Species*, Darwin shows that the wondrously complex adaptations of orchids, so intricately fashioned to aid fertilization by insects, are all jury-rigged from the ordinary parts of flowers, not built to the optimum specifications of an engineer's blueprint. "The use of each trifling detail of structure," Darwin writes, "is far from a barren search to those who believe in natural selection" (1888 ed., p. 286).

Throughout all these arguments, Darwin also

We must look for imperfections and oddities, because any perfection in organic design or ecology obliterates the paths of history and might have been created as we find it

---

showed his keen appreciation for the other great principle of historical science—the importance of proper taxonomies. In a profession more observational and comparative than experimental, the ordering of diverse objects into sensible categories becomes a sine qua non of causal interpretation. A taxonomy is not a mindless allocation of objective entities into self-evident pigeonholes, but a theory of causal ordering. Proper taxonomies require two separate insights: the identification and segregation of the basic phenomenon itself, and the division of its diverse manifestations into subcategories that reflect process and cause. Consider, for example, Steno's *Prodromus* (1669). This founding document of geology is, fundamentally, a new taxonomy (see Gould 1983). Steno identifies solid objects enclosed in other solids as the basic phenomenon (a stunningly original and peculiar choice in the light of ordering principles generally accepted in his time); he then makes a fundamental division into objects hard before surrounded (fossils) and those introduced without initial solidity into a rigid matrix (crystals in geodes, for example). Using these divisions, Steno could identify the organic origins of fossils and the temporal nature of strata—the cornerstones of historical geology.

Darwin's *Different Forms of Flowers on Plants of the Same Species* (1877) is a fine illustration of how taxonomy informs history. The basic recognition of the phenomenon itself as a unitary puzzle poses a historical question. The work then becomes a long argument about subdivisions by function (heterostyly to assure cross-fertilization, cleistogamy to allow some advantageous selfing while retaining other forms of flowers for occasional crossing, for example), and about ancestral states and the paths of potential transformation.

The *Origin of Species* achieves its conceptual power by using all these forms of historical argument: uniformitarianism in extrapolating the observed results of artificial selection by breeders and farmers; inference of history from temporal ordering of coexisting phenomena (in constructing, for example, a sequence leading from variation within a population, to small-scale geographic differentiation of races, to separate species, to the origin of major groups and key innovations in morphology); and, most often and to such diverse effect, the panda principle of imperfection (vestigial organs, odd biogeographic distributions made sensible only as products of history, adaptations as contrivances jury-rigged from parts available).

I do not know whether Darwin operated by conscious design to construct his multivolumed treatise on historical method. Since great thinkers so often work by what our vernacular calls intuition (though the process involves no intrinsic mystery, as logical reconstruction by later intellectual biographers can attest), conscious intent is no criterion of outcome. Still, I like to think that the last paragraph of Darwin's last book records his own perception of connection—for he closes his last treatise on worms by remembering his first on corals, thereby linking both his humble subjects and his criteria of history:

It may be doubted whether there are many other animals which have played so important a part in the history of the world, as have these lowly organized creatures [worms]. Some other animals, however, still more lowly organized, namely corals, have done far more conspicuous work in having constructed innumerable reefs and islands in the great oceans.

Using the panda principle

Historical science is still widely misunderstood, underappreciated, or denigrated. Most children first meet science in their formal education by learning about a powerful mode of reasoning called "*the* scientific method." Beyond a few platitudes about objectivity and willingness to change one's mind, students learn a restricted stereotype about observation, simplification to tease apart controlling variables, crucial experiment, and prediction with repetition as a test. These classic "billiard ball" models of simple physical systems grant no uniqueness to time and object—indeed, they remove any special character as a confusing variable—lest repeatability under common conditions be compromised. Thus, when students later confront history, where complex events occur but once in detailed glory, they can only conclude that such a subject must be less than science. And when they approach taxonomic diversity, or phylogenetic history, or biogeography—where experiment and repetition have limited application to systems in toto—they can only conclude that something beneath science, something merely "descriptive," lies before them.

These historical subjects, placed into a curriculum of science, therefore become degraded by their failure to match a supposedly universal ideal. They become, in our metaphors, the "soft" (as opposed to "hard") sciences, the "merely descriptive" (as opposed to "rigorously experimental"). Every year Nobel prizes are announced to front-page fanfare, and no one who works with the complex, unrepeated phenomena of history can win— for the prizes only recognize science as designated by the stereotype. (I'm not bitching—since it makes for a much more pleasant profession—only making a social comment.) Plate tectonics revolutionized our view of the

earth, but its authors remain anonymous to the public eye; molecular phylogeny finally begins to unravel the complexities of genealogy, and its accomplishments rank as mere narrative. Harvard organizes its Core Curriculum and breaks conceptual ground by dividing sciences into the two major styles of experimental-predictive and historical, rather than, traditionally, by discipline. But guess which domain becomes "Science A" and which "Science B"?

In a perverse way, the best illustration of this failure to understand the special character of history can be found in writings by opponents of science, who use clever rhetoric to argue against evolution because it doesn't work like their simplistic view of physics. In his book *Algeny* (1983), for example (see Gould 1985b for a general critique), Jeremy Rifkin dismisses Darwin because evolution can't be turned into a controlled laboratory experiment: "To qualify as a science, Darwin's theory should be provable by means of the scientific method. In other words, its hypotheses should be capable of being tested experimentally" (p. 117). Rifkin then cites Dobzhansky's statement about history as though it represents a fatal confession: "Dobzhansky laments the fact that 'evolutionary happenings are unique, unrepeatable, and irreversible. It is as impossible to turn a vertebrate into a fish as it is to effect the reverse transformation.' Dobzhansky is chagrined" (p. 118). But Dobzhansky in this passage is neither sheepish nor troubled; he is simply commenting upon the nature of history. Rifkin concludes nonetheless: "Embarrassing, to say the least. Here is a body of thought, incapable of being scientifically tested. . . . If not based on scientific observation, then evolution must be a matter of personal faith" (pp. 118–19).

When creationist lawyer Wendell Bird took my deposition in pretrial hearings on the constitutionality of Louisiana's "creation-science" law (mercifully tossed out without trial by a federal judge), he spent an inordinate amount of time (and verbal trickery) trying to make me admit that a suspension of natural law—a miracle— might fall within the purview of science. At the close of this lamentable episode, I was astounded when he asked "Are you familiar with the term singularistic?" and then tried to argue that complex historical events (singularistic in that sense), as unrepeatable, are somehow akin to miracles (singularities) and therefore make such historical sciences as evolution either as good or as bad as the Genesis-literalism of so-called creation-science!

These arguments about history are red herrings and we would do well to suppress them by acknowledging the *different* strengths of historical science, lest the simplistic stereotype be turned, as in these cases, against all of science.

The "lesser" status of historical science may be rejected on two grounds. First, it is not true that standard techniques of controlled experimentation, prediction, and repeatability cannot be applied to complex histories. Uniqueness exists in toto, but "nomothetic undertones" (as I like to call them) can always be factored out. Each mass extinction has its endlessly fascinating particularities (trilobites died in one, dinosaurs in another), but a common theme of extraterrestrial impact (Alvarez et al. 1980) may trigger a set of such events, even perhaps on a regular cycle (Raup and Sepkoski 1984). Nature, moreover, presents us with experiments aplenty, imperfectly controlled compared with the best laboratory standards, but having other virtues (temporal extent, for example) not attainable with human designs.

Second, as argued earlier, Darwin labored for a lifetime to meet history head on, and to establish rigorous methods for inference about its singularities. History, by Darwin's methods, is knowable in principle (though not fully recoverable in every case, given the limits of evidence), testable, and different. We do not attempt to predict the future (I could already retire in comfort if someone paid me a dollar for each rendition of my "paleontologists don't predict the future" homily following the inevitable question from the floor at all presentations to nonscientists—"Well, where is human evolution going anyway?"). But we can postdict about the past—and do so all the time in historical science's most common use of repeatability (every new iridium anomaly at the Cretaceous-Tertiary boundary is a repeated postdicted affirmation of Alvarez's conjecture about impact, based upon just three sites in the original article).

Finally, history's richness drives us to different methods of testing, but testability (via postdiction) is our method as well. Huxley and Darwin maintained interestingly different attitudes toward testing in history. Huxley, beguiled by the stereotype, always sought a crucial observation or experiment (the destruction of theory by a "nasty, ugly, little fact" of his famous aphorism). Darwin, so keenly aware of both the strengths and limits of history, argued that iterated pattern, based on types of evidence so numerous and so diverse that no other coordinating interpretation could stand—even though any item, taken separately, could not provide conclusive proof—must be the criterion for evolutionary inference. (The great philosopher of science William Whewell had called this historical method "consilience of inductions.") Huxley sought the elusive crucial experiment; Darwin strove for attainable consilience. Di Gregorio's recent treatise on Huxley's scientific style contains an interesting discussion of this difference (1984). (Ironically, Whewell, a conservative churchman, later banned Darwin's *Origin* from the library of Trinity College, Cambridge, where he was master. What greater blow than the proper use of one's own arguments in an alien context.)

In an essay of 1860, for example, Huxley wrote: "but there is no positive evidence, at present, that any group of animals has, by variation and selective breeding, given rise to another group which was, even in the least degree, infertile with the first. Mr. Darwin is perfectly aware of this weak point, and brings forward a multitude of ingenious and important arguments to diminish the force of the objection" (quoted by di Gregorio, p. 61). Note particularly Huxley's subtle misunderstanding of Darwin's methodology. What Huxley views as a set of indirect arguments, presented faute de mieux in the absence of experimental proof, *is* Darwin's consilience, positively developed as the proper method of historical inference. Darwin complained of just this misunderstanding in a letter to Hooker in 1861: "change

of species cannot be directly proved . . . the doctrine must sink or swim according as it groups and explains phenomena. It is really curious how few judge it in this way, which is clearly the right way" (di Gregorio, p. 62). And, more specifically, in his *Variation of Animals and Plants Under Domestication* (1868):

Now this hypothesis [natural selection] may be tested—and this seems to me the only fair and legitimate manner of considering the whole question—by trying whether it explains several large and independent classes of facts; such as the geological succession of organic beings, their distribution in past and present times, and their mutual affinities and homologies. If the principle of natural selection does explain these and other large bodies of facts, it ought to be received. [I, 657]

Historical scientists try to import bodily an oversimplified caricature of "hard" science

Despite the ready availability of these powerful, yet different, modes of inference, historical scientists have often been beguiled by the stereotype of direct experimental proof, and have wallowed in a curious kind of self-hate in trying to ape, where not appropriate, supposedly universal procedures of *the* scientific method. Many of the persistent debates within evolutionary biology are best viewed as a series of attempts to divest evolution of history under the delusion that scientific rigor gains thereby—with responses by defenders that history cannot be factored away, and that good science can be done just splendidly with it.

In extreme versions, for example, the welcome and powerful movement of "equilibrium" biogeography and ecology, which developed in the 1960s and peaked in the 1970s, not only denied history, but viewed its singularities as impediments to real science. Equilibrium models avoid history by explaining current situations as active balances maintained between competing forces now operating. Such models apply a reverse panda principle by identifying nonequilibria as signs of history—situations not yet balanced and therefore in a relevant state of history. Equilibria, when reached, are timeless, changing only when the measurable inputs alter, and not by any historically bound "evolution" of the system.

Ironically, the founding document of this movement was written by two fine historical scientists who understood proper limits, and who used their models to identify nomothetic undertones of a valued history (MacArthur and Wilson 1967). They also explicitly discussed the interactions of history and equilibrium, and the long-term evolutionary adjustments that continue, albeit at slower rates, within systems at equilibrium: "The equilibrium model has the virtues of making testable predictions which were not immediately obvious and of making the individual vagaries of island history seem somewhat less important in understanding the diversity of the island's species. Of course the history of islands remains crucial to the understanding of the taxonomic composition of species" (1967, p. 64). But in the hands of singleminded and less thoughtful supporters, equilibrium ecology moved from suggestive simplification (or search for repeated undertones) to a hard substantive claim spearheading a crusade for bringing "real" science into an antiquated domain of descriptive natural history. The campaign quickly stalled, however; nature fights back effectively.

The most common denial of history made by self-styled Darwinian evolutionists resides in claims for optimality—conventionally for the mechanics of morphology, more recently for behavior and ecology. Again, optimality theory has its place and uses (primarily in designating ideals for assessing natural departures). Committed votaries think that they are celebrating evolution by showing how inexorably and fine the mills of natural selection can grind; in fact, they are attempting to abrogate Darwin's most important criterion of history—the panda principle of imperfection and oddity as signs of previous lifestyles and affinities genealogically pre-

Evolution is a bush, not a ladder

served. Under certain conditions of minimal constraint, we may legitimately seek optimality (animal color patterns, often less subject than morphology to developmental covariance, represent one promising domain—see Cott 1940 for the classic statement). Usually, history and complexity must assert themselves prominently.

The sad tale goes on and on. Historical scientists, who should take legitimate pride in their different ways, try to import bodily an oversimplified caricature of "hard" science, or simply bow to pronouncements of professions with higher status. Some accepted Kelvin's last and most restrictive dates for a young earth, though fossils and strata spoke differently; many more foreswore their own data when physicists proclaimed that continents cannot move laterally. Charles Spearman misused factor analysis to identify intelligence as a measurable physical thing in the head, and then said of psychology that "this Cinderella among the sciences has made a bold bid for the level of triumphant physics itself" (see Gould 1981, chap. 6).

But the great historical scientists have always treasured both their rigorous, testable methods and their singular data. D'Arcy Thompson, whose own vision of optimal form, impressed directly upon organisms by physical forces, must rank among the most ahistorical of approaches to evolution (see Gould 1971), nonetheless knew that a retrievable history pervaded all objects—and that the panda principle can recover it. He wrote in his incomparable prose (1942):

Immediate use and old inheritance are blended in Nature's handiwork as in our own. In the marble columns and architraves of a Greek temple we still trace the timbers of its wooden prototype, and see beyond these the tree-trunks of a primeval sacred grove; roof and eaves of a pagoda recall the sagging mats which roofed an earlier edifice; Anglo-Saxon land-tenure influences the planning of our streets and the cliff-dwelling and cave-dwelling linger on in the construction of our homes! So we see enduring traces of the past in the living organism—landmarks which have lasted on through altered function and altered needs. [pp. 1020–21]

The triumph of homology

Louis Agassiz chose an enigmatic name for his institution—and for good cause. He called it the Museum of Comparative Zoology (I am sitting in its oldest section as I write this article) in order to emphasize that the sciences of organic diversity do not usually seek identity in repeated experiment, but work by comparing the similarities among objects of nature as given. Kind, extent, and amount of similarity provide the primary data of historical science.

As a problem, recognized since Aristotle, natural similarities come in two basic, largely contradictory styles. We cannot simply measure and tabulate; we must factor and divide. Similarities may be homologies, shared by simple reason of descent and history, or analogies, actively developed (independently, but to similar form and effect) as evolutionary responses to common situations.

Systematics (the science of classifying organisms) is the analysis of similarity in order to exclude analogy and recognize homology. Such an epigram may sacrifice a bit of subtlety for crisp epitome, but it does capture the first goal of historical science. Homologous similarity is the product of history; analogy, as independent tuning to current circumstance, obscures the paths of history.

The major brouhaha about cladistics (see Hennig 1966; Eldredge and Cracraft 1980; any issue of *Systematic Zoology* for the past decade, or, for self-serving misuse by yet another opponent of science, Bethell 1985) has unfolded in needless acrimony because cladistics has not been properly recognized, even by some of its strongest champions, as a "pure" method for defining historical order and rigidly excluding all other causes of similarity. Cladistics allies objects in branching hierarchies defined only by relative times of genealogical connection. Closest, or "sister-group," pairs share a unique historical connection (a common ancestor yielding them as its only descendants). The system then connects sister-group pairs into ever-more inclusive groups sharing the same genealogical uniqueness (common ancestry that includes absolutely all descendant branches and absolutely no other groups). Cladistics is the science of ordering by genealogical connection, *and nothing else*. As such, it is the quintessential expression of history's primacy above any other cause or expression of similarity—and, on this basis alone, should be received with pleasure by evolutionists.

Nonetheless, several of the most forceful cladists never grasped this central point clearly; they buried their subject in frightful terminology and such exaggerated or extended claims that they antagonized many key systematists and never won the general approbation they deserved. In an almost perverse interpretation (literal, not ethical), some supporters, the self-styled "transformed" or "pattern" cladists, have actually negated the central strength of their method as a science of history by claiming—based on a curiously simplistic reading of Karl Popper—that science should eschew all talk of "process" (or cause) and work only with recoverable "pattern" (or the branching order of cladograms). Pattern cladists are not anti-evolution (as misportrayed by Bethell 1985); rather, as a result of narrow commitment to an extreme

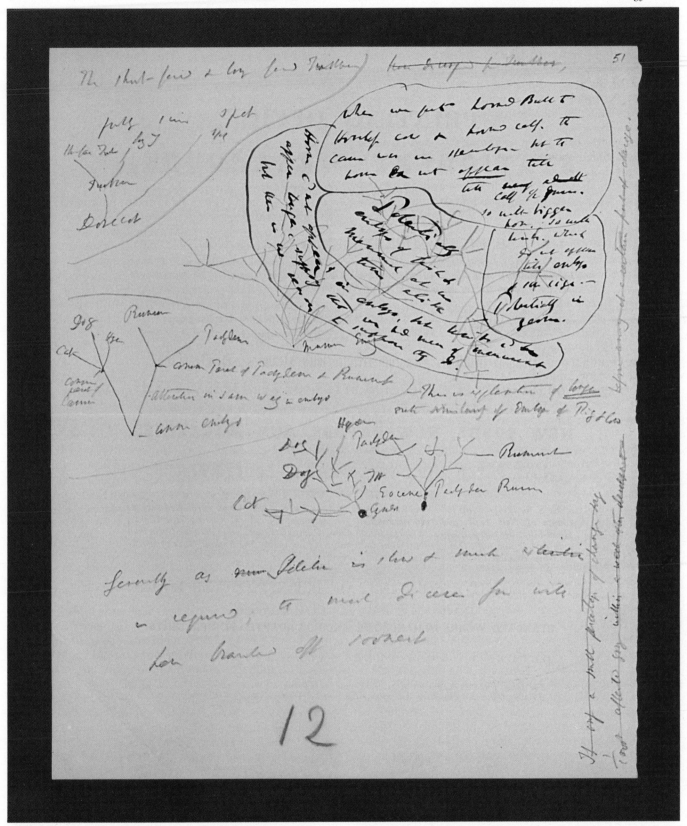

This unpublished page from Darwin's file "Embryology" shows Darwin constructing congruent trees of relationship, based in one case on developmental similarities, and in the other on time of divergence. The page, which dates from the mid-1850s, though by no means his first use of trees, is one of his most sophisticated. In the upper left, Darwin shows the descent of two breeds of tumbler pigeons from the common pigeon. Below, he arranges carnivores on one great branch and ungulates (divided into pachyderms and ruminants) on the other, which unite at a base labeled "common embryo." The central tree relates the same groups, but grows through geological time, rather than developmental space; here Darwin writes "Eocene Pachyderm Ruminant." On the right, he discusses the inheritance, expression, timing, and alteration of characters in development. (Summary by S. Rachootin; photograph of DAR 205.6:51 reproduced by permission of the Syndics of the Cambridge University Library.)

view of empiricism and falsification, they choose to ignore evolution as unprovable talk about process, and to concentrate on recovery of branching pattern alone. All well and good; it can be done with cladistic methods. But unless we wish to abandon a basic commitment to cause and natural law, branching order must arise for a reason (by a process, if you will). And that process is history, however history be made. In other words, the clearest method ever developed for discerning history has been twisted by some supporters to a caricatured empiricism that denies its subject. What an irony. We need to recover cladistics—a pure method for specifying those histories that develop by divergent and irreversible branching—from some of its own most vociferous champions.

Cladistics, as a methodology, is conceptually sound. In biological systematics, it has often proved inconclusive in practice (leading to a false suspicion that the method itself might be logically flawed) because taxonomists have worked, generally faute de mieux, or by simple weight of tradition, with the inappropriate data of morphology. Homology can often be recovered from morphology, but the forms of organisms often include an inextricable mixture of similarities retained by history (homology) and independently evolved in the light of common function (analogy). Morphology is not the best source of data for unraveling history. Thoughtful systematists have known this ever since Darwin, but have proceeded as best they could in the absence of anything better.

In principle, the recovery of homology only requires a source of information with two properties: sufficiently numerous and sufficiently independent items to preclude, on grounds of mathematical probability alone, any independent origin in two separate lineages. The "items" of morphology are too few and too bound in complex webs of developmental correlation to yield this required independence. Yet the discoveries and techniques of molecular biology have now provided an appropriate source for recovering homology—a lovely example of science at its unified best, as a profession firmly in the camp of repetition and experiment provides singular data for history. Molecular phylogenies work not because DNA is "better," more real, or more basic than morphology, but simply because the items of a DNA program are sufficiently numerous and independent to ensure that degrees of simple matching accurately measure homology. The most successful technique of molecular phylogeny has not relied upon sophisticated sequencing, but rather the "crudest" brute-force matching of all single-copy DNA—for such a method uses a maximum number of independent items (Sibley and Ahlquist 1981, 1984, and 1985).

Consider just one elegant example of the triumph of homology (as determined by molecular methods)—a case that should stand as sufficient illustration of the primacy of history as both a basic cause of order and a starting point for all further analyses. The songbirds of Australia have posed a classic dilemma for systematists because they seem to present a biogeographic and taxonomic pattern so different from the mammals. The marsupials of Australia form a coherent group bound by homology and well matched to the geographic isolation

of their home. But the songbirds include a large set of creatures apparently related to several Eurasian and American groups: warblers, thrushes, nuthatches, and flycatchers, for example. This classification has forced the improbable proposal that songbirds migrated to Australia in several distinct waves. Even though birds fly and mammals do not, the discordance of pattern between the two groups has been troubling, since most birds are not nearly so mobile as their metaphorical status suggests. But Sibley and Ahlquist (1985) have now completed the molecular phylogeny (true homology) of Australian songbirds—and they form a coherent group of homologues, just like the marsupials. The similarities to Eurasian and American taxa are all convergences—independently evolved analogies—not homologies. Marsupials, of course, have also evolved some astounding convergences upon placentals (marsupial "wolves," "mice," and "moles"), but systematists have not been confounded because all marsupials retain signatures of homology in their pouches and epipubic bones. We now realize that the apparent and deeply troubling discordance between Australian birds and mammals arose only because the birds did not retain such morphological signatures of homology, and systematists were therefore fooled by the striking analogies.

Once we map homologies properly, we can finally begin to ask interesting biological questions about function and development—that is, we can use morphology for its intrinsic sources of enlightenment, and not as an inherently flawed measure of genealogical relationships. The historical flow of protoplasm through branching systems of genealogy *is* the material reality that structures biology. But analogies have a different (and vital) meaning that the resolution of history finally permits us to appreciate. They are functional themes that stand, almost like a set of Platonic forms, in a domain for biomechanics, functional morphology, and an entire set of nonhistorical disciplines. "Wrenness," "mouseness," and "wormness" may not have the generality of forms so basic that they have rarely been confused with homologies (bilateral symmetry, for example), but they still represent iterated themes, standards of design external to history. The material reality of history—phylogeny—flows through them again and again, forming a set of contrasts that define the fascination of biology: homology and analogy, history and immanence, movement and stability.

If the primacy of history is evolution's lesson for other sciences, then we should explore the consequences of valuing history as a source of law and similarity, rather than dismissing it as narrative unworthy of the name science. I argued in the first section that Darwin cut through a tangle of unnecessary complexity by proposing "just history" as a disconcertingly simple answer in domains where science had sought a deeper, rational immanence (as for the ordering of taxonomy). I wonder how far we might extend this insight. Consider our relentless search for human universals and our excitement at the prospect that we may thereby unlock something at the core of our being. So Jung proposes archetypes of the human psyche in assessing the similarities of mythic systems across cultures, while others invoke brain structure and natural selection as a source of uncannily complex repetitions among human groups long separate. Yet evolution is a bush, not a ladder. *Homo sapiens* had a discrete and recent origin, presumably as an isolated local population, crowned with inordinate later success. Many of these similarities may therefore be simple homologies of a contingent history, not deep immanences of the soul. Such an offbeat idea might provide an astonishingly simple solution to some of the oldest dilemmas born of the Socratic injunction—know thyself.

Finally, history seems to be extending its influence to ever-widening domains. Soon we may no longer be able even to maintain the basic division of two scientific styles discussed in this paper (which only advocates equal treatment for the equally scientific, but different, historical sciences). The latest researches in cosmology are suggesting that the laws of nature themselves, those supposed exemplars of timeless immanence, may also be contingent results of history! Had the universe passed through a different (and possible) history during its first few moments after the big bang, nature's laws might have developed differently. Thus everything, ultimately, may be a product of history—and we will need to understand, appreciate, and use the principles of historical science throughout our entire domain of zealous research.

This has been an unconventional discharge of an appointed duty—to write a centennial essay on the impact of evolution. I have not written a traditional review. I have not chronicled the major advances and discoveries of evolution. I have tried, instead, to suggest that evolution's essential impact upon the practice of science has been methodological—validating the historical style as equally worthy and developing for it a rigorous methodology, outlined by Darwin himself in his most distinctive (but largely untouted) contribution, and continually refined to the kind of ultimate triumph for homology that molecular phylogeny can provide.

I have presented nothing really new, only a plea for appreciating something so basic that we often fail to sense its value. With a bow to that overquoted line from T. S. Eliot, I only ask you to return to a place well known and see it for the first time.

References

Alvarez, L. W., W. Alvarez, F. Asaro, and H. V. Michel. 1980. Extraterrestrial cause for the Cretaceous-Tertiary extinction. *Science* 208:1095–1108.

Bethell, T. 1985. Agnostic evolutionists: The taxonomic case against Darwin. *Harper's*. Feb., pp. 49–61.

Carr, E. H. 1961. *What Is History?* Vintage Books.

Chambers, R. 1844. *Vestiges of the Natural History of Creation*. London: John Churchill. (Published anonymously.)

Cott, H. B. 1940. *Adaptive Coloration in Animals*. Methuen.

Darwin, C. 1842. *The Structure and Distribution of Coral Reefs*. London: Smith, Elder.

———. 1859. *On the Origin of Species*. London: John Murray.

———. 1862. *The Various Contrivances by which Orchids Are Fertilized by Insects*. London: John Murray.

———. 1868. *Variation of Animals and Plants Under Domestication*. London: John Murray.

———. 1877. *Different Forms of Flowers on Plants of the Same Species*. London: John Murray.

———. 1881. *The Formation of Vegetable Mould, Through the Action of Worms*. London: John Murray.

di Gregorio, M. A. 1984. *T. H. Huxley's Place in Natural Science*. Yale Univ. Press.

Eldredge, N., and J. Cracraft. 1980. *Phylogenetic Patterns and the Evolutionary Process*. Columbia Univ. Press.

Eldredge, N., and S. J. Gould. 1972. Punctuated equilibria: An alternative to phyletic gradualism. In *Models in Paleobiology*, ed. T. J. M. Schopf, pp. 82–115. Freeman, Cooper, and Co.

Ghiselin, M. 1969. *The Triumph of the Darwinian Method*. Univ. of California Press.

Gingerich, P. D. 1983. Rates of evolution: Effects of time and temporal scaling. *Science* 222:159.

Gould, S. J. 1965. Is uniformitarianism necessary. *Am. J. Sci.* 263:223–28.

———. 1971. D'Arcy Thompson and the science of form. *New Lit. Hist.* 2 (2):229–58.

———. 1980. *The Panda's Thumb*. W. W. Norton.

———. 1981. *The Mismeasure of Man*. W. W. Norton.

———. 1982a. Darwinism and the expansion of evolutionary theory. *Science* 216:380–87.

———. 1982b. The meaning of punctuated equilibrium and its role in validating a hierarchical approach to macroevolution. In *Perspectives on Evolution*, ed. R. Milkman, pp. 83–104. Sunderland, Mass.: Sinauer Assoc.

———. 1983. *Hen's Teeth and Horse's Toes*. W. W. Norton.

———. 1984. Smooth curve of evolutionary rate: A psychological and mathematical artifact. *Science* 226:994–95.

———. 1985a. On the origin of specious critics. *Discover*. Jan., pp. 34–42.

———. 1985b. The paradox of the first tier: An agenda for paleobiology. *Paleobiology* 11:2–12.

Hennig, W. 1966. *Phylogenetic Systematics*. Univ. of Illinois Press.

Hooykaas, R. 1963. *The Principle of Uniformity in Geology, Biology, and Theology*. Leiden: E. J. Brill.

Huxley, T. H. 1893. The origin of species [1860]. In *Collected Essays*, vol. 2: *Darwiniana*, pp. 22–79. Appleton.

Kitcher, P. 1985. Darwin's achievement. In *Reason and Rationality in Science*, ed. N. Rescher, pp. 123–85. Pittsburgh: Stud. Phil. Sci.

Lamarck, J. B. 1809. *Philosophie zoologique*. Trans. 1984 by H. Elliot as *Zoological Philosophy*. Univ. of Chicago Press.

Lyell, C. 1842. *Principles of Geology*. 6th ed., 3 vols. Boston: Hilliard, Gray, and Co.

MacArthur, R. H., and E. O. Wilson. 1967. *The Theory of Island Biogeography*. Princeton Univ. Press.

Oken, L. 1805–11. *Lehrbuch der Naturphilosophie*. Jena: F. Frommand.

Raup, D. M., and J. J. Sepkoski, Jr. 1984. Periodicity of extinctions in the geologic past. *PNAS* 81:801–05.

Rifkin, J. 1983. *Algeny*. Viking Press.

Rudwick, M. J. S. 1972. *The Meaning of Fossils*. London: MacDonald.

Ruse, M. 1979. *The Darwinian Revolution: Science Red in Tooth and Claw*. Univ. of Chicago Press.

Sibley, C. G., and J. E. Ahlquist. 1981. The phylogeny and relationships of the ratite birds as indicated by DNA-DNA hybridization. In *Evolution Today*, ed. G. G. E. Scudder and J. L. Reveal, pp. 301–35. Proc. Second Intl. Cong. Syst. Evol. Biol.

———. 1984. The phylogeny of the hominoid primates, as indicated by DNA-DNA hybridization. *J. Mol. Evol.* 20:2–15.

———. 1985. The phylogeny and classification of the Australo-Papuan passerine birds. *The Emu* 85:1–14.

Sober, E. 1984. *The Nature of Selection*. MIT Press.

Steno, N. 1669. *De solido intra solidum naturaliter contento dissertationis prodromus*. Trans. 1916 by J. G. Winter as *The Prodromus of Nicolaus Steno's Dissertation*. Macmillan.

Swainson, W. 1835. *Classification of Quadrupeds*. London: Dr. Lardner's Cabinet Cyclopedia.

Thompson, D. W. 1942. *Growth and Form*. Macmillan.

Vrba, E. S., and N. Eldredge. 1984. Individuals, hierarchies and processes: Towards a more complete evolutionary theory. *Paleobiology* 10:146–71.

Bert Hölldobler
Edward O. Wilson

The Evolution of Communal Nest-Weaving in Ants

Steps that may have led to a complicated form of cooperation in weaver ants can be inferred from less advanced behavior in other species

One of the most remarkable social phenomena among animals is the use of larval silk by weaver ants of the genus *Oecophylla* to construct nests. The ants are relatively large, with bodies ranging up to 8 mm in length, and exclusively arboreal. The workers create natural enclosures for their nests by first pulling leaves together (Fig. 1) and then binding them into place with thousands of strands of larval silk woven into sheets. In order for this unusual procedure to succeed, the larvae must cooperate by surrendering their silk on cue, instead of saving it for the construction of their own cocoons. The workers bring nearly mature larvae to the building sites and employ them as living shuttles, moving them back and forth as they expel threads of silk from their labial glands.

The construction of communal silk nests has clearly contributed to the success of the *Oecophylla* weaver ants. It permits colonies to attain populations of a half million or more, in spite of the large size of the

The authors take pleasure in dedicating this article to Caryl P. Haskins, fellow myrmecologist and distinguished scientist and administrator, on the occasion of his seventy-fifth birthday and his retirement from the chairmanship of the Board of Editors of American Scientist.

Bert Hölldobler is Alexander Agassiz Professor of Zoology at Harvard University. After completing his Dr. rer. nat. at the University of Würzburg in 1965 and his Dr. habil. at the University of Frankfurt in 1969, he served at the latter institution as privatdocent and professor until 1973, when he joined the Harvard faculty. Edward O. Wilson is Frank B. Baird Jr. Professor of Science and Curator in Entomology of the Museum of Comparative Zoology, Harvard University. He received his Ph.D. from Harvard in 1955, held a Junior Fellowship in the Society of Fellows during 1953–56, and has served on the faculty continuously since that time. Address: Museum of Comparative Zoology, Harvard University, Cambridge, MA 02138.

workers, because the ants are freed from the spatial limitations imposed on species that must live in beetles' burrows, leaf axils (the area between the stems of leaves and the parent branch), and other preformed vegetative cavities. This advance, along with the complex recruitment system that permits each colony to dominate up to several trees at the same time, has helped the weaver ants to become among the most abundant and successful social insects of the Old World tropics (Hölldobler and Wilson 1977a, b, 1978; Hölldobler 1979). A single species, *O. longinoda*, occurs across most of the forested portions of tropical Africa, while a second, closely related species, *O. smaragdina*, ranges from India to Queensland, Australia, and the Solomon Islands. The genus is ancient even by venerable insect standards: two species are known from Baltic amber of Oligocene age, about 30 million years old (Wheeler 1914). *O. leakeyi*, described from a fossil colony of Miocene age (approximately 15 million years old) found in Kenya, possessed a physical caste system very similar to that of the two living forms (Wilson and Taylor 1964).

Our recent studies, building on those of other authors, have revealed an unexpectedly precise and stereotyped relation between the adult workers and the larvae. The larvae contribute all their silk to meet the colony's needs instead of their own. They produce large quantities of the material from enlarged silk glands early in the final instar rather than at its end, thus differing from cocoon-spinning ant species, and they never attempt to construct cocoons of their own (Wilson and Hölldobler 1980). The workers have taken over almost all the spinning movements from the larvae, turning them into passive dispensers of silk.

It would seem that close attention to the exceptional properties of *Oecophylla* nest-weaving could shed new light on how cooperation and altruism operate in ant colonies, and especially on how larvae can function as an auxiliary caste. In addition, a second, equally interesting question is presented by the *Oecophylla* case: How could such extreme behavior have evolved in the first place? As is the case with the insect wing, the vertebrate eye, and other biological prodigies, it is hard to conceive how something so complicated and efficient in performance might be built from preexisting structures and processes. Fortunately, other phyletic lines of ants have evolved communal nest-weaving independently and to variably lesser degrees than *Oecophylla*, raising the prospect of reconstructing the intermediate steps leading to the extreme behavior of weaver ants. These lines are all within the Formicinae, the subfamily to which *Oecophylla* belongs. They include all the members of the small Neotropical genus *Dendromyrmex*, the two Neotropical species *Camponotus* (*Myrmobrachys*) *senex* and *C.* (*M.*) *formiciformis*, which are aberrant members of a large cosmopolitan genus, and various members of the large and diverse Old World tropical genus *Polyrhachis*.

Two additional but doubtful cases have been reported outside the Formicinae. According to Baroni Urbani (1978), silk is used in the earthen nests of some Cuban species of *Leptothorax*, a genus of the subfamily Myrmicinae. However, the author was uncertain whether the material is obtained from larvae or from an extraneous source such as spider webs. Since no other myrmicine is known to produce silk under any circumstances, the latter alternative seems the more probable.

Similarly, the use of silk to build nests was postulated for the Javan ant *Technomyrmex bicolor textor*, a member of the subfamily Dolichoderinae, in an early paper by Jacobson and Forel (1909). But again, the evidence is from casual field observations only, and the conclusion is rendered unlikely by the fact that no other dolichoderines are known to produce silk.

During the past ten years we have studied the behavior of both living species of *Oecophylla* in much greater detail than earlier entomologists, and have extended our investigations to two of the other, poorly known nest-weaving genera, *Dendromyrmex* and *Polyrhachis*. This article brings together the new information that resulted from this research and some parallel findings of other authors, in a preliminary characterization of the stages through which the separate evolving lines appear to have passed.

In piecing together our data, we have utilized a now-standard concept in organismic and evolutionary biology, the phylogenetic grade. The four genera of formicine ants we

have considered are sufficiently distinct from each other on anatomical evidence as to make it almost certain that the communal nest-weaving displayed was in each case independently evolved. Thus it is proper to speak of the varying degrees of cooperative behavior and larval involvement not as the actual steps that led to the behavior of *Oecophylla* but as grades, or successively more advanced combinations of traits, through which autonomous evolving lines are likely to pass. Other combinations are possible, even though not now found in living species, and they might be the ones that were actually traversed by extreme forms such as *Oecophylla*. However, by examining the behavior of as many species and phyletic lines as possible, biologists are sometimes able to expose consistent trends and patterns that lend convincing weight to particular evolutionary reconstructions. This technique is especially promising in the case of insects, with several million living species to sample. Within this vast array there are more than 10,000 species of ants, most of which have never been

studied, making patterns of ant behavior exceptionally susceptible to the kind of analysis we have undertaken and are continuing to pursue on communal nest-building.

The highest grade of cooperation

The studies conducted on *Oecophylla* prior to our own were reviewed by Wilson (1971) and Hemmingsen (1973). In essence, nest-weaving with larval silk was discovered in *O. smaragdina* independently by H. N. Ridley in India and W. Saville-Kent in Australia, and was subsequently described at greater length in a famous paper by Doflein (1905). Increasingly detailed accounts of the behavior of *O. longinoda*, essentially similar to that of *O. smaragdina*, were provided by Ledoux (1950), Chauvin (1952), Sudd (1963), and Hölldobler and Wilson (1977a).

The sequence of behaviors by which the nests are constructed can be summarized as follows. Individual workers explore promising sites within the colony's territory, pulling at the edges and tips of leaves. When

Figure 1. To make a nest out of leaves and larval silk, worker ants of the species *Oecophylla smaragdina*, the Australian green tree ant, first choose a pliable leaf. They then form a row and pull in unison, as shown in the photograph, until they force two leaves to touch or one leaf to curl up on itself. (All photographs are by Bert Hölldobler unless otherwise indicated.)

Figure 4. Once leaves have been pulled together to form a nest, the workers hold them in place with larval silk. A simple form of weaving is practiced by the workers of an Australian *Polyrhachis* species similar to *doddi*; at the top, one worker holds a larva above the surface, allowing it to perform most of the weaving movements. The most sophisticated type of weaving has been developed by *O. smaragdina*; in the bottom photograph, *O. smaragdina* workers perform almost all the movements while the larvae serve principally as passive shuttles.

Intermediate steps

The existence of communal nest-weaving in *Polyrhachis* was discovered in the Asiatic species *Polyrhachis* (*Myrmhopla*) *dives* by Jacobson (Jacobson and Wasmann 1905). However, few details of the behavior of these ants have been available until a recent study by Hölldobler, reported here for the first time.

A species of *Polyrhachis* (*Cyrtomyrma*), tentatively classified near *doddi*, was observed in the vicinity of Port Douglas, Queensland, where its colonies are relatively abundant. The ants construct nests among the leaves and twigs of a wide variety of bushes and trees (Fig. 5). Most of the units are built between two opposing leaves, but often only one leaf serves as a base or else the unit is entirely constructed of silk and is well apart from the nearest leaves.

Polyrhachis ants have never been observed to make chains of their own bodies or to line up in rows in the manner routine for *Oecophylla*. Occasionally a single *Polyrhachis* worker pulls and slightly bends the tip or edge of a leaf, but ordinarily the leaves are left in their natural position and walls of silk and debris are built between them.

The weaving of *Polyrhachis* also differs markedly from that of *Oecophylla*. The spinning larvae are considerably larger and appear to be at or near the end of the terminal instar (Fig. 4). The workers hold them gently from above, somewhere along the forward half of their body, and allow the larvae to perform all of the spinning movements. In laying silk on the nest wall, the larvae use a version of the cocoon-spinning movements previously observed in the larvae of *Formica* and other formicine ants. Like these more "typical" species, which do not engage in communal nest-building, *Polyrhachis* larvae begin by protruding and retracting the head relative to the body segments while bending the forward part of the body downward. Approximately this much movement is also seen in *Oecophylla* larvae prior to their being touched to the surface of a leaf.

The *Polyrhachis* larvae are much more active, however, executing most of the spinning cycle in a sequence very similar to that displayed by cocoon-spinning formicines. Each larva begins with a period of bending and stretching, then returns to its original position through a series of arcs directed alternately to the left and right; in sum, its head traces a rough figure eight. Because the larvae are held by the workers, the movements of their bodies are restricted. They cannot complete the "looping-the-loop" and axial rotary movements described by Wallis (1960), by which larvae of other formicine ants move around inside the cocoon to complete its construction. In fact, the *Polyrhachis* larvae do not build cocoons. They pupate in the naked state, having contributed all their expelled silk to the communal nest. In this regard they fall closer to the advanced *Oecophylla* grade than to the primitive *Dendromyrmex* one, discussed below.

Polyrhachis ants are also intermediate between *Oecophylla* and *Dendromyrmex* in another important respect. The *Polyrhachis* workers do not move the larvae constantly like living shuttles as in *Oecophylla*, nor do they hold the larvae in one position for long periods of time or leave them to spin on their own as in *Dendromyrmex*. Rather, each spinning larva is held by a worker in one spot or moved slowly forward or to the side for a variable period of time (range 1–26 sec, mean 8 sec, SD 7.1 sec, n = 29). After each such brief episode the larva is lifted up and carried to another spot inside the nest, where it is permitted to repeat the stereotyped spinning movements. While the larva is engaged in spinning, the worker touches the substrate, the silk, and the front half of the larva's body with its antennae. However, these antennal movements are less stereotyped than in *Oecophylla*.

The product of this coordinated activity is an irregular, wide-meshed network of silk extending throughout the nest. The construction usually begins with the attachment of the silk to the edge of a leaf or stem. As the spinning proceeds, some workers bring up small particles of soil and bark, wood chips, or dried leaf material that the ants have gathered on the ground below. They attach the detritus to the silk, often pushing particles into place with the front of their heads, and then make the larvae spin additional silk around the particles to secure them more tightly to the wall of the nest. In this way a sturdy outside shell is built, consist-

ing in the end of several layers of silk reinforced by solid particles sealed into the fabric. The ants also weave an inside layer of pure silk, which covers the inner face of the outer wall and the surfaces of the supporting leaves and twigs. Reminiscent of wallpaper, this sheath is thin, very finely meshed, and tightly applied so as to follow the contours of the supporting surface closely. When viewed from inside, the nest of the *Polyrhachis* ant resembles a large communal cocoon (Fig. 5).

A very brief description of the weaving behavior of *Polyrhachis*

(Myrmhopla) simplex by Ofer (1970) suggests that this Israeli species constructs nests in a manner similar to that observed in the Queensland species. The genus *Polyrhachis* is very diverse and widespread, ranging from Africa to tropical Asia and the Solomon Islands. Many of the species spin communal nests, apparently of differing degrees of complexity, and further study of their behavior should prove very rewarding.

A second intermediate grade is represented by *Camponotus (Myrmobrachys) senex*, which occurs in moist forested areas of South and Central

Figure 5. The nest of the Australian *Polyrhachis* species (*top*) is at an intermediate level of complexity, consisting of sheets of silk woven between leaves and twigs and reinforced by soil and dead vegetable particles. The interior of this type of nest (*bottom*) has a layer of silk tightly molded to the supporting leaf surfaces.

America. It is one of only two representatives of the very large and cosmopolitan genus *Camponotus* known to incorporate larval silk in nest construction (although admittedly very little information is available about most species of this genus), and in this respect must be regarded as an evolutionarily advanced form. The most complete account of the biology of *C. senex* to date is that of Schremmer (1972, 1979a, b).

Unlike the other weaver ants, *C. senex* constructs its nest almost entirely of larval silk. The interior of the nest is a complex three-dimensional maze of many small chambers and connecting passageways. Leaves are often covered by the silken sheets, but they then die and shrivel, and thereafter serve as no more than internal supports. Like the Australian *Polyrhachis*, *C. senex* workers add small fragments of dead wood and dried leaves to the sheets of silk along the outer surface. The detritus is especially thick on the roof, where it serves to protect the nest from direct sunlight and rain.

As Schremmer stressed, chains of worker ants and other cooperative maneuvers among workers of the kind that characterize *Oecophylla* do not occur in *C. senex*. The larvae employed in spinning are relatively large and most likely are near the end of the final instar. Although they contribute substantial amounts of silk collectively, they still spin individual cocoons—in contrast to both *Oecophylla* and the Australian *Polyrhachis*. Workers carrying spinning larvae

can be most readily seen on the lower surfaces of the nest, where walls are thin and nest-building unusually active. During Schremmer's observations they were limited to the interior surface of the wall and consequently could be viewed only through the nascent sheets of silk. Although numerous workers were deployed on the outer surface of the same area at the same time, and were more or less evenly distributed and walked slowly about, they did not carry larvae and had no visible effect on the workers inside. Their function remains a mystery. They could in fact be serving simply as guards.

Although Schremmer himself chose not to analyze the weaving behavior of *C. senex* in any depth, we have been able to make out some important details from a frame-by-frame analysis of his excellent film (Schremmer 1972). In essence, *C. senex* appears to be very similar to the Australian *Polyrhachis* in this aspect of their behavior. Workers carry the larvae about slowly, pausing to hold them at strategic spots for extended periods. They do not contribute much to the contact between the heads of the larvae and the surface of the nest. Instead, again as in *Polyrhachis*, the larvae perform strong stretching and bending movements, with some lateral turning as well. When held over a promising bit of substrate, larvae appear to bring the head down repeatedly while expelling silk. We saw one larva perform six "figure eight" movements in succession, each time touching its

head to the same spot in what appeared to be typical weaving movements. The duration of the contact between its head and the substrate was measured in five of these cycles; the range was 0.4–1.5 sec and averaged 0.8 sec. During the spinning movements the workers play their antennae widely over the front part of the body of the larva and the adjacent substrate.

The nest-weaving of *C. senex*, then, is the same as that of the Australian *Polyrhachis*. The only relevant difference between the two is that *C. senex* larvae construct individual cocoons and *Polyrhachis* larvae do not.

The simplest type of weaving

A recent study of the tree ants *Dendromyrmex chartifex* and *D. fabricii* has revealed a form of communal silk-weaving that is the most elementary conceivable (Wilson 1981). The seven species of *Dendromyrmex* are concentrated in Brazil, but at least two species (*chartifex* and *fabricii*) range into Central America. The small colonies of these ants build oblong carton nests on the leaves of a variety of tree species in the rain forest (Weber 1944).

The structure of the nests is reinforced with continuous sheets of larval silk (Fig. 6). When the nest's walls are deliberately torn to test their strength, it can be seen that the silk helps hold the carton together securely. Unlike *Oecophylla* larvae, those of *Dendromyrmex* contribute silk only at the end of the final instar, when they are fully grown and ready to pupate. Moreover, only part of the silk is used to make the nest. Although a few larvae become naked pupae, most enclose their own bodies with cocoons of variable thickness. Workers holding spinning larvae remain still while the larvae perform the weaving movements; in *Oecophylla*, the larvae are still and the workers move. Often the larvae add silk to the nest when lying on the surface unattended by workers. Overall, their nest-building movements differ from those of cocoon-spinning only by a relatively small change in orientation. And, not surprisingly, this facultative communal spinning results in a smaller contribution to the structure of the nest than is the case in *Oecophylla* and other advanced weaver ants.

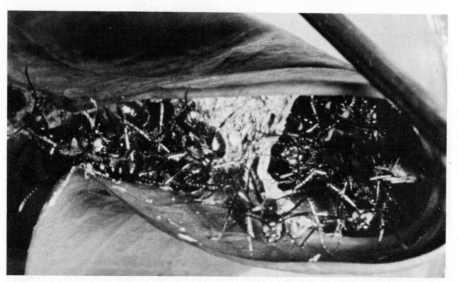

Figure 6. *Dendromyrmex chartifex*, of Central and South America, makes the simplest type of woven nest, a carton-like structure of chewed vegetable fibers reinforced with larval silk. (Photograph from Wilson 1981.)

Anatomical changes

The behavior of communally spinning ant larvae is clearly cooperative and altruistic in nature. If general notions about the process of evolution are correct, we should expect to find some anatomical changes correlated with the behavioral modifications that produce this cooperation. Also, the degree of change in the two kinds of traits should be correlated to some extent. And finally, the alterations should be most marked in the labial glands, which produce the silk, and in the external spinning apparatus of the larva.

These predictions have generally been confirmed. *Oecophylla*, which has the most advanced cooperative behavior, also has the most modified external spinning appara-

tus. The labial glands of the spinning larvae of *Oecophylla* and *Polyrhachis* are in fact much larger in proportion to the size of the larva's body than is the case in other formicine ant species whose larvae spin only individual cocoons (Karawajew 1929; Wilson and Hölldobler 1980). On the other hand, *C. senex* larvae do not have larger labial glands than those of other *Camponotus* larvae. Schrem-

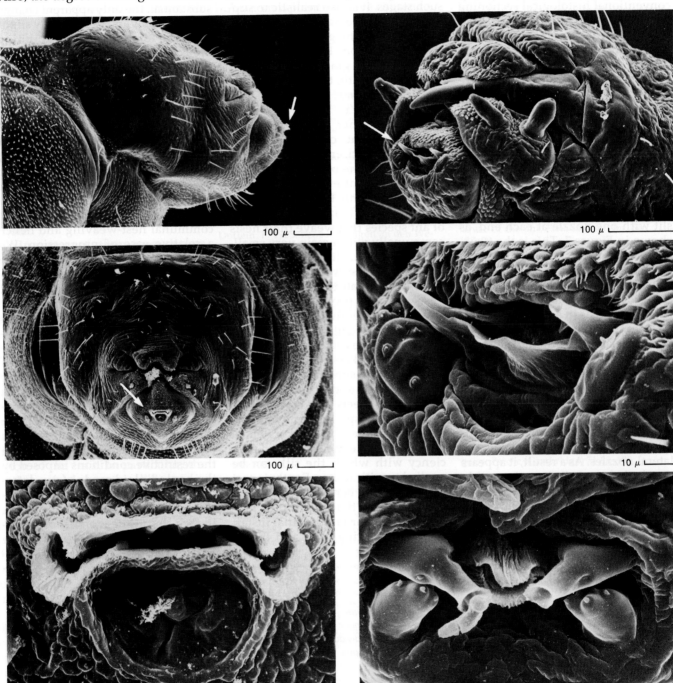

Figure 7. Scanning electron micrographs reveal adaptations in the spinning apparatus of ant larvae in *Oecophylla*. At the left, the head of an *O. longinoda* larva is shown from the side (*top*) and front (*middle*); the arrows indicate the slit-shaped opening of the silk glands, which is modified substantially from the more primitive forms at the right. The reduced lateral nozzles in *O. longinoda* and the larger central nozzle are clearly visible at the bottom left. In *Nothomyrmecia macrops*, a living Australian ant thought to be similar to the earliest formicines, there is no central nozzle and the lateral nozzles are much more prominent; the arrow at the top right points to the area enlarged at the middle right. The silk-gland opening of the Australian weaver ant, a species of *Polyrhachis* (*bottom right*), is similar in structure to that in *Nothomyrmecia*. (Micrographs by Ed Seling.)

Untangling the Evolution of the Web

A spider's web leaves no trace in the fossil record. How, then, can the evolutionary history of webs be deciphered?

William A. Shear

Figure 1. *Linyphia triangularis*, a close relative of spiders that weave geometric orb webs, constructs not an orb but an aerial sheet with scaffolding. Linyphiid sheet webs pose an intriguing problem for evolutionary biologists. The orb web, once considered the pinnacle of spider-web evolution, may have evolved from a single origin, or it may have appeared during the evolution of at least two lines of spiders, only to be lost or replaced in some families. New research suggests that web architecture, and web-building behavior, may have evolved in a com-

The delicate tracery of a spider's web must be one of the most unlikely candidates for fossilization in all of nature. Only a few threads of ancient spider silk have been preserved, and these in relatively recent deposits of amber; we know of no complete fossil spider webs.

Arachnologists who would reconstruct the evolutionary history of the web must examine, then, the wonderful record found in the garden, grassland and forest: the webs built by living spiders. By comparing these webs and integrating the information with evolutionary trees based on spider anatomy, it should be possible to discern the course of web evolution. After all, ephemeral though the web may be, the spider is nearly unique among animals in leaving a detailed record of its behavior.

And what a varied record it is! The careful exploration of tropical rain forests brings to light new species and new webs every year. There are orbs: two-dimensional, point-symmetrical arrays, a strong silk frame enclosing a series of radiating lines. Then there are other aerial webs, so unlike the orb: suspended sheets and three-dimensional space-filling webs, or cobwebs, that lack any obvious organization.

On the ground, spiders living in burrows or under objects weave silken collars or sheets that extend outward from the mouth of a burrow. Still others have abandoned their burrows entirely and construct short silk tubes on tree bark, camouflaged by bits of bark and lichens, and sometimes closed by hinged doors at either end. Some construct webs that they use more as weapons than traps, throwing them at passing insects or holding them under tension to release them when prey blunders into a thread.

How does a scientist interpret the story written in the spider's web? The answer has changed substantially over the past century. A few years after Darwin proposed his theory of evolution by natural selection in 1859, biologists turned this powerful analytical tool on the Araneae, the order of spiders. In 1895, the British arachnologist and biogeographer Reginald Innes Pocock proposed a scheme describing the course of evolution in spider webs. His work was followed by other proposals, including, most notably, those of William Bristowe in 1930, and B. J. Kaston in 1964.

These scenarios were essentially static, based on observations and descriptions of finished webs. Students of animal behavior, however, have added a new and dynamic dimension to the study of web evolution by carefully examining the actual process of web construction. It turns out that some spiders previously thought to be unrelated share patterns of web-building behavior, which is tightly controlled by a spider's genetic program, even though their webs may look radically different to us.

Meanwhile systematists have been rearranging spider classification itself as new tools (such as the scanning electron microscope) for observing anatomical features have become available. Detailed studies of tropical and Southern Hemisphere spiders have revealed undetected relationships among groups. The earlier hypotheses about web evolution no longer, in many cases, line up with the evidence about the evolution of the spiders themselves.

The web, though a wonderfully detailed record, is now known to tell only part of its own ancestral story. In the full account, as it continues to unravel, is seen all the richness of the new evolutionary biology that has grown from the natural-history studies of Darwin's century. Information from anatomy, systematics, ecology and ethology can be eclectically combined to produce and test new hypotheses. The result, in the case of the spider web, has been a picture far less simple and linear than the old taxonomy. The familiar garden spider's orb web, for example, may have been tried and then abandoned by some species; for all its magnificence, the orb may be not the pinnacle of web evolution but an intermediate form.

Spiders and Their Silk

Arachnologists now recognize three main groups of spiders, differentiated by anatomical and behavioral traits

William A. Shear is Charles Patterson Distinguished Professor and chairman of the Biology Department at Hampden-Sydney College and a research associate in the Department of Entomology, American Museum of Natural History. He received his doctorate in 1971 from Harvard University. He has published more than 100 articles and chapters on the systematics and evolution of arachnids and myriapods. Recently his research has focused on the fossil evidence for early terrestrial ecosystems, and he wrote about this subject with Jane Gray for American Scientist in the September–October 1992 issue. Address: Department of Biology, Hampden-Sydney College, Hampden-Sydney, VA 23943. Internet: bills@tiger.hsc.edu.

plex, nonlinear fashion separately from spider anatomy, the traditional basis for classifying spiders and discerning their evolutionary history.

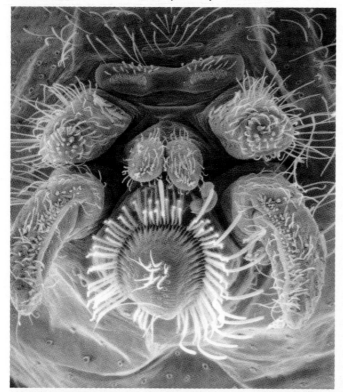

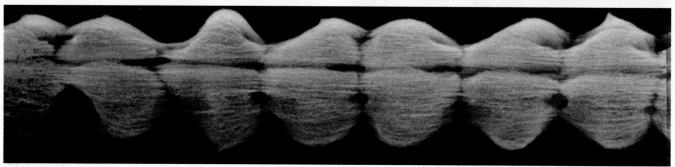

Figure 2. Presence or absence of a cribellum, a broad median plate that replaces one of four pairs of spinnerets (silk-spinning organs) in many spider families, was used by early taxonomists to divide araneomorph spiders into two ancestral groups. *Oecobius*'s cribellum is evident above its spinnerets and anal tubercle in the upper-left image, produced by a scanning electron microscope. The underside of another spider, *Uroctea (upper right),* displays only the spinnerets and tubercle. Cribellate spiders produce a woolly, puffy silk *(bottom micrograph)* that functions as an effective insect trap. Surprisingly, many araneomorphs have lost the cribellum. It is no longer considered the basis for separating families of spiders; in fact, *Oecobius* and *Uroctea* are now known to be closely related. (Micrographs courtesy of Charles Griswold, California Academy of Sciences *(top)* and Brent Opell, Virginia Polytechnic Institute and State University.)

and some general differences in their webs. The most primitive of these, the Mesothelae, are known from living examples found from Japan south to Indonesia, and from fossils as old as 300 million years. They differ from all other spiders in having an obviously segmented abdomen, and in having their spinning organs, or spinnerets, located near the middle of the abdomen, rather than at the posterior end.

Tarantulas, now becoming popular as pets, belong to a second group, the Mygalomorphae. Although the mygalomophs resemble the mesotheles in some characteristics, their spinnerets are at the end of the abdomen, and there are nev-

er more than six. Mygalomorphs are common and diverse in the tropics; most species are sedentary, some almost never leaving their burrows.

The third group of spiders, the Araneomorphae, include most North American spiders and are the "true spiders" best known to most of us. Unlike the mesotheles and mygalomorphs, their fangs point toward each other at right angles to the long axis of the body, and most have only a single pair of lungs.

Their silk glands and use of silk unambiguously define all these groups as spiders. Not all living spiders spin webs, but since 1950 web-building species have been found in almost all

the families of spiders once thought of as wandering hunters. It now seems very likely that all spiders who actively hunt their prey, or who use little or no silk in prey capture, are descendants of web builders. We also know that modern-looking, functional spinnerets were characteristic of spiders that lived 375 million years ago. So a fundamental problem in the study of web and spider evolution is the origin of silk itself.

A little more than a century ago, Henry McCook, one of the earliest American observers of spider behavior, proposed a hypothesis for the origin of silk that today remains the best-supported idea. McCook suggested that primitive proto-

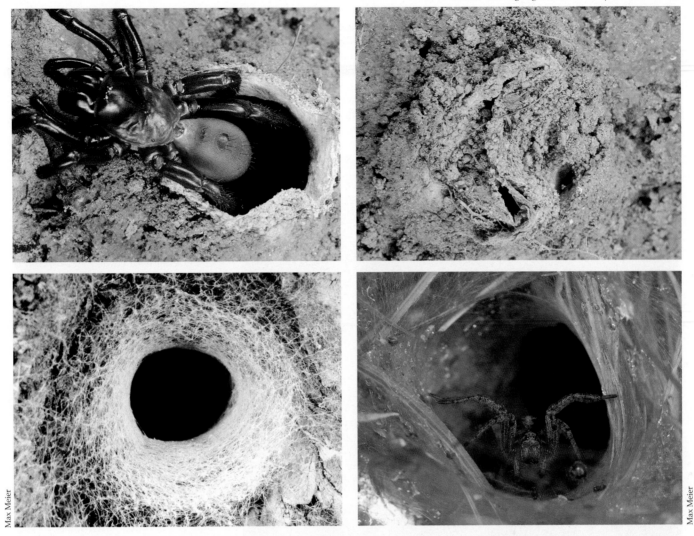

Max Meier

Max Meier

Figure 3. Early stages of web evolution are thought to be represented by silk-lined burrows, sometimes with triplines to extend the spider's sensory area. Next a silk collar is constructed as an extension of the lining, followed by the sheet webs developed by such spiders as the araneomorph family Agelenidae, which incorporates a retreat for the spider. Primitive web-building is seen among mygalomorphs (*Atypoides unicolor, upper two photographs*) that build silk-lined burrows and cover them with trap doors. A transitional stage is evident in the silk collar extended by *Amaurobius ferox*, an araneomorph (*lower left*). Finally, an agelenid, *Agelena labyrinthica*, weaves a ground-level sheet with a retreat. **(Upper photographs courtesy of Fred Coyle, Western Carolina University.)**

spiders, like modern centipedes, trailed excretory matter from kidneylike glands at the bases of the legs as they walked about. The chemical trails so laid down would have been useful in finding mates and returning to burrows or hiding places under stones.

Eventually, McCook reasoned, the excretory function of these many coxal glands was taken over entirely by a few in the anterior (forward) part of the body, leaving those in the abdomen to function entirely in trail-making. The trail of excretory material was replaced by longer-lasting protein—silk—and some of the abdominal appendages became transformed into spinnerets.

Even today, most spiders continuously trail out a dragline of silk as they move about, and the silk lines of females can be followed by males in search of mates. Studies of development have shown that the spinnerets do indeed originate from the rudiments of abdominal appendages.

There remain two alternative views, one of them originating with Pocock and Bristowe, who thought that the original function of silk was to protect eggs. In this hypothesis, silk was first produced from the mouth region and smeared over the egg bundles. Unfortunately, this scheme rests partly on the argument that the gum produced from the jaws of one spider, *Scytodes thoracica*, is a primitive "pre-silk," and it turns out that *Scytodes thoracica* is not a primitive spider. There is also the problem of transferring the production of silk from the mouth region to the abdomen.

The second alternative has been offered recently by Arthur Decae of the National Museum of Natural History in Leiden, Holland, who suggested in 1984 that spiders developed silk even before they became terrestrial. The function of the silk would have been to keep burrows in marine mud from collapsing or being filled with sediment, and the silk might have served as a sort of gill for later (and entirely hypothetical) amphibious spiders who would periodically be submerged by tides. Decae did not speculate on how silk might actually have originated.

Anatomy and Evolution

A functional explanation for the origins of silk and the spinning habit may be impossible to achieve, but the evolution of silk-spinning *organs* has been studied, and debated, extensively. Revealing evidence has come from the histology of silk glands—the details of their cellular construction—and from the embryological development of the spinnerets themselves. Histological evidence allows us to draw connections, or homologies, between silk glands in different spider groups, and embryology shows clearly that the spinnerets are paired abdominal appendages, with the silk issuing from modified setae, or hairs. So much information is available on the anatomy of the spinning apparatus, in fact, that the traditional view of web evolution rests heavily on a classification derived from the form and position of spinnerets.

As mentioned above, the spinnerets are located near the middle of the abdomen in the primitive spiders, the mesotheles, but at the end of the abdomen in the other two families. With the evolution of the araneomorphs there came a further development: The frontmost pair of spinnerets (presumably inherited from a mesothele-like ancestor) became a broad, median plate called a cribellum. The cribellum is covered with minute tubules, each capable of producing an extremely fine silk fiber, and the araneomorphs that have this plate are called cribellate spiders.

Using special bristles on the last pair of legs, the cribellate spiders tease these fibers into a woolly ribbon that is laid on core fibers produced by other, less modified, spinnerets. The result is an effective insect trap. The tiny cribellate fibers not only entangle bristles and hairs on insects, but also may adhere by means of electrostatic attraction to even the smoothest surface. At least two families of cribellate spiders make orb webs, in which the catching spiral consists of this kind of silk.

Surprisingly, many groups of araneomorphs have lost the cribellum. In their webs insects do not adhere to the silk but are simply impeded by it—long enough for the web's owner to capture them. Among the groups in which the cribellum has evidently been lost—called ecribellates—members of one superfamily, Araneoidea, have substituted a new kind of sticky silk. This silk is actually wet. Special adhesive glands add a liquid cement to ordinary silk fibers as they emerge from the spinnerets. The adhesive is extremely sticky, but it loses this property when it dries out, so that most spiders making webs including this kind of silk (called viscid silk) must periodically renew the sticky threads. Among the araneoids are many makers of intricate orbs, including the common garden spider.

The role of the cribellum in the evolution of spider webs has been the subject of considerable debate. Without the cribellum, the mesotheles and the mygalomorphs never achieved the orb web. Mesothele webs are generally silk linings for their burrows. Mygalomorphs make various webs, including aerial sheets, but they have not evolved orbs or space-filling cobwebs. Only araneomorphs make orbs and cobwebs, and only some araneomorphs have cribella. The cribellum poses one of evolutionary biology's most intriguing ques-

Max Meier

Figure 4. Orb web of an ecribellate araneoid, the garden spider *Araneus diadematus* (top), differs only in detail from the orb of a cribellate uloborid spider, the New Zealand species *Piha waitkerensis*. The leg movements used by both during web construction are very similar, making it likely that there is a close evolutionary relation between them, despite their anatomical differences. The orb web may, however, have arisen in both groups by convergent evolution—a common adaptation to similar environmental pressures. (Bottom photograph courtesy of Brent Opell.)

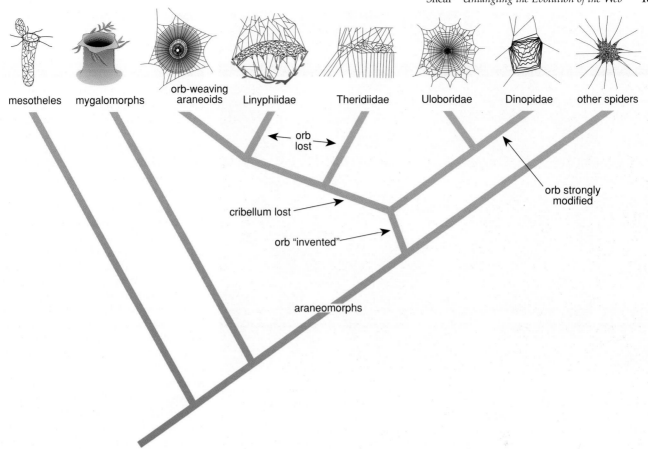

Figure 5. Phylogenetic tree shows how some families of spiders may have arrived at similar and different web designs. The Mesothelae are generally believed to have evolved first from the common ancestor of all spiders, followed by the Mygalomorphae and the "true spiders," the Araneomorphae. The monophyletic hypothesis of orb-web origin (which is incorporated into this diagram) holds that the orb web was invented by an araneomorph, the common ancestor of araneoid and uloborid spiders, that had a cribellum. The cribellum was acquired by a spider that was the common ancestor of all araneomorphs, including the araneoid superfamily and the uloborids. The araneoids lost the cribellum, and some araneoid families later lost the orb. Among the uloborids and their close relatives, the dinopids, are many species that have modified the orb.

tions: What is the relation between the evolution of anatomy and of behavior?

The spider taxonomy of the late 19th century, which relied heavily on anatomical distinctions, proposed an answer. Pocock considered the cribellum a stable anatomical trait, important enough to divide the araneomorphs into two major groups, the cribellates and the ecribellates. This distinction was maintained by Bristowe and Kaston, but a Finnish taxonomist, Pekka Lehtinen, pointed out in 1967 that there are many pairs of spider taxa that are very similar, except that one has a cribellum and the other does not. In the new scheme of spider phylogeny a single family or genus sometimes includes both cribellate and ecribellate spiders.

A case in point involves the two spider genera *Uroctea* and *Oecobius*. Lehtinen pointed out that these two genera of anatomically peculiar spiders, then segregated in their separate families, are virtually identical except that *Oecobius*

species have a cribellum and *Uroctea* species do not. The similarities extend to many fine details of their anatomy and behavior. In 1970, I studied *Oecobius* and found that upon sexual maturity males lose the calimistrum, or comb, used to process cribellate silk, and that in both sexes of many species, the cribellum appears degenerate.

Today it is accepted that the two genera are each other's closest relatives and that they belong in the same family— *Uroctea* was derived from an *Oecobius*-like ancestor through the loss of the cribellum and calimistrum. The main outcome of this change in our view of anatomical evolution has been that we can no longer be certain that similar webs woven by cribellate and ecribellate spiders were independent developments.

This is not to say that convergent and parallel evolution have no place in the scheme. It seems clear that when, for example, mygalomorph and araneomorph spiders make similar advanced

webs, the design is unlikely to be attributable to descent from a common ancestor. Indeed, the division of spiders into the three main lines of evolution (mesothele, mygalomorph and araneomorph) probably took place when the founders of all three lines were still burrow-dwellers. Earlier schemes of web evolution erred, in my judgment, not only in assuming that there was extensive convergence between the "distinct" cribellate and ecribellate lines, but also in arranging web types in lockstep linear sequences. A phylogenetic diagram of web evolution *(Figure 5)* shows instead a reticulate pattern in which different lines of spiders (which would include both cribellate and ecribellate species) may have followed different pathways to the same web design.

Interestingly, the orb web, the most extensively studied of all webs, probably originated only once, though from what precursor is still not clear. I shall return to the evolutionary problem of the orb later.

Figure 6. Uloborid orb may have evolved along the lines that web-making follows during the lives of cribellate spiders of the New Guinean genus *Fecenia*. *Fecenia* species make near-orbs (shown is *Fecenia angustata*'s). *Fecenia ochracea* alters the form of the web from a sheet to a near-orb over its lifetime, incorporating a retreat. To achieve an orb from a *Fecenia* web it is only necessary to dispense with the retreat and complete the circle. (Photograph courtesy of H. W. Levi, Harvard University.)

Figure 7. *Synotaxus turbinatus*'s web resembles a fishnet, incorporating regularly arranged threads. As a close relative of the cobweb-weaving theridiid spiders, and thus part of the superfamily Araneoidea, which includes most of the orb-weaving groups, *Synotaxus* would be expected to arrange its threads in an orb. The fact that it does not supports the idea that the orb developed in separate families by convergent evolution, not from a single origin. (Photograph by Jonathan Coddington, courtesy of H. W. Levi.)

Tubes, Trap Doors and Triplines

My own consideration of web types and their distribution among spider taxa has led me to propose a nonlinear evolutionary scheme. It appears probable that several web types are the product of convergent evolution—that is, that the same web has evolved in unrelated species that have adapted to similar environmental circumstances. Convergence provides an explanation of the appearance of similar web types in different families: These webs may be as much a product of a spider family's ecology as of its phylogenetic relationships.

The simplest and probably the oldest spider's web is simply a silk lining for the retreat in which the animal spends much of its time. Many living spiders in all three major groups modify crevices or holes in their environment in this way. Others actively dig burrows in soft soil or sand and line them with silk. The retreat or burrow may be closed by an elaborately hinged door or by a collapsible collar that is an extension of the silk lining.

The design of such webs, and the behavior of the spiders that construct them, is an intriguing story in itself and has been the subject of extensive study by Fred Coyle of Western Carolina University. Coyle has shown that making a collapsible collar for a trap door to a burrow can improve predatory efficiency and still give a modicum of protection. He has also demonstrated that vibrations carried by the ground are the most important sensory cue used by these spiders to aim and time their lunges at prey.

These vibrations suggest the reason for a modification: the lines of silk that many spiders extend from their burrows. Coyle has hypothesized that these lines could be used by the myopic spider as triplines to extend its sensory area. The importance of an extended sensory area is shown by the fact that many burrowers that build trap doors and rarely exit their tunnels incorporate twigs and leaves, or even tabs of tough silk, in their entrances in a radiating pattern. Prey touching the twigs, leaves or tabs is attacked.

A burrow with silk lines extending from the entrance is found, for example, in *Liphistius batuensis*, a mesothele. The lines are not produced accidentally, but are deliberately laid out and later reinforced. Tweaking them with a stick induces the spider to strike. In addition to this primitive spider, such arrangements also turn up among a variety of mygalomorphs and araneomorphs.

The Sheet-Web Weavers

Most living spiders do not make their webs in burrows. Indeed, by far the most common web is the agelenid sheet web, which appears to represent the next stage in web evolution. These webs, made by members of the araneomorph family Agelenidae, include a tubular retreat reminiscent of the burrow. From this retreat extends a dense, horizontal sheet of silk. It seems likely that this sheet might have developed in at least two ways, perhaps by extension from an original turret, or by the addition of a complex of cross lines to the original triplines. In any case, the sheet not only signals the presence of prey over a wide area, it also impedes the movements of the prey and gives the spi-

der more time to reach a victim. Perhaps 90 percent of living spiders make sheet webs with a retreat.

Many agelenids also produce an extensive, irregular tangle of threads above the sheet. As B. J. Kaston pointed out in his reexamination of the evolution of web-building in 1964, the tangle confers multiple advantages. It can make the sheet structurally more rigid so that it better conducts vibrations made by crawling prey. It has also been observed to intercept and knock down low-flying insects so that they fall on the sheet and can be captured. A number of families related to the agelenids also construct these sheet-plus-retreat webs near the ground. Species in several mygalomorph families have adopted this way of life, and a few have achieved aerial sheet webs. But

relatively few of them have added the aerial tangle, and no known mygalomorph species have gone past the sheet-web stage of web design.

Agelenids and their relatives run over the top surface of the sheet, but other families of sheet-web weavers hang from beneath it. This may provide some additional protection from predators. In both groups of sheet-weavers, some have dispensed with the retreat entirely and have moved the sheet up into vegetation, where the abundant supply of flying insects may be exploited.

The distribution of sheet webs among spider families follows ecological constraints more closely than family relationships, making a strong case for convergence. The mygalomorph spiders of the genus *Euagrus* make retreat-plus-sheet webs difficult to dis-

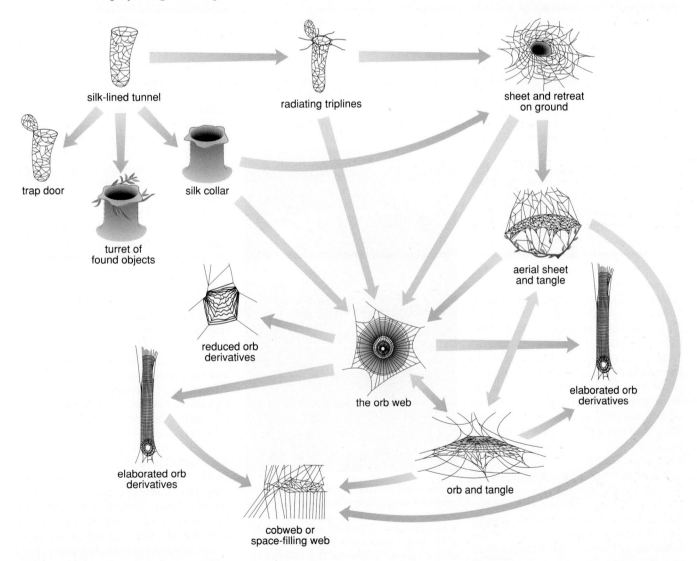

Figure 8. Hypothetical pathways of spider-web evolution form a tangled web of their own, with the question of the orb's origin, and its role as a possible precursor to other webs, at the center. In several cases it is not clear which web is ancestral; it is possible that some aerial sheet webs preceded the orb web, whereas others developed from the orb. Pathways that are less likely are indicated by light-orange arrows; for some of them there is no direct evidence.

tinguish from those of the araneomorph agelenids; these spiders also replace the agelenids ecologically in the deserts of the southern United States and in Mexico. In New Zealand, Ray Forster of the Otago Museum has described a number of families of spiders that make aerial sheets, just as do other, unrelated families in the Northern Hemisphere. Such examples suggest that the common possession of an unspecialized type of web cannot be used to argue for relationships between spider groups, since the resemblance of the webs may be a result either of the retention of a primitive ancestral pattern common to many spiders, or of convergent evolution.

Enter the Orb

At the center of the phylogenetic Gordian knot of web evolution lies the orb. The geometry of the orb webs fashioned by cribellate and ecribellate araneomorphs is nearly identical, yet the spiders that make them seem quite different. Are the resemblances among the orb webs the results of common ancestry, or are they the product of convergent evolution? The evidence is mixed, and the controversy in this case may not soon be resolved.

The debate over the orb takes the shape of two proposed scenarios. One view, called the "monophyletic hypothesis," states that all orb webs have a common origin; that is, all orb-weaving spiders descend from a common ancestor. This proposal is the older of the two, having first been suggested by Teodor Thorell in 1886.

But the alternative, the convergence hypothesis, has been dominant in scenarios of web-building evolution for more than 50 years, because the dominant thinking has been that all ecribellate spiders were related to one another and formed a group not necessarily close to the cribellates. This distinction meant that any resemblances between the cribellates and ecribellates, including web form, had to result from convergent evolution.

When arachnologists realized that having a cribellum was the primitive condition for all araneomorphs, it became clear that the ecribellate spiders do not form a single group but originated many times by multiple losses of the cribellum. The monophyletic hypothesis again surfaced, and has been persuasively argued in recent years by Jonathan Coddington of the Smithson-

ian Institution. According to Coddington, the cribellate orb weavers and the ecribellate ones had a common ancestor, a cribellate that was probably also an orb weaver.

Much of the evidence marshaled by Coddington comes from careful observations of leg movements that take place in the process of web construction. These observations, reported by William Eberhard of the University of Costa Rica, have had a significant effect not only on the debate over web evolution but also on the views of how behavior ought to be considered in classifying spiders. Eberhard has established that most of the spiders that weave geometrical orbs follow a highly stereotyped sequence of leg movements, even if the orb web has become modified to the point that it no longer resembles a "typical" orb.

Orb weavers are found among five families of an ecribellate superfamily of spiders mentioned above, the Araneoidea, and a cribellate family, the Uloboridae. The studies of Eberhard and Yael Lubin of the Ben Gurion University of the Negev have found that the uloborids' leg movements are identical to

Figure 9. Reduced orb webs reflect ingenious predatory strategies in which activity appears to be substituted for silk. *Mastophora* plays out a single line of silk with a large glue blob at the end, luring male moths to the blob with a volatile substance that mimics the female pheromone of the desired species. Reduced orbs may be considered evolutionary outgrowths of the orb. (Photograph by Mark Stowe, University of Florida.)

those of all five araneoid families of ecribellate orb weavers. Of course, the silk in the webs is different; the uloborids use cribellate silk for the catching spiral, whereas the garden spider and other araneoids use viscid silk.

The origin of the uloborid orb web is not difficult to understand. In 1966, Raja Szlep of the Hebrew University of Jerusalem described the web of a cribellate sheet-weaver, *Titanoeca albomaculata*. She found that around the periphery of its small sheet, *Titanoeca* weaves regularly arranged arrays of threads surprisingly similar in appearance to partial orbs. The regular sectors had numbers of radiating lines, and across these were laid adhesive cribellate threads in a back-and-forth pattern which, if continued entirely around the web, could be visualized as a spiral. *Titanoeca*'s web suggested a route to the orb.

Later, similar behavior was observed by Michael Robinson of the Smithsonian Institution and Lubin. Working in New Guinea on the cribellate *Fecenia ochracea*, a member of a family possibly related to *Titanoeca*'s, Robinson and Lubin discovered that in the course of its lifetime, an individual of this species alters its web form from a typical sheet to what could only be described as a near-orb, the whole web consisting of a *Titanoeca*-like sector, with the radiating lines converging on a retreat. Since then, other examples of sheets with regularly arranged sectors have been found, but only among cribellate araneomorphs. To achieve the orb, it seems only necessary to dispense with the retreat of a *Fecenia*-like web and complete the circle of radial lines.

We are still left puzzling over the question of where the Araneoidea got the orb web. The monophyletic hypothesis urges that the orb developed among certain cribellates that are the common ancestors of araneoid and uloborid orb weavers. In this view the uloborids and the araneoids are each other's closest relatives, differing in that the araneoids have since lost the cribellum. The convergent hypothesis, on the other hand, maintains that the araneoids lost their cribella, and then independently invented the orb web. In this view, uloborids and araneoids are not each other's closest relatives.

The origin of the orb may, then, lie in a *Titanoeca-Fecenia* scenario carried to its geometric conclusion by a common ancestor, followed by the loss of the cribellum among the araneoids. Or a similar scenario may have been played out in similar environments by separate ancestors. The riddle can only be solved by independent evidence. If a close relationship between uloborids and araneoids can be demonstrated by means independent of web features, the monophyletic hypothesis is supported. If, on the other hand, it turns out that the uloborids' closest relatives are some other group of cribellates that do not make orbs, the convergent hypothesis is supported.

Coddington has indeed found independent evidence supporting the monophyletic hypothesis. He has found at least two anatomical characters, or stable traits, that occur only in uloborids and araneoids, and that therefore were also probably inherited from a common ancestor. Both groups have specialized silk glands connected to the posterior lateral spinnerets, and both have a special muscularized valve in the anterior lateral spinnerets. Admittedly, only a few spider families have been examined for these characters, but of those looked at so far, the specializations occur only in Uloboridae and Araneoidea. In addition, Coddington has listed nine detailed behavioral characters connected with orb web construction. An examination of the list suggests that if there was a common cribellate ancestor of uloborids and araneoids, it too wove an orb web and passed down several common behavioral traits to both its descendant groups.

The strongest argument so far advanced for the convergent-evolution hypothesis is the fact that two large families in the Araneoidea, the Linyphiidae and the Theridiidae, do not make orbs, and their behavior, as currently understood, includes no trace of the characteristic silk-handling movements discovered by Eberhard. If there is no evidence that these araneoids descended from an orb-making ancestor, then the common behavioral traits of cribellate and ecribellate orb weavers might be better explained as adaptation to common environments, rather than expressions of a shared inheritance.

The members of the Linyphiidae make aerial sheets. It could be argued that these webs are derived from horizontal orbs with added elements, but similarly persuasive is a scenario in which, following the traditional linear-phylogeny argument, the linyphiid sheets are ancestral to orbs.

Figure 10. *Cyrtophora* **has elaborated rather than reduced the orb. Above and below the horizontal orb, the spider has added an extensive tangle of threads. The addition of a tangle, so common in aerial sheet webs, suggests a pathway from which aerial sheets might have evolved from orb webs. It is possible that the orb arose from a sheet web, but that some species have gone from weaving orbs to making a new form of sheet web.**

Max Meier

The crucial evidence is missing; linyphiid web-building is not well enough understood. If the special silk-handling movements of orb weavers are indeed completely absent in the Linyphiidae, their ancestors were probably never orb weavers, and the convergent hypothesis is supported because of the close relationship of the linyphiids to the five orb-weaving families.

The second non-orb-weaving family in the Araneoidea, the cobweb weavers of the Theridiidae, is one of the largest spider families in numbers of species. The arguments just stated for linyphiids apply to theridiids as well, but there is an additional point. Eberhard has discovered a theridiid spider, *Synotaxus* (now considered not a theridiid at all but placed in its own family, Synotaxidae), that makes a web with regularly arranged threads. The problem this presents for the monophyletic hypothesis is that the web is not an orb, but a totally different design using rectangular modules that make the whole web resemble a fisherman's net. If the monophyletic hypothesis is correct, one might reasonably predict that such a web would resemble an orb; the fact that it does not supports the idea of convergence.

Beyond the Orb
Among all six of the orb-weaving families I have just mentioned, there are some members whose webs do not closely resemble orbs, nor are they sheets. A few make no webs at all. They can be recognized as family members by their anatomical features and, if webs are still present, by the characteristic movements they use to handle silk. The webs and behavior of these spiders suggest that the orb itself is ancestral to various ingenious modified webs, some of them more complex and others elegant for their simplicity and efficiency.

The most interesting may be the reduced webs. Web construction is of course only part of a spider's approach to predation and self-protection. Among some species activity has been substituted for silk. In the Uloboridae, Brent Opell of Virginia Tech has studied members of the genera *Hyptiotes* and *Miagrammopes*. These spiders hold their webs under tension, releasing them when they come in contact with prey. With the use of this strategem an insect can be effectively entangled with a simpler web. In *Hyptiotes*, the triangle spider, the web consists of only a pie-shaped sector of the orb, and in *Miagrammopes* only one or a few single sticky threads make up the entire web.

Jonathan Coddington has recently found that *Dinopis*, a cribellate spider of the family Dinopidae (neither a uloborid nor an araneoid), exhibits the characteristic movements of an orb builder. *Dinopis*, however, makes a small, rectangular web best described as the outer half of a pie-shaped sector of an orb, and is famous for its habit of throwing itself and its web at pedestrian and flying prey. It is tempting to propose that muscular movement is cheaper than silk, and thus action has replaced the elaborate web. Recent work by Opell provides indirect support for this idea.

Mark Stowe of the University of Florida argues persuasively that another route to web reduction lies through prey specialization. His field studies have shown that certain araneoids catch mostly moths and have evolved an especially sticky glue for their webs for that purpose. The "super glue" is needed because moths can escape most spider webs by shedding the scales that cover their bodies, leaving the scales stuck to the web. The presence of this glue has allowed the New Guinean spider *Pasilobus* to reduce its web to only a few threads, which, however, remain extraordinarily effective in catching moths.

Stowe and Eberhard have also been able to document that some moth specialists have carried specialization to an extreme: They catch prey of only one or a few species, and only the males of those species. To do so they produce a volatile substance that mimics the female pheromones of the moth species they catch. In the case of the moth-attracting spiders of the genus *Mastophora*, the web is reduced to a single line, with a large glue blob at the end, which is manipulated by the spider. Only their anatomy connects these spiders to the orb weavers.

Robert Jackson of Canterbury University, New Zealand, and I have independently suggested that yet another route to web reduction involves the habit of kleptoparasitism, in which one spider species lives in the web of another and steals prey. Some species of the genus *Argyrodes* still make their own webs and catch insects for themselves, but others invade the webs of unrelated species and cut out part of the host web, replacing it with their own threads. From this base, they make forays into the host web to steal food. Still other species go beyond

stealing prey; they kill and eat their host spiders, making web-building virtually superfluous. All known members of the large ecribellate family Mimetidae make no webs of their own. By imitating the struggles of prey in others' webs, they entice other spiders near enough to be killed and eaten.

Finally, there appear to be orb weavers that have added structures to their webs. Lubin has found that some uloborids construct a second orb beneath the first and pull it into a cone. *Cyrtophora*, an ecribellate orb weaver, adds an extensive tangle of threads above and below its horizontal orb. These strong lines probably provide *Cyrtophora*, which sits on its web during the day, a modicum of additional protection from parasites and predators, and may also increase predatory efficiency by knocking down flying insects, which then fall on the horizontal orb. Similar tangles are found above and below many linyphiid sheet webs, which have also recently been found to incorporate viscid silk. The tangle itself, without the horizontal orb, resembles a cobweb. Discovering the characteristic movements of orb weavers in the makers of such webs would reinforce the hypothesis that these spiders originated from orb-weaving ancestors.

Despite the almost complete lack of a fossil record for spiders' webs, and only a very sketchy one for spiders themselves, information from anatomy, systematics, ethology and ecology has been combined to produce hypotheses about the course that evolution has taken in forming this fascinating arachnid artifact. Predictions from these hypotheses can be checked by careful studies of web types old and new, and the results of these observations will lead to further refinements. There are still many closely woven threads to be untangled in the story of web evolution.

Bibliography

Coddington, J. A. 1990. Cladistics and spider classification: araneomorph phylogeny and the monophyly of orbweavers (Araneae: Araneomorphae; Orbiculariae). *Acta Zoologica Fennica* 190:75–87.

Coddington, J., and C. Sobrevila. 1987. Web manipulation and two stereotyped attack behaviors in the ogre-faced spider *Deinopis spinosus* Marx (Araneae, Deinopidae). *Journal of Arachnology* 15:213–225.

Coyle, F. A., and N. D. Ketner. 1990. Observations on the prey and prey capture behavior of the funnelweb mygalomorph spider genus *Ischnothele* (Araneae, Dipluridae). *Bulletin of the British Arachnological Society* 8:87–104.

Eberhard, W. G. 1975. The 'inverted ladder' orb web of *Scoloderus* sp. and the intermediate web of *Eustala* (?) sp. Araneae: Araneidae. *Journal of Natural History* 9:93–106.

Eberhard, W. G. 1980. The natural history and behavior of the bolas spider *Mastophora dizzydeani* sp. N. (Araneidae).

Eberhard, W. G. 1982. Behavioral characters for the higher classification of orb–weaving spiders. *Evolution* (36(5):1067–1095.

Eberhard, W. G. 1990. Function and phylogeny of spider webs. *Annual Review of Ecological Systems* 21:341–372.

Eberhard, W. G. 1990. Early stages of orb construction by *Philoponella vicinia, Leucauge mariana,* and *Nephila clavipes* (Araneae, Uloboridae and Tetragnathidea), and their phylogenetic implications. *Journal of Arachnology* 18:205–234.

Jackson, R. R., and M. E. A. Whitehouse. 1986. The biology of New Zealand and Queensland pirate spiders (Araneae, Mimetidae): Aggressive mimicry, araneophagy and prey specialization. *Journal of Zoology, London (A),* 210:279–303.

Kaston, B. J. 1964. The evolution of spider webs. *American Zoologist* 4:191–207.

Kullmann, E. J. 1972. The convergent development of orb-webs in cribellate and ecribellate spiders. *American Zoologist* 12:395–405.

Levi, H. W. 1978. Orb-webs and phylogeny of orb-weavers. *Symposium Zoological Society of London* 42:1–15.

Lubin, Y. D., B. D. Opell, W. G. Eberhard and H. W. Levi. 1982. Orb plus cone-webs in Uloboridae (Araneae), with a description of a new genus and four new species. *Psyche* 89(1-2):29–64.

Opell, B. D. 1990. Material investment and prey capture potential of reduced spider webs. *Behavioral Ecology and Sociobiology* 26:375–381.

Platnick, N. I., and W. J. Gertsch. 1976. The suborders of spiders: A cladistic analysis (Arachnida, Araneae). *American Museum Novitates* 2607:1–15.

Reed, C. F., P. N. Witt, M. B. Scarboro and D. B. Peakall. 1970. Experience and the orb web. *Developmental Psychobiology* 3(4):251–265.

Robinson, M. H., and B. Robinson. 1975. Evolution beyond the orb web: The web of the araneid spider *Pasilobus* sp., its structure, operation and construction. *Zoological Journal of the Linnean Society* 56(4):301–314.

Robinson, M. H., and Y. D. Lubin. 1979. Specialists and generalists: The ecology and behavior of some web-building spiders from Papua New Guinea. II. *Psechrus argentatus* and *Fecenia* sp. (Araneae: Psechridae). *Pacific Insects* 21(2-3):133–164.

Shear, W. A., ed. 1986. *Spiders: Webs, Behavior, and Evolution*. Stanford, Calif.: Stanford University Press.

Szlep, R. 1966. Evolution of the web spinning activities: The web spinning in *Titanoeca albomaculata* luc. (Araneae, Amaurobidae). *Israel Journal of Zoology* 15:83–88.

Vollrath, F. 1992. Spider webs and silks. *Scientific American* 266(3):70–76.

Witt, P. N., and C. F. Reed. 1965. Spider-web building. *Science* 149(3689):1190–1197.

Why Do Bowerbirds Build Bowers?

Females prefer to visit courtship areas that provide easy avenues of escape, thereby protecting them from forced copulations

Gerald Borgia

Male bowerbirds of Australia and New Guinea clear and decorate courts and build bowers at display sites where they mate. Bowerbird species, however, differ in several characteristics, including the type and color of court decorations and the form of a bower, if one is even built. Moreover, some male bowerbirds possess bright crest and body plumages, and others do not. Charles Darwin's observations of satin bowerbirds—in the Blue Mountains of Australia during his round-the-world journey on the HMS *Beagle*—contributed to the then-controversial central element of his theory of sexual selection called female choice. The highly sculptured structure of a bower and a male's use of brightly colored decorations suggested to Darwin that female bowerbirds might shop for the most attractive bower, thereby directing the evolution of these display traits.

Nevertheless, several other mechanisms could have driven the evolution of bowers. The so-called good-genes model, for instance, suggests that male-display traits, including bowers, might indicate a male's vigor and, ultimately, his quality as a sire. That is, more vigorous males might have better bowers. A bower could even directly benefit a female, perhaps protecting her from threats, including predators that might attack her during mating or

males that might try to force her to copulate. Bower building could even arise from an arbitrary or pre-existing female preference, such as an attraction to nest-like structures.

I have used Darwin's method of comparisons of related species to reconstruct the evolution of bower building. My work on several species of bowerbirds confirms the existence of female preferences for males with well-built and highly decorated bowers. The origins of bower building, however, can be best explained as a trait that attracts females because of the protection it provides them from forced copulation by bower owners.

Evaluating Bower-Building Hypotheses
Picking one model of bower-building evolution over another proves difficult because of several problems. One cannot always reconstruct what happened long ago, especially for display behavior that leaves no fossil record. Moreover, bower building may have evolved over a period of time, and different stages of its evolution may have served different functions. Although experiments can show the plausibility of a particular evolutionary process, understanding the origins of traits can best be accomplished by careful comparisons between species whose relationships are known.

Such an analysis depends on accurate and detailed descriptions of bowers and how they are used in courtship in modern species. Obtaining detailed quantitative information on courtship and mating through direct observation proves nearly impossible, because bowers are separated widely, the mating period may last several months and a large proportion of males do not or rarely copulate. Remote-controlled cameras aimed at bowers where males perform their dis-

plays and mate, however, have allowed intensive monitoring of more than 30 bowers for a single species through an entire mating season. That information has provided a direct measurement of male attractiveness and detailed information on how males and females use a bower during courtship. For most of the species that my colleagues and I have studied, we were the first to see these bowerbirds perform successful courtships, which ended in copulation.

By comparing mitochondrial DNA sequences, my colleagues, Robert Kusmierski and Ross Crozier, and I have developed a highly reliable bowerbird phylogeny, which shows evolutionary relationships among species. Our phylogeny indicates, in contrast to some earlier speculation, that all 18 species of bowerbirds evolved from a single ancestral species. Three species branched off from others long ago, and they employ the predominant avian pattern of monogamy with both parents caring for their offspring. All other bowerbirds are polygynous (males mate with more than one female), and they create elaborately decorated display courts. All but two of the polygynous species build bowers. The second major divergence developed between species that build avenue bowers—two vertical stick walls, separated by a central avenue—and those that build maypole bowers—sticks woven around a sapling to create a decorated pillar. It appears that bower building evolved once and then diverged into two types of bowers. The two species that do not build bowers, toothbill and Archbold's bowerbirds, apparently lost bower-building behavior, but they do clear and decorate display courts. Comparisons of bowers and relationships among living species suggest that a

Gerald Borgia is professor of zoology at the University of Maryland. He holds a long interest in mate choice, which he studied in insects for his doctoral research at the University of Michigan. He began studying bowerbirds 15 years ago, while he was a postdoctoral fellow at the University of Melbourne. This year he and his students will return to his original bowerbird study site for further investigations of how male display affects mate choice. Address: Department of Zoology, University of Maryland, College Park, MD 20742.

R. Brown, Vireo

Figure 1. Male bowerbirds build bowers, where they court and mate with females. Some species, such as this satin bowerbird, build avenue bowers, made of two freestanding stick walls. Others build maypole bowers, in which sticks are placed around a central sapling (*Figure 3*). In addition, a male may decorate his bower with a variety of objects, including the pieces of blue plastic shown here. This complex behavior of building and decorating prompts a fundamental question: How did it evolve?

decorated sapling—similar to a simple maypole bower—may represent the ancestral bower type.

Several criteria can be used to evaluate hypotheses for bower evolution based on mating behavior and the evolutionary relationships among species. To the extent that these criteria are met, we can identify the likely initial causes of bower building. First, the proposed function of incipient bowers should be consistent with the design of the supposed ancestral bower. That is, the bower type that appears most consistent with the ancestral bower type should be capable of functioning in accordance with the hypothesized cause of bower origins. Second, the proposed function of the earliest bowers should be consistent with the design of modern bower types. The persistence of bower building among the polygynous species suggests that ancestral functions may remain important. If a consistent function exists for modern bowers, it would be a likely candidate for the ancestral function. Third, species that do not build bowers should possess alternative solutions to the prob-

lem solved by a bower. These species should possess compensatory behaviors, which work in the absence of a bower to protect females from forced copulations by the courting male.

Avenue-Bower Builders

The group of avenue-bower builders consists of three genera and eight species, including the satin bowerbird. Satins inhabit rain forests along the

Figure 2. Inside an avenue bower, a female satin bowerbird observes a male's courtship display. A male flits back and forth across the avenue opening, flicks his wings, mimics the calls of other birds and performs other displays. When a male runs to the rear entrance of the avenue to mate with a female, she either waits to copulate or departs through the front opening.

W. Peckover, Vireo

S. Pruett-Jones, Vireo

Figure 3. Macgregor's bowerbird *(left)* **constructs the simplest maypole bower** *(right)*. **A male selects a thin sapling, stacks sticks around it and covers a surrounding display court with a compressed-moss mat. In addition, he decorates the court.**

eastern fringe of Australia. A male aligns his bower along a north-south line, with a display court at the north end. He decorates his display court with blue, yellow and white objects including feathers, flowers, leaves, snail shells and, where available, plastic and paper, over a background of yellow straw. The male trims leaves from above the court, and the northern orientation causes the sun to illuminate the decorated site, perhaps making it more attractive. Males of several species destroy each other's bowers and steal decorations.

A visiting female usually lands in cover south of the bower and then moves rapidly into the avenue between the two stick walls. On the display court, a male makes vocalizations, including guttural chortles and squeaks that progress into a typical call sequence: initial mechanical buzzing followed by mimicking a kookaburra, a Lewin's honeyeater and less frequently a crow. During the buzzing, a male moves swiftly across the northern bower entrance and rapidly flicks one or both wings. When he begins mimicking other birds, he stops at one side of the bower entrance, puffs up his body feathers, holds his wings at his side,

faces the female with a small decoration—usually a yellow leaf—in his mouth and performs a series of knee bends. After that, he usually moves away from the bower, makes several harsh calls and then returns to the bower for more displaying.

In courtships that lead to copulation, a female in the bower avenue crouches deeply as courtship progresses, and a slight lifting of her tail signals her willingness for mating. A male circles around to the opposite end of the bower and mounts her for a three-second copulation. After mating, a female shakes and flaps her wings in or near the bower for a few minutes before leaving. Although a female may visit several bowers, she usually mates with only one male. The average courtship lasts about four minutes. If a male moves to the southern end of a bower before the female is ready, she escapes through the northern exit.

Females exert strong preferences in mating, and only a small proportion of males achieve most of the matings. Males with high quality bowers—with symmetrical walls formed from thin, densely packed sticks—and many decorations on their courts mate most often. Although only nine percent of

satin courtships lead to copulation, the most attractive males mate in 25 percent of their courtships. The higher rate of courtship success by specific males, the significant effect of small decorations and the fine details of bowers on mating success, and the changes in a female's behavior that indicate her readiness to mate after she arrives at the male bower indicate that a female makes her mating decisions after she arrives at the court.

Maypole-Bower Builders

The other major group of bower builders make maypole bowers—a central "pole" surrounded by a circular display court. Some maypole builders cover part of the display court with a hut-like structure. The simplest structure, however, comes from Macgregor's bowerbird, which lives at high elevations in the mountains of central and eastern New Guinea. It decorates a sapling with sticks and moss. This bower may be most similar to the ancestral one for all bowerbird species.

A male Macgregor's bowerbird selects a thin sapling, usually from three to six centimeters in diameter, and surrounds it with horizontal piles of sticks, which increase the pole's diameter to about 25 centimeters. He covers the lower part of the maypole and the court floor with a fine compressed-moss mat that rises up to form a circular rim about 40 centimeters from the pole. He decorates the court with small objects including seeds, and he hangs regurgitated fruit pulp near the ends of the maypole sticks. On the court's rim and nearby logs, he adds woody black fungi.

We have observed young male Macgregor's bowerbirds clearing courts around the naked trunks of small trees. A selected tree's diameter usually approaches that of a fully developed maypole, which is much larger than the saplings selected by an adult male. This shows a functional correspondence between the trees used by young males and the size of maypoles built by adults.

A female arriving for courtship on an adult male's bower usually lands on the maypole and then hops down to the court. The male moves to the opposite side of the maypole with his chest close to it. He calls, and as the female moves around the maypole, he makes a counter move to keep the maypole between them. Calling, moving and counter moving go on for one to two minutes. Then the female stops moving, and the male expands his bright orange head plume and shakes his head from side to side, giving the female a view of rapid orange flashes on alternating sides of the maypole. While shaking his head, the male moves toward the female to copulate. In some cases, the male may charge the female without prolonged head shaking, causing her to escape around the opposite side of the maypole.

Bowerbirds without Bowers

Two species of bowerbirds, toothbill and Archbold's, clear and decorate courts, but they have lost their bower-building behavior. The unique mating tactics of each of these species suggest functional alternatives to bower building.

Archbold's bowerbird has lost bower-building behavior. Instead, it clears a display court that is about four meters long and 2.5 meters wide, and it covers the court with a thick mat of ferns. A male decorates his court with beetle wings, dark fruit, King of Saxon (a bird of paradise) head plumes and snail shells, and he places smaller decorations in piles near a court's edge and on limbs that overhang a court. A male also drapes orchid vines on numerous

Figure 4. Macgregor's bowerbirds dodge around the maypole during courtship. A male calls, a female moves around the pole and the male makes a countermove to keep the pole between them. If a female wants to mate, she stops moving, allowing a male to approach her.

overhanging limbs, making a set of curtains that crisscross and nearly touch the display court.

Archbold's bowerbird courtship begins with a male chasing a court-visiting female. He flies and hops low, close to the court surface, beneath the vine curtains. After repeated chases, a female stops moving, apparently signaling the male that he may approach her. The male then presses his body close to the fern mat and moves toward the female. With his head near the ground, the male faces the female and makes a chattering call, during which he moves his head rapidly with slight side-to-side movements and occasionally jerks up his head and tail. If the female remains stationary after that frontal display, the male moves behind her, staying near the ground, and then rises rapidly to perform a brief copulation. The low position of the male held throughout courtship, in part assured by the low-hanging vine curtains, reduces his opportunity for forced copulations by jumping on the female.

Male toothbills clear courts that are about two meters in diameter. A court encompasses several small trees, and their bases are cleaned meticulously. Unlike other bowerbirds, a toothbill decorates its court with large objects: fresh leaves turned upside down so that their light undersides are showing. Although not visible from adjacent

courts, the courts of different males are aggregated, in a so-called *lek*, and are often less than 30 meters apart allowing them to interact through loud calls. Dominant males interrupt the calls of males on adjacent courts. In addition, toothbills spend little time on the ground and far less time on their courts than do other species. Males at the center of an aggregation—the birds that preliminary studies show to be dominant in vocal interactions—have the highest mating success.

Figure 5. Toothbill bowerbirds clear courts but do not build bowers. A male decorates his court with large leaves, which he turns upside down. The rectangular pieces of paper on this court were added in an experiment to test the colors of artificial objects males would used. (Photograph courtesy of the author.)

Figure 6. Archbold's bowerbirds also build bowerless courts. A male approaches a court-visiting female with low-level flights or hops. He also makes calls and jerks his head from side to side. The female either remains stationary for mating or escapes.

During courting, a female arrives on the court and stands still, as if waiting for a male. After little or no display, the male aggressively mounts the female. The longest observed courtship lasted just 3.8 seconds. Toothbill copulation, however, lasts longer and appears violent compared with the brief and cooperative mating of other bowerbirds. During mating, a male toothbill makes low buzzing calls and beats his wings. After mating, a female leaves immediately. The use of exceptionally loud calls and large decorations and evidence of a female preference for central males on leks suggests that toothbill females may assess males before arriving on the court. If a female chooses a mate before arriving on a court,

there is little need for a bower's protection from forced copulation.

Why Build Bowers?

Although avenue and maypole bowers differ in form, observations of courtship behavior at bowers show that both provide a barrier that protects a visiting female from forced copulations by a courting male. Both avenue- and maypole-building males perform prolonged and active courtship displays. A male watches a female until she signals her readiness for mating, then he moves behind her to copulate. A female not prepared to mate can escape while an approaching male moves around a barrier created by a bower wall or maypole. A bower also allows a female to observe court decorations from close range with a reduced threat of forced copulation. The freedom from forced copulation offered by the bower may explain the high degree of elaboration of the decorated ground that has evolved in this group, including the use of small decorations on a ground court.

The two species that build courts without bowers offer alternative solutions to the problem of restricted mate choice because of forced copulation. Toothbill females select desirable mates before arriving on a court, so they do not need the protection of a bower. The low position of an Archbold's bowerbird male while courting allows a female to escape an unwanted copulation.

Males of many species gain reproductive success through forced copulation, so why would male bowerbirds build a structure that limits their opportunities for this behavior? The pro-

tection of a bower probably attracts females and increases their visitation, which more than compensates a male for losing forced copulations. Given that female bowerbirds choose the courts they visit, they should prefer the ones that provide protection from forced copulations. A female that freely chooses her mate should also be less likely to mate with another male. That combination of behaviors provides bower-building males with increased visitation by females and a high chance of being a female's only mate.

In some bowerbird species, males attack visiting females during courtship, and a bower might protect a female from such a threat. A maypole bower could serve that purpose. At avenue bowers, on the other hand, a male faces a female during courtship, so the bower offers no protection. Moreover, a female confined inside avenue-bower walls makes a susceptible target; she could only escape by moving backward, because the walls prevent her from moving sideways or turning around. If bowers served originally as protection against aggressive attacks from courting males, the evolution of the avenue bowers would require the loss of that function and replacement with others. Although the prostrate position taken by a courting male Archbold's may provide protection from attacks, no such behavior has been observed in toothbills. Overall, it seems unlikely that bowers initially evolved to protect females from attacks by males.

The good-genes hypothesis gains support from some observations, including the tendency of females to choose vigorous males and the intense, athletic displays of males in species with widely separated bowers. These characteristics, however, may derive from the origins of male courtship "dances" and vocal behavior rather than from bower building. In some modern species, a female might assess a male's genetic quality from his ability to maintain his bower in the face of destruction by rivals, but such a process seems unlikely early in the evolution of bower building, when only a few males had bowers. Assessing male quality by his bower probably arose as a secondary function after bower building evolved.

The so-called runaway model suggests that female preferences and male traits evolve together, driven by a mating advantage gained by males that possess

B. Chudleigh, Vireo

Figure 7. Great bowerbirds build avenue bowers and decorate them with green objects and shells. A male great bowerbird organizes the decorations strategically, such that they contrast with his lilac crest during his display. The combination of decorations and behavior may enhance a female's interest.

an extreme version of a trait, such as bower building. No evidence, though, suggests that males with large bowers mate more. In addition, recent versions of the runaway model expect high costs for a trait that confers a strong mating advantage. In an intense study of satin bowerbirds, I found no evidence of high cost, despite strong effects of bower quality on male mating success.

Several other hypotheses also lack support. The predation hypothesis seems unlikely, in part, because no example of predation on females or males appeared during more than 100,000 hours of monitoring bowerbird display courts in 10 species. That result proves especially relevant given that males of most species are not protected from predators during courtship. In addition, neither major bower type protects a female from behind, where a predator or a marauding male seeking forced copulation might approach.

Evolution of Bower Building

Determining the evolution of many traits requires an explanation of how incipient stages could be used. Our work suggests that the first bowers consisted of a sapling on a display court. If that is correct, why might a female prefer a male that has a sapling on his display court over ones with other attractive attributes that might benefit her or her offspring? Maybe females simply sought protected courts. A court with a natural barrier, such as a sapling, could separate a female from a courting male and allow her to closely observe the male's display and decorations without committing to mating.

By placing sticks around a sapling a male would be less constrained by sapling size and location. He could utilize a much wider range of saplings and ones in particularly suitable locations, by enhancing the diameter of a maypole to an appropriate size with a stick covering. In addition, the soft edge created by a stick maypole allows males and females to observe each other and anticipate each other's moves, which would be more difficult around a tree of equal diameter. In this scenario, stick-built bowers would have been an improvement on a previous practice of using natural barriers on courts. It might have begun with the rearrangement of fallen sticks that were already present to enhance a court's protective qualities and led to a simple maypole bower.

Once the tendency toward stick-built bowers evolved, two trends could emerge. First, bower form could diversify to serve other functions. Second, free-standing stick barriers would allow males even more freedom in selecting bower sites and in concentrating decorations in advantageous locations. The transition to avenue bowers required losing the use of a sapling as a bower support and the addition of a different barrier. The two-walled barrier oriented a female toward parts of a court where a male could concentrate his decorations on a well-lit stage and arrange the decorations to his best advantage. Many avenue-building males separate decoration types in zones around a bower in an apparently functional manner. Male avenue-building great bowerbirds, for example, place green objects beneath the spot where they display their lilac crest. The decorations are a complementary color to a male's crest and probably increase the contrast of his display. The hut-like cover on some maypole-bower courts also orients a female to a male's display.

No one knows whether bower building or decorating came first, but it appears that the development of complex bowers may have strongly influenced the use of decorations and the evolution of male plumage. The late E. Thomas Gilliard of the American Museum of Natural History argued that the degree of male head-crest elaboration correlates inversely with bower size and the number of decorations in maypole bowers. He suggested that plumage characteristics were transferred to the bower and its decorations, but offered no explanation for the transfer.

Observations of how bower shape constrains a male's display may reveal a relationship between plumage, bowers and display areas. Around the simple maypole of Macgregor's bowerbirds or the large bowerless court of Archbold's bowerbirds, decorations are spread widely around the bower. The males of both species possess well-developed crests, which they use actively during displays. That behavior contrasts with most avenue builders, which have either a reduced or no crest and build more complex bowers that orient a female toward a more limited area where decorations can be concentrated and kept in her view. In most polygynous avian species, including bowerbirds with simple or no bowers, the position of males and females varies during displays, and for females to see bright colors males must carry bright plumage. In bowerbirds with more complex bowers, which focus a female's attention on concentrations of decorations, costly bright plumage may be replaced by strategically located arrays of decorations.

The combination of analyzing bowerbird behavior and constructing a phylogeny produces an unexpectedly coherent picture of bower function, despite the diversity in structural form. All types of bowerbird behavior indicate that females seek protection from unwanted mating. No other hypothesis proves consistent with current bower function, the function of a presumed ancestral bower and novel behavior in derived bowerless species. The significance of protection from a courting male suggests an important role for models that predict direct benefits that females might gain from elaborate male traits.

Acknowledgments

This research was supported by the National Science Foundation and the University of Maryland. The New South Wales and Queensland National Parks, The Australian Bird and Bat Banding Scheme, and the PNG Wildlife and Conservation Department provided permits. R. Crozier, J. Dimuda, G. Harrington, I., J., N. and M. Hayes, J. Lauridsen, M. J. Littlejohn, J. Kikkawa, M. Raga, J. Hook, and M. and J. Turnbull provided important support. C. Depkin, D. Bond, K. Collis, R. Condit, A. Day, J. Helms, I. Kaatz, C. Loffredo, J. Morales and D. Sejkora participated as team leaders and/or co-investigators. More than 100 volunteers provided excellent field assistance.

Bibliography

Borgia, G. 1995. Threat reduction as a cause of differences in bower architecture, bower decoration and male display in two closely related bowerbirds *Chlamydera nuchalis* and *C. maculata*. *Emu* 95:1–12.

Borgia, G. 1985. Bowers as markers of male quality. Test of a hypothesis. *Animal Behavior* 35:266–271.

Borgia, G., and U. Mueller. 1992. Bower destruction, decoration stealing, and female choice in the spotted bowerbird (*Chlamydera maculate*). *Emu* 92:11–18.

Kusmierski, R., G. Borgia, R. Crozier and B. Chan. 1993. Molecular information on bowerbird phylogeny and the evolution of exaggerated male characters. *Journal of Evolutionary Biology* 6:737–752.

Hunter-Gatherers of the New World

Kim Hill
A. Magdalena Hurtado

Foraging peoples live by hunting, fishing, and collecting wild plants and insects. In the last 20 years anthropological research among foragers, or hunter-gatherers, as they are often called, has become increasingly important, for three reasons. First, the opportunity to study humans as foragers has been disappearing at an extremely rapid rate in the last 100 years. Because of worldwide economic trends, the transition from foraging to other forms of subsistence may be irreversible. Thus, we are likely to be the last generation to witness our fellow humans living in a way that was typical of most of human history.

Second, modern foragers live in relatively small groups (usually 15 to 100 individuals) in which subsistence activities produce immediate results, and a limited number of behavioral options are open to band members. This means that the ability to study direct links between ecological or social variables and behavioral patterns is generally greater in foraging societies than in more complex human settings.

Third, because our hominid ancestors spent all but the last 10,000 years (less than one percent of the time span of hominid history) living in small groups that subsisted on wild resources, we should be able to learn a great deal about human history and the evolution of human traits by studying modern foragers. Virtually all modern human anatomical traits evolved when foraging was universal, and the human nervous system and physiological mechanisms that generate behavior also evolved while humans lived by hunting and gathering.

Stimulated by these factors, and strongly influenced by the seminal work of Richard Lee on the !Kung San as well as a provocative conference on "Man the Hunter" (Lee and DeVore 1968), a number of ecologically oriented research projects on modern foragers have been carried out in the past two decades: the Harvard Kalahari project and other !Kung San studies (for example, Lee and DeVore 1976; Tanaka 1980), research in Australia and Indonesia (Meehan 1977; Jones 1980; Griffin and Estokio-

> *Observations of the Ache, a foraging people in Paraguay, indicate that no single pattern of behavior is typical of the hunter-gatherer way of life*

Griffin 1985), research in the Arctic and sub-Arctic (Binford 1978; Smith 1985), the Harvard Pygmy Project and other pygmy studies (Hart 1978; Harako 1981; Bailey and Peacock 1988), and the Utah hunter-gatherer project on Ache and Hiwi foragers, as well as the Utah-UCLA Hadza project (O'Connell et al. 1988; Hawkes et al. 1989).

The most important lesson that can be derived from these and numerous other forager studies is that very few "typical" hunter-gatherer patterns emerge. Instead, groups vary in almost every parameter that has been measured: composition of diet, food-sharing, men's and women's work patterns, subsistence strategies, childcare, settlement patterns, marriage systems, and fertility and mortality. The nonspecialist anthropological audience has been slow to appreciate the importance of this variability. Instead, the !Kung San studies of Lee are often cited as the typical hunter-gatherer pattern. This is almost certainly due to the outstanding quality of Lee's work and to the supposed appropriateness of the African savanna as an ecological context for understanding earlier hominids. Indeed, the !Kung study remains a cornerstone of modern forager research, but we must learn to build upon it if we are to develop our understanding of human evolutionary history.

One of the most useful approaches currently employed in forager studies is that of behavioral ecology, or the study of behavior from an evolutionary perspective. This approach assumes that behavioral patterns are generally adaptive and that variations are due to differences in the costs and benefits to fitness in each environmental and social context.

Much of the work presented in this paper was stimulated by four main issues in hunter-gatherer studies. First, what are the likely causes and consequences of the major dietary and technological shifts observed in the archaeological record, and what can these tell us about the diet of our hominid ancestors? Next, how and why is food shared among current foragers, and what are the implications of these patterns for the evolution of group living, settlement patterns, and the sexual division of labor in hominids? Third, what can activity profiles of modern foragers tell us about our past? Is the foraging way of life one of ease and leisure or a difficult one requiring hard work to survive? Fourth, in the area of demographics, what are the basic trends in fertility and

Kim Hill is an assistant professor in the Dept. of Anthropology and a member of the Evolution and Human Behavior program at the University of Michigan. He received his B.A. in biology and his Ph.D. in anthropology from the University of Utah. A. Magdalena Hurtado is a research scientist at the Instituto Venezolano de Investigaciones Científicas and in the Dept. of Anthropology at Michigan. She received her B.A. in anthropology from SUNY-Purchase and her Ph.D. in anthropology from Utah. Address: Dept. of Anthropology, 1054 L.S.A., University of Michigan, Ann Arbor, MI 48109-1382.

mortality that characterize modern foragers?

In addition to these issues, most researchers are now concerned with the relationship between ecological constraints and different behavioral patterns. An understanding of this relationship should make it possible to build models that specify how independent variables will affect the behavior observed in any human community. Only through this uniformitarian approach can we hope to know what our ancestors did in the unobservable past.

This paper reports some of the results obtained in eight years of work with Ache foragers of Paraguay and describes preliminary findings from an ongoing project with the Hiwi foragers of the Venezuelan savanna. In all these studies we emphasize the variation through time and across groups of individuals, in order to derive likely explanations for differences observed between foraging groups.

The Ache of Paraguay

The Ache are a native population of Paraguay who until recently were full-time nomadic foragers (Clastres 1972; Melia et al. 1973). They consist of four independent groups; we have studied primarily the northernmost group, the last to make permanent peaceful contact (in the 1970s). This group was made up of 10 to 15 small bands that had no specific territories but roamed over an area of about 18,500 km^2 in eastern Paraguay. Each band had a smaller home range, but adults generally knew the entire area covered by the group. Recall by informants suggests the median size of a band was 48 people; the range on a given day was from 3 to 160.

The region is mainly neotropical, semideciduous evergreen forest, with a tree canopy about 20 m high and undergrowth more dense than that observed in many other primary tropical forests of South America. Since 1975, much of the area has been cut for agriculture and cattle pasture, but sizable pieces of primary forest (about 2,400 km^2) still exist near the Ache settlements. Transects measuring mammalian densities suggest a crude biomass of only about 400 kg/km^2 for the most commonly encountered mammals. This is about half the crude biomass measured for the same species in Barro Colorado, Panama (Eisenberg and Thorington 1973), and most forests of South America that have been studied show considerably higher species diversity.

Before contact, Ache bands foraged in this area and moved campsite frequently while hunting and gathering. Now they live primarily at agricultural mission stations but still spend about 25% to 35% of their time on overnight trips back in the forest, foraging for subsistence. We have monitored the Ache diet almost from the point of first contact to the present and are able to

reconstruct the traditional diet to some extent by means of observations on forest trips (Hill 1983).

The data from short-term trips (with a range of 4 to 15 days) suggest that as foragers the Ache eat an astounding 3,700 calories per person per day (Hill et al. 1984). (By contrast, active adult Americans consume about 2,700 calories per day.) When the Ache are living in the forest, an average of 56% of their calories come from mammalian meat (ranging from 46% to 66%, depending on the season), with honey making up 18% (range 6–30%) and plants and insects providing an average of 26% (range 15–49%).

On a foraging trip, camp members rise early, eat whatever is left over from the previous day, and set out in search of food. Men lead the way, carrying only bows and arrows, and women and children follow, the women carrying young children and the family's possessions in a woven basket. Some men walk with their wives and carry children on their shoulders. Ache foragers do not walk on trails but break a new path through the forest each day. Usually the leaders set out in the direction of an area known or thought to contain important food resources.

After walking together for about an hour, the two sexes separate, with men walking further and more rapidly in search of game, and women and children slowly progressing in the general direction the men have set out. Men generally eat very little during the day, but women and children sometimes collect and eat fruits and insects while men hunt, and women often process palm trunks for their starchy fiber near the end of the day. This snacking usually accounts for less than 5% of all food consumed (Hill et al. 1984).

All camp members come together again at the end of the day, when they clear a small camp in the underbrush, build fires, and prepare and share food extensively. Evening is considered the most pleasant time, with band members enjoying their only large meal of the day, and joking and singing in the night. While in the forest the Ache sleep on the ground or on palm-leaf mats in a small circle. They build palm-leaf huts to sleep in only if it begins to rain. The next morning the band moves on again in search of food unless there is heavy rainfall throughout the day.

A model for food choice

The tropical forests of Paraguay are believed to contain several hundred species of edible mammals, birds, reptiles, amphibians, and fish, but the Ache have been observed to exploit only about 50 of them. Similarly, the forest holds hundreds of edible fruits and insects, yet the Ache exploit only about 40 of these. Over 98% of the total calories in the diet we observed between 1980 and 1983 were supplied by only 17 different resources. What can account for the fact that the Ache seem to ignore many edible resources in their environment?

The optimal diet model (MacArthur and Pianka 1966; Emlen 1966) was developed to predict which of an array of resources will be exploited if organisms attempt to maximize rates of food acquisition. In order to maximize the overall rate of return, foragers should attempt to obtain a resource only when the expected return rate is higher than what they can obtain on average if they

Figure 1. The Ache are a people of eastern Paraguay; most lived as nomadic foragers until the 1970s. Now based at agricultural mission stations, they continue to make short trips into the primary forest, where they move camp each day, gathering insects and plants and hunting mammals for meat. (Photo by A. M. Hurtado.)

Figure 2. When in the forest the Ache eat well, taking in as much as 3,700 calories per day. The fruit *Rheedia brazilense (above),* which ripens in January, is a favorite food, as is *Cebus apella,* the capuchin monkey *(right).* (Photo above by K. Hill; photo at right by K. Hawkes.)

ignore it and continue to search for other resources.

Between 1980 and 1986, using Ache data, we were able to test the prediction that each resource exploited should be characterized by a higher caloric value than that observed for foraging overall. Our data show that all 16 items observed to be exploited in 1980, and 25 of 26 items exploited in 1981–83, are characterized by higher return rates than a forager could expect to obtain if he ignored the item and continued foraging (Hawkes et al. 1982; Hill et al. 1987). Whereas return rates for a whole day of foraging worked out to about 1,250 cal/h for men and about 1,090 cal/h for women, the average rate for any particular food obtained was, for men, about 3,500 cal per hour of foraging, and for women about 2,800 cal per hour of foraging. Although we were unable to measure resources *not* taken by the Ache, experience suggests that many would indeed be characterized by low returns (small fruits, birds, insects, reptiles). Thus, Ache foragers apparently do behave as if they chose to exploit only those resources that would increase their overall rate of food acquisition.

The model may be particularly useful for understanding subsistence changes that occur through time, or as a result of changes in technology. For example, in 1980 a few Ache hunters acquired shotguns, which raised their overall return rate from 910 cal/h (with bow and arrow) to 2,360 cal/h. Because some of the game taken by Ache men is characterized by return rates below the new 2,360 cal/h but above the rate of bow-and-arrow hunting, shotgun hunters should ignore some low-return animals that are taken by bow hunters. This prediction was generally met by observations in the field (Hill and Hawkes 1983). Most notably, shotgun hunters spent less than 2% of their time pursuing capuchin monkeys (with a return rate of 1,215 cal/h), whereas bow hunters spent over 13% of their time chasing capuchin monkeys *on the same foraging trips.* Several times shotgun hunters were observed to leave monkey hunts and continue searching for other, more profitable game.

Continued work with the Ache has pointed out both the utility and some shortcomings of models derived

from optimal foraging theory (Hill et al. 1987; Hill 1988). Limitations that we have noted are, first, that some items that must be processed extensively may be exploited even when the resulting return rates are low, if processing normally takes place during times that foraging is not possible. Second, a short-term risk of imminent starvation or a reduction in the variance of daily food intake may lead to foraging behavior not predicted by simple models. Third, the biological value of foods is probably not reducible to calories when food types differ greatly.

A balance of nutrients is likely to be especially important in decisions in which the forager faces a choice between foods high in carbohydrates (such as plants) and those high in protein and lipids (animals and insects). From observations on the Ache, it appears that the sexual division of labor and the foraging strategy of males in general can be predicted by optimal foraging models only if the higher value of foods rich in proteins and lipids over carbohydrates is taken into account (Hill et al. 1987). The most important lesson from the Ache studies, however, is that simple models based on the assumption that individuals will attempt to maximize their rate of food acquisition while foraging are indeed useful for predicting subsistence patterns.

Food-sharing

The extensive sharing of food has been reported for many foraging peoples, but until recently there have been no quantitative studies that allow us to determine exactly what the sharing pattern looks like and how it varies from one foraging group to another. Understanding this variation is crucial, because food-sharing has been postulated to be critical in shaping the unique character of human sociality (e.g., Washburn and Lancaster 1968; Isaac 1978; Lancaster 1978; Kithara-Frisch 1982; Zihlman 1983).

The Ache share food throughout the band. Women who share vegetable items are usually praised, and young children are taught that stinginess is the worst trait a person can have. All hunters, regardless of status or hunting success, give up their kills to be distributed by others, and they almost never eat from their own kill. Nevertheless, there is some interesting variation in the way different resources are shared.

Although about 75% of all food consumed in an Ache band is acquired by a person outside the consumer's nuclear family, different resources are not all shared to the same extent (Kaplan et al. 1984). Game items are shared most, with more than 90% of the meat a hunter acquires being consumed by individuals not in his immediate family. Honey is shared somewhat less, and plant and insect foods least (Kaplan and Hill 1985).

Statistical analyses show that wives, children, and siblings receive no more of the meat or honey acquired by a man (their husband, father, or brother) than would be predicted by random chance if all the food were simply divided up among band members. This clearly contradicts a common assumption that food is always shared preferentially with close kin. However, the husband, children, and siblings of a woman were found to consume more of the food she collected than would be predicted by chance (Kaplan et al. 1984).

Further analyses suggest food-sharing among the Ache may serve to reduce daily variance in consumption. Food-sharing should be most beneficial if, by their own foraging, individuals acquire more than they can eat on some days and nothing on other days—a pattern that, indeed, characterizes Ache men's hunting returns (Hill and Hawkes 1983; Kaplan et al. 1989). If sharing is a strategy to reduce daily variance in food intake, and if different types of food (e.g., vegetables and meat) are not interchangeable, we should find a correlation between the daily variance in acquisition of a type of food and the extent to which it is shared. Our data from the Ache confirm this relationship. Moreover, the absolute reduction in variance of daily food intake is high enough to be biologically significant. During the time of our study, the average nuclear family reduced its daily variance from 13,243 cal to 4,863 cal by sharing food. A simple model suggests that food-sharing may lead to an 80% increase in nutritional status (Kaplan and Hill 1985).

Differences in observed patterns of food-sharing across modern foragers may therefore be partially due to differences in the daily variance of major food types acquired. This hypothesis leads to useful predictions: for example, among foragers that are able to store food, we might expect very little sharing, because storage reduces variance in the availability of food. Additionally, the data suggest that food-sharing and associated social patterns may not have arisen in hominid history until our ancestors began to use subsistence strategies that produced a high daily variance in food acquisition.

The division of labor

One of the major issues in the study of modern foragers has been just what determines how much time they spend in different activities. Some researchers see foragers as members of an original "affluent society" (e.g., Sahlins 1972), in which work effort is low because "needs" are few and easily met. Others (for example, Hawkes et al. 1985) have questioned this generalization. Early quantitative work with the !Kung San of Africa (Lee 1968) tended to support the low-effort model (which was partially derived from !Kung data) and has led more recent workers to monitor carefully how much time is spent in subsistence work and what other activities are important throughout the day. In addition, because a marked division of labor along sexual lines is an important characteristic distinguishing humans from other primates, it is of interest to describe the range of activities specifically for men and for women.

Our collected data—some 63 days' worth of focal

Figure 3. The divison of labor along sexual lines is clearly marked among the Ache, with men spending almost 7 hours per day in hunting and in processing both meat and other foods. Resting, socializing, and light work together account for about another 4.5 hours per day. (Photo by K. Hill.)

mates may result in female competition for the best mates if genetic variation among males affects offspring survival and if males of high genetic quality limit female reproductive success. Although there is little information with which to examine the influence of high-quality sires on the operation of sexual selection, there are a number of studies, which we will now discuss, that examine sexual selection in species in which males provide immediate, material services.

Contributions to offspring

Both males and females can provide for offspring in a variety of ways. Females invest directly in their offspring both through the material investment in eggs and zygotes as well as through maternal care. The better-known examples of male investment concern direct paternal care of offspring; this has been observed in a large number of species in a variety of animal groups (Ridley 1978). However, males can also contribute indirectly to offspring by providing benefits to their mates, both before and after mating. Examples include the nutritional benefits of "courtship feeding," which is observed in certain birds (Nisbet 1973) and insects, such as the hangingfly *Hylobittacus apicalis* shown in Figure 1 (Thornhill 1976). Protection of the mate is another example of such benefits (Gwynne 1984b; Thornhill 1984). The nature of selection leading to the evolution of benefit-providing males is still poorly understood (Alexander and Borgia 1979; Knowlton 1982). But regardless of the evolved function of the phenomenon, with the evolution of benefit-providing males there may be a change in the action of sexual selection on the sexes.

Contributions by one sex that affect the number and survival of offspring potentially limit the reproduction of the opposite sex. Thus the relative investment of the sexes in these sorts of reproductive efforts should determine the extent to which each sex competes for the opposite sex. This hypothesis can be tested by comparing species or populations with differing investment patterns or by directly manipulating resources that limit the reproduction of the population. Examples we will review use these methods.

There have been few attempts to estimate the extent of sexual selection on the sexes in nature. However, variance in the reproductive success of the sexes has been estimated for *Drosophila melanogaster* (Bateman 1948), a damselfly (Finke 1982), red-winged blackbirds (Payne 1979), and red deer (Clutton-Brock et al. 1982). In these species, the parental contribution by the male is smaller than that of the female; as predicted, all species show greater sexual selection on the males.

Observed sexual differences, the consequences of sexual selection, typically serve as evidence for the relative intensity of sexual selection on the sexes in the evolutionary past. In most species, females provide a large amount of parental contribution and males little, and it is primarily the males that show secondary sexual traits of morphology and behavior which function in competition for mates. It is also well known that sexual differences are greatly reduced under monogamy. This is as expected, because both sexes of monogamous species engage in similar levels of parental care. However, for this comparative test to succeed, a sex reversal in the courtship and competitive roles should be observed in species in which males provide a greater portion of the total contribution affecting offspring number and survival.

Parental care provided only by the male is found throughout the vertebrates, particularly in frogs and toads, fishes, and birds, and is likely to represent a limiting resource for female reproduction. In certain seahorses and pipefishes (Syngnathidae), males care for eggs in a specialized brood pouch (Breder and Rosen 1966). In these fishes, male parental care appears to limit female reproductive success, and females are larger and more brightly colored than males, as well as being more competitive in courtship (Williams 1966, 1975; Ridley 1978). In frogs of the genus *Colostethus,* it is the male in one species and the female in another that provide parental care by carrying tadpoles on their backs. In both species, as predicted by theory, it is the sex emancipated from parental duties that defends long-term mating territories and has a higher frequency of competitive encounters (Wells 1980). In species of birds in which males provide most of the parental care, the roles in courtship behavior are reversed; females compete for mates and sometimes are the larger or more brightly colored sex (reviewed by Ridley 1978).

Exclusive paternal care of eggs or larvae is restricted to about 100 species of insects, all of which are within the order Hemiptera, or true bugs (Smith 1980). In the giant water bugs (Hemiptera: Belostomatidae), females adhere eggs to the wing covers of their mates, and the males aerate the eggs near the water surface and protect them from predators. For *Abedus herberti,* Smith (1979) provides evidence that male back space is a limiting resource for females and that male parental care is essential for offspring survival; females actively approach males during courtship, and males reject certain females as mates.

Although direct investment in offspring through parental care is uncommon within the insects, indirect paternal contributions, with males supplying the females with nutrition or other services such as guarding, is widespread in a number of taxa.

The guarding of the female by the male after mating is usually thought of as functioning to prevent other males from inseminating the female (Parker 1970). However, an alternative hypothesis is that guarding evolved in the context of supplying protection for the female and that it thereby enhances male reproductive success (Gwynne 1984b; Thornhill 1984). Male guarding is known to benefit the female in several species: in damselflies (*Calopteryx maculata*), guarding by males after mating allows females to oviposit undisturbed by other males (Waage 1979); in waterstriders (*Gerris remigis*), harassment of guarded females by other males is similarly reduced, allowing females much longer periods during which to forage for food (Wilcox 1984). At present there is no information for these species concerning whether certain males protect females better than others, which would lead to female competition for more protective males. If female competition occurs, selection should favor mate choice by males.

Mate choice by males has been observed in species in which males provide protection or other services to females. In brentid weevils (*Brentus anchorago*) males prefer large females as mates and are known to assist their mates in competition for oviposition sites by driving away nearby ovipositing females (Johnson and Hubbell 1984). Male lovebugs (*Plecia nearctica*), so named for their two- to three-day-long periods of copulation, also prefer to mate with large females (Hieber and Cohen 1983). Lengthy copulation in this species may be beneficial for females in that copulating pairs actually fly faster than unattached lovebugs (Sharp et al. 1974). Similar benefits may be obtained by paired amphipod crustaceans (*Gammarus pulex*); pairs in which males are larger than females have a superior swimming performance that minimizes the risk of being washed downstream (Adams and Greenwood 1983). Perhaps these sorts of services supplied by male crustaceans explain the presence of male choice of mates seen in certain groups (e.g., Schuster 1981).

Males can supply nutrition in several ways. Our research has dealt with courtship feeding, where food items such as prey or nutritious sperm packages (spermatophores) are eaten by females, and we discuss this behavior in detail below for male katydids and scorpionflies. There are also more subtle forms of contribution; in several insect species spermatophores or other ejaculatory nutrients are passed into the female's genital tract at mating (Thornhill 1976; Gwynne 1983; Thornhill and Alcock 1983). A number of researchers have done some interesting work on a similar phenomenon in crustaceans. Electrophoretic studies of proteins in the ovaries and the male accessory gland of a stomatopod shrimp (*Squilla holoschista*) strongly suggest that a specialized protein from the male's accessory glands is transferred with the ejaculate into the female's gonopore and then is translocated intact into the developing eggs; females of this species usually initiate mating and will mate repeatedly (Deecaraman and Subramoniam 1983). In a detailed study of another

Figure 2. A female katydid (*Requena verticalis*) just after mating shows the large spermatophore that has been attached by the male to the base of her ovipositor (*top*). The female grasps the nutritious spermatophylax (*middle*) and eats it (*bottom*), leaving the sperm ampulla portion of the spermatophore in place. After insemination, the ampulla is also eaten. The nutrients in the spermatophore represent a considerable material contribution by the male in the reproductive success of the female. (Photos by Bert Wells.)

Figure 3. While this female Mormon cricket (*Anabrus simplex*) is atop the male, the male apparently weighs his potential mate and will reject her if she is too light. Males select mates among females that compete for access to them, preferring females that are larger and therefore more fecund. This represents a reversal of the sex roles much more commonly found in nature. (Photo by Darryl Gwynne.)

stomatopod, *Pseudosquilla ciliata*, Hatziolos and Caldwell (1983) report a reversal in sex roles, with females courting males that appear reluctant to mate; in the absence of obvious male parental contribution, these researchers cite work with insects in suggesting that male *Pseudosquilla* may provide valuable nutrients in the ejaculate.

Studies with several butterfly species have used radiolabeling to show that male-produced proteins are incorporated into developing eggs as well as into somatic tissues of females (e.g., Boggs and Gilbert 1979). Lepidopteran spermatophores potentially represent a large contribution by the male (up to 10% of body weight), and proteins ingested by females are likely to represent a limiting resource for egg production in these insects that feed on nectar as adults (Rutowski 1982). Preliminary experiments by Rutowski (pers. com.) with alfalfa butterflies (*Colias* spp.) indicate that females receiving larger spermatophores lay more eggs. Consistent with theory, there is evidence of males choosing females and of competition by females for males. For example, in the checkered white butterfly (*Pieris protodice*), males prefer young, large (and thus more fecund) females to older, small-

er individuals (Rutowski 1982). And in *Colias*, certain females were observed to solicit courtship by pursuing males; these females may have had reduced protein supplies, as they were shown to have small, depleted spermatophores in their genital tracts (Rutowski et al. 1981). Although there is variation between species of butterflies in the size of the male spermatophore, this variation apparently does not result in large differences between species in the male contribution; a review of the reproductive behavior of several butterfly species did not show consistent differences in courtship when species with small spermatophores were compared to those with large spermatophores (Rutowski et al. 1983). However, as shown by Marshall (1982) and confirmed by our studies described below, spermatophore size is not always a useful measure of the importance of the male nutrient contribution.

Reversal of courtship roles in katydids

Katydids (Orthoptera: Tettigoniidae) are similar to butterflies in that males transfer spermatophores to their mates, and, as shown by radiolabeling, spermatophore nutrients are

used in egg production (Bowen et al. 1984). In contrast to the mated female butterfly, the katydid female ingests the spermatophore by eating it (Fig. 2). The spermatophore consists of an ampulla which contains the ejaculate and a sperm-free mass termed the spermatophylax (Gwynne 1983). Immediately after mating, the female first eats the spermatophylax; while this is being consumed, insemination takes place, after which the empty sperm ampulla is also eaten (Gwynne et al. 1984). However, the katydid spermatophylax appears not to function as protection of the ejaculate from female feeding. The spermatophylax of the katydid *Requena verticalis* is more than twice the size necessary to allow the transfer both of the spermatozoa and of substances that induce a four-day nonreceptive period in females (Gwynne, unpubl.).

Spermatophore nutrients are important to the reproductive success of the female katydid. Laboratory experiments have shown that as consumption of spermatophylax increases, both the size and the number of eggs that females subsequently lay also increase (Gwynne 1984c). Furthermore, the increase in the size of eggs appears to be determined only by male-provided nutriment; an increase in protein in the general diet increases egg number but does not affect egg size (Gwynne, unpubl.).

The size of the spermatophore produced by male katydids varies from less than 3% of male body weight in some species to 40% in others (Gwynne 1983). Differences in the size of the male contribution conform with the predictions of sex-difference theory: in two species of katydids that make very large investment in each spermatophore (25% or more of male weight) and that have been examined in detail—the Mormon cricket (*Anabrus simplex*) and an undescribed species (*Metaballus* sp.) from Western Australia—there is a complete reversal in sex roles, with females competing aggressively for access to males that produce calling sounds, and males selecting mates, preferring large, fecund females (Gwynne 1981, 1984a, 1985). Figure 3 illustrates this reversal in the Mormon cricket. There is no evidence of such a reversal of courtship roles in species with smaller spermatophores; in these species males compete for mating territories and females select mates (Gwynne 1983).

It is evident, however, that a complete estimate of the contribution to offspring requires more than a simple measure of the relative contribution by the sexes to offspring such as the weight of the spermatophore relative to the weight of a clutch of eggs. Both species of katydids showing a role reversal in courtship behavior also had populations that showed no evidence of the reversed roles. For the Mormon cricket, the simple measure of relative contribution did not show a higher contribution by males at the sites of role reversal (Gwynne 1984a). However, these sites had very high population densities, with individuals of both sexes competing vigorously for food in the form of dead arthropods and certain plants. These observations suggest the hypothesis that the limited food supplies at these sites resulted in few spermatophores being produced and that spermatophore nutrients were thus a limiting resource for female reproduction. Food did not appear to be scarce at sites of low population density where the reversal in courtship roles was not observed. Support for the hypothesis that food is a limiting resource at high-density sites came from dissections of the reproductive accessory glands that produce the spermatophore in a sample of males from each of the sites. Only the few calling males at sites of high density had glands large enough to produce a spermatophore, whereas most males at the low-density site had enlarged glands. This difference between the males at the high- and low-density sites was not a result of a higher number of matings by males at the high-density site.

Differences between individuals from the two sites indicate that sexual selection on females at high-density sites was intense compared to the low-density site. (Sexual selection is measured by variance in mating success; see Wade and Arnold 1980.) Some females were very successful at obtaining spermatophores. These tended to be large females that were preferred by males as mates. The evolutionary consequences of the apparently greater sexual selection on high-density females was not only aggressive female behavior in competition for calling mates but also a

larger female body size at this site relative to males. This sexual dimorphism was not seen at sites of low density.

Variation in the expression of sexual differences within the same katydid species suggested that behavior might be flexible; that is, females become competitive and males choosy when they encounter certain environments. This hypothesis was examined using the undescribed *Metaballus* species of katydid from Western Australia, which is similar to the Mormon cricket in that only certain populations show female competition for mates and show males that reject smaller, less fecund females. In this species, discriminating males call females by producing a broken "zipping" song from deep in the vegetation. Sites where courtship roles are reversed consist of mainly the zipping male song, whereas at sites of male competition, males produce continuous songs that appear to be louder. An experiment was conducted which involved shifting a number of males and females from a site where role reversal was not observed to one in which it was noted. The behavior of the males that were moved to the role-reversed site changed to resemble that of the local males: their song changed from a continuous to a zipping song, the duration of courtship increased (possibly to assess the quality of their mates), and they even rejected females as mates. Thus, sexual differences in behavior are plastic; courtship roles of the sexes appear to be dependent on the environment encountered.

For the Mormon cricket, it is likely that the relative contributions of the sexes is the factor controlling sexual selection. A comparison of the weights of spermatophores and egg clutches is undoubtedly a poor estimate of relative contribution by the sexes; spermatophores seem to be important to the reproduction of females at both sites (Gwynne 1984a). However, if food supplies limit spermatophore production at sites where reversals in courtship roles are observed, and if females cannot obtain spermatophore nutrients from other food sources, then spermatophores are likely to have a greater influence on female fecundity and thereby are more valuable to female reproduction at these sites. Thus, the total

contribution from the males at these sites is probably larger than that of the females.

Nuptial feeding in scorpionflies

Most of the evidence supporting the hypothesis that the relative contribution of the sexes to offspring is an important factor controlling the extent of sexual selection has been derived from comparisons between or, in the katydid work, within species. In contrast, studies were conducted in which the relative contribution of males was manipulated to determine its effect on the extent of sexual selection (Thornhill 1981, 1986). This research has focused on scorpionflies of the genus *Panorpa* (Panorpidae), in which males use either dead arthropods or nutritious products of salivary glands to feed their mates.

Males must feed on arthropod carrion, for which they compete through aggression, before they can secrete a salivary mass. Males in possession of a nuptial gift release pheromone that attracts conspecific females from some distance. Females can obtain food without male assistance, but doing so is risky because of exposure to predation by web-building spiders. Movement in the habitat required to find dead arthropods exposes females and males to spider predation, and dead arthropods unattended by males are frequently found in active spider webs. The gift-giving behavior of males is an important contribution, because dead arthropods needed by females to produce eggs are limited both in the absolute sense and in terms of the risks in obtaining them.

In a series of experiments, individually marked male and female *Panorpa latipennis* were placed in field enclosures, and variances in mating success of the sexes were determined in order to estimate the relative intensity of sexual selection. Dead crickets taped to vegetation represented the resource that males defended from other males and to which females were attracted. In one experiment, three treatments were established in which equal numbers of males and females were added to each enclosure and the number of dead crickets varied—two, four, or six crickets per enclosure. As predicted, competition among males was greatest in the enclosures with two crickets; the intensity of sexual selection, calculated by variance in male mating success, was greatest in this treatment and was lowest in the treatment with six crickets.

Variance in female mating success was low and was not significant across cricket abundances over the seven days of the experiment. Sexual selection on females probably often arises from female-female competition for the best mates regardless of the number of mates. *Panorpa* females prefer males that provide large, fresh nuptial gifts of dead arthropods over males that provide salivary masses, and males only secrete saliva when they cannot compete successfully for dead arthropods (Thornhill 1981, 1984). This female mating preference is adaptive in that females mating with arthropod-providing males lay more eggs than females mating with saliva-providing males. Thus an accurate measure of sexual selection on female *Panorpa* would include the variation in egg output by females in relation to the resource provided by the mates of females. This information is not available at present.

However, the results on males from this experiment clearly support the hypothesis that sexual selection is determined by the relative contribution of the sexes. As food is a limiting resource for reproduction by female scorpionflies, the total contribution of food by males in enclosures with more crickets was greater than in enclosures with fewer crickets, and the intensity of sexual selection on males declined as males contributed relatively more.

Such studies of the factors controlling the operation of sexual selection are important for two major reasons. The first is simply that sexual selection has been such an important factor in the evolution of life. Sexual selection seems inevitable in species with two sexes, because, as Bateman (1948) first pointed out, the relatively few large female gametes will be the object of sexual competition among the males, whose upper limit to reproductive success is set by the number of ova fertilized rather than by production of the relatively small, energetically cheap sperm. The role of sexual selection in the history of life can best be explored when such controlling factors are fully understood. The second reason is related to the first: the difference in the operation of sexual selection on the sexes may ultimately account for all sexual differences. Only sexual selection acts differently on the sexes per se (Trivers 1972). Natural selection may act on and may even magnify sexual differences in behavior and morphology, but probably only after these differences already exist as a result of the disparate action of sexual selection.

The insight of Williams (1966) and Trivers (1972) is that the relative contribution of materials and services by the sexes in providing for the next generation is the most important factor controlling the operation of sexual selection. In insects, contributions supplied by males to their mates include not only the paternal care of young, a well-studied phenomenon in vertebrates, but also other services such as courtship feeding, subtle forms of nutrient transfer via the reproductive tract, and "beneficial" guarding of mates.

References

Adams, J., and P. J. Greenwood. 1983. Why are males bigger than females in precopula pairs of *Gammarus pulex*? *Behav. Ecol. Sociobiol.* 13:239–41.

Alexander, R. D., and G. Borgia. 1979. On the origin and basis of the male-female phenomenon. In *Sexual Selection and Reproductive Competition in the Insects*, ed. M. S. Blum and N. A. Blum, pp. 417–40. Academic.

Bateman, A. J. 1948. Intrasexual selection in *Drosophila*. *Heredity* 2:349–68.

Boggs, C. L., and L. E. Gilbert. 1979. Male contribution to egg production in butterflies: Evidence for transfer of nutrients at mating. *Science* 206:83–84.

Bowen, B. J., C. G. Codd, and D. T. Gwynne. 1984. The katydid spermatophore (Orthoptera: Tettigoniidae): Male nutrient investment and its fate in the mated female. *Aust. J. Zool.* 32:23–31.

Bradbury, J. W., and S. L. Vehrencamp. 1977. Social organization and foraging in emballonurid bats. III. Mating systems. *Behav. Ecol. Sociobiol.* 2:1–17.

Breder, C. M., and D. E. Rosen. 1966. *Modes of Reproduction in Fishes*. Nat. Hist. Press.

Clutton-Brock, T. H., F. E. Guinness, and S. D. Albon. 1982. *Red Deer: Behavior and Ecology of Two Sexes*. Univ. of Chicago Press.

Darwin, C. 1874. *The Descent of Man and Selection in Relation to Sex*, 2nd ed. New York: A. L. Burt.

Deecaraman, M., and T. Subramoniam. 1983. Mating and its effect on female reproductive physiology with special reference to the fate of male accessory sex gland secretion in the stomatopod, *Squilla holoschista*. *Mar. Biol.* 77:161–70.

Emlen, S. T., and L. W. Oring. 1977. Ecology, sexual selection, and the evolution of mating systems. *Science* 197:215–22.

Finke, O. M. 1982. Lifetime mating success in a natural population of the damselfly *Enallagma hageni* (Walsh) (Odonata: Coenagrionidae). *Behav. Ecol. Sociobiol.* 10:293–302.

Gwynne, D. T. 1981. Sexual difference theory: Mormon crickets show role reversal in mate choice. *Science* 213:779–80.

———. 1983. Male nutritional investment and the evolution of sexual differences in the Tettigonidae and other Orthoptera. In *Orthopteran Mating Systems: Sexual Competition in a Diverse Group of Insects*, ed. D. T. Gwynne and G. K. Morris, pp. 337–66. Westview.

———. 1984a. Sexual selection and sexual differences in Mormon crickets (Orthoptera: Tettigoniidae, *Anabrus simplex*). *Evolution* 38:1011–22.

———. 1984b. Male mating effort, confidence of paternity, and insect sperm competition. In Smith 1984, pp. 117–49.

———. 1984c. Courtship feeding increases female reproductive success in bushcrickets. *Nature* 307:361–63.

———. 1985. Role-reversal in katydids: Habitat influences reproductive behavior (Orthoptera: Tettigoniidae, *Metaballus* sp.). *Behav. Ecol. Sociobiol.* 16:355–61.

Gwynne, D. T., B. J. Bowen, and C. G. Codd. 1984. The function of the katydid spermatophore and its role in fecundity and insemination (Orthoptera: Tettigoniidae). *Aust. J. Zool.* 32:15–22.

Hatziolos, M. E., and R. Caldwell. 1983. Role-reversal in the stomatopod *Pseudosquilla ciliata* (Crustacea). *Anim. Behav.* 31:1077–87.

Hieber, C. S., and J. A. Cohen. 1983. Sexual selection in the lovebug, *Plecia nearctica*: The role of male choice. *Evolution* 37:987–92.

Johnson, L. K., and S. P. Hubbell. 1984. Male choice: Experimental demonstration in a brentid weevil. *Behav. Ecol. Sociobiol.* 15:183–88.

Knowlton, N. 1982. Parental care and sex role reversal. In *Current Problems in Sociobiology*, ed. King's College Sociobiology Group, pp. 203–22. Cambridge Univ. Press.

Marshall, L. D. 1982. Male nutrient investment in the Lepidoptera: What nutrients should males invest? *Am. Nat.* 120:273–79.

Nisbet, I. C. T. 1973. Courtship-feeding, egg size, and breeding success in common terns. *Nature* 241:141–42.

Parker, G. A. 1970. Sperm competition and its evolutionary consequences in the insects. *Biol. Rev. Cambridge Philos. Soc.* 45:525–67.

Payne, R. B. 1979. Sexual selection and intersexual differences in variance of breeding success. *Am. Nat.* 114:447–66.

Ridley, M. 1978. Paternal care. *Anim. Behav.* 26:904–32.

Rutowski, R. L. 1982. Mate choice and lepidopteran mating behavior. *Fla. Ent.* 65:72–82.

Rutowski, R. L., C. E. Long, and R. S. Vetter. 1981. Courtship solicitation by *Colias* females. *Am. Midl. Nat.* 105:334–40.

Rutowski, R. L., M. Newton, and J. Schaefer. 1983. Interspecific variation in the size of the nutrient investment made by male butterflies during copulation. *Evolution* 37:708–13.

Schuster, S. M. 1981. Sexual selection in the Socorro Isopod *Thermosphaeroma thermophilum* (Cole) (Crustacea: Peracarida). *Anim. Behav.* 29:698–707.

Sharp, J. L., N. C. Leppala, D. R. Bennett, W. K. Turner, and E. W. Hamilton. 1974. Flight ability of *Plecia nearctica* in the laboratory. *Ann. Ent. Soc. Am.* 67:735–38.

Smith, R. L. 1979. Paternity assurance and altered roles in the mating behaviour of a giant water bug, *Abedus herberti* (Heteroptera: Belostomatidae). *Anim. Behav.* 27:716–25.

———. 1980. Evolution of exclusive postcopulatory paternal care in the insects. *Fla. Ent.* 63:65–78.

———, ed. 1984. *Sperm Competition and the Evolution of Animal Mating Systems*. Academic.

Thornhill, R. 1976. Sexual selection and paternal investment in insects. *Am. Nat.* 110:153–63.

———. 1980. Sexual selection in the black-tipped hangingfly. *Sci. Am.* 242:162–72.

———. 1981. *Panorpa* (Mecoptera: Panorpidae) scorpionflies: Systems for understanding resource-defense polygyny and alternative male reproductive effort. *Ann. Rev. Ecol. Syst.* 12:355–86.

———. 1983. Cryptic female choice in the scorpionfly *Harpobittacus nigriceps* and its implications. *Am. Nat.* 122:765–88.

———. 1984. Alternative hypotheses for traits believed to have evolved in the context of sperm competition. In Smith 1984, pp. 151–78.

———. 1986. Relative parental contribution of the sexes to offspring and the operation of sexual selection. In *The Evolution of Behavior*, ed. M. Nitecki and J. Kitchell, pp. 10–35. Oxford Univ. Press.

Thornhill, R., and J. Alcock. 1983. *The Evolution of Insect Mating Systems*. Harvard Univ. Press.

Trivers, R. L. 1972. Parental investment and sexual selection. In *Sexual Selection and the Descent of Man, 1871–1971*, ed. B. Campbell, pp. 136–79. Aldine.

Waage, J. K. 1979. Adaptive significance of postcopulatory guarding of mates and non-mates by *Calopteryx maculata* (Odonata). *Behav. Ecol. Sociobiol.* 6:147–54.

Wade, M. J., and S. J. Arnold. 1980. The intensity of sexual selection in relation to male sexual behaviour, female choice, and sperm precedence. *Anim. Behav.* 28:446–61.

Wells, K. D. 1980. Social behavior and communication of a dendrobatid frog (*Colostethus trinitatis*). *Herpetologica* 36:189–99.

Wilcox, R. S. 1984. Male copulatory guarding enhances female foraging in a water strider. *Behav. Ecol. Sociobiol.* 15:171–74.

Williams, G. C. 1966. *Adaptation and Natural Selection*. Princeton Univ. Press.

———. 1975. *Sex and Evolution*. Princeton Univ. Press.

Animal Genitalia and Female Choice

William G. Eberhard

When I was a senior in college I took a course in ichthyology and learned to enjoy thumbing through taxonomic drawings, which displayed the fascinating theme-and-variations patterns that are so common in nature. The various species in a genus were basically similar, but each had a set of seemingly senseless and often surprising and aesthetically pleasing differences. Later that year I became interested in spiders, and I can still remember my disappointment upon finding that similar drawings of whole spiders did not accompany papers on spider taxonomy. Instead, illustrations in spider papers were limited to male and female genitalia, which are generally extremely complex structures lacking the elegant sweep of fish profiles. Even closely related species of spiders can usually be distinguished by the genitalia alone.

This was my first encounter with a major pattern in animal evolution: among closely related species that employ internal fertilization, the genitalia—especially male genitalia—often show the clearest and most reliable morphological differences. For some reason, the genitalia of most spiders have evolved rapidly, becoming distinct even in recently diverged lines. In contrast, animals that employ external fertilization, such as most fish, do not have species-specific genital morphology.

These trends are widespread. Groups in which intromittent genitalia (for placing gametes inside the mate) are often useful for distinguishing species include flatworms, nematodes, oligochaete worms, insects, spiders, millipedes, sharks and rays, some lizards, snakes, mites, opilionids, crustaceans, molluscs, and mammals (including rodents, bats, armadillos, and primates). In contrast, groups that employ external fertilization all lack species-specific genitalia; they include echinoderms, most polychaete worms, hemichordates, brachiopods, sipunculid worms, frogs, birds, a few insects, and most

Rapid evolutionary divergence of male genitalia may be explained by the ability of females to choose the paternity of their offspring

William Eberhard is a member of the staff of the Smithsonian Tropical Research Institute and a professor at the University of Costa Rica. He received both undergraduate and graduate degrees from Harvard University. His research interests include the behavior and ecology of web-spinning spiders, functional morphology of beetle horns and earwig forceps, evolutionary interactions between subcellular organelles and plasmids and the cells that contain them, and the evolution of animal genitalia. Address: Escuela de Biología, Universidad de Costa Rica, Ciudad Universitaria, Costa Rica.

fish. In such cases, both males and females have only a simple opening through which gametes are released. Even within groups that have recently switched from external to internal fertilization, for example guppies, whose males use a modified anal fin to introduce sperm into the female, the intromittant organs are often useful for distinguishing species.

Rapid and divergent genital evolution also occurs in species in which the male, rather than penetrating the female himself, introduces a spermatophore, or package of his sperm, into the female. In many octopuses, squids, scorpions, some pseudoscorpions, some snails and slugs, some arrow worms, and pogonophoran worms, it is the spermatophore, rather than male genital structures, that is morphologically complex and species-specific.

Why do male mating structures possess such a bewildering diversity of forms? Surely the transfer of a small mass of gametes does not require the elaborate genital structures carried by the males of many groups. Two explanations were proposed some time ago: lock-and-key and pleiotropy. Neither is particularly convincing. According to the lock-and-key hypothesis, females have evolved under selection favoring those individuals that avoided wasting eggs by having them fertilized by sperm of other species. Elaborate, species-specific female genitalia (locks) admit only the genitalia of conspecific males (keys), enabling females to avoid mistakes in fertilization.

Originally proposed nearly 150 years ago for insects (see Nichols 1986), the lock-and-key idea fell into disrepute when it was established that locks are too easily picked. Studies of groups in which females have complex genitalia showed that the female genitalia could not exclude the genitalia of males of closely related species (for a summary of evidence see Shapiro and Porter 1989). The lock-and-key hypothesis is inapplicable in many other groups in which the female genitalia are soft and mechanically incapable of excluding incorrect keys while the male genitalia or spermatophores are nevertheless species-specific in form (flatworms, nematodes, arrow worms, annelid worms, sharks and rays, guppies and their relatives, snakes, lizards, snails and slugs, octopuses and squids, and many insects).

A species-isolation function, whether mechanical or otherwise, is improbable for several reasons. In some

groups with species-specific genitalia, males and females exchange species-specific signals during courtship, and probably seldom if ever reach the point of making genital contact with members of other species. For example, some female moths attract males with species-specific blends of pheromones, and the males, after finding the females, court them with additional species-specific pheromones before beginning to copulate. Yet the male moths have species-specific genitalia (Baker and Cardé 1979). Species-specific genitalia have even evolved in situations in which mistaken, cross-specific matings are essentially impossible, for example in island-dwelling species isolated from all close relatives, or parasitic species that mate on hosts which never harbor more than one species of the parasite. In some of these groups, such as the pinworms of primates, male genitalia provide the best morphological characteristics known to distinguish closely related species (Inglis 1961).

The alternative hypothesis, pleiotropy, is no more satisfying. It holds that genital characteristics are chance effects of genes that code primarily for other characteristics, such as adaptations to the environment. But this idea fails to explain why incidental effects should consistently occur on genitalia and not elsewhere. Nor does it explain why incidental effects fail to occur in species employing external rather than internal fertilization. It also cannot account for the genital morphology of a number of groups, such as spiders and guppies, in which organs (e.g., a pedipalp or anal fin) other than the primary male genitalia acquire the function of introducing sperm into the female; these other organs consistently become subject to the putative "incidental" effects while the primary genitalia do not.

So, until recently, a pattern widespread in animal evolution was left without a plausible explanation. Recently, however, a resurgence of interest in Darwin's ideas on sexual selection and advances in evolutionary theory to encompass male-female conflicts have stimulated new hypotheses.

Two mechanisms proposed by Darwin are potentially involved: male-male competition and female choice. Recent hypotheses are that male genitalia function to remove or otherwise supersede sperm introduced in previous matings of the female (Waage 1979; Smith 1984), and that male genitalia are often used as "internal courtship devices"—inducing the female to use a male's sperm—and thus are under sexual selection by female

Figure 1. The male of this pair of *Altica* beetles performs complex courtship behavior after copulation has begun. Within the first two minutes of copulation, he inflates inside the female a sac which emerges from the tip of the brown, cylindrical basal portion of his genitalia *(visible in the top photo)*, and passes his sperm into the female. During the rest of the approximately 20-minute copulation, he periodically thrusts forward as far as he can *(middle photo)* in a more or less stereotyped pattern. The thrusts do not move his genitalia deeper into the female's genital tract; rather they stretch the walls of the entire basal portion of the tract as it is displaced forward within her body. Between bouts of thrusting the male often rubs his rear tarsi gently but persistently near the tip of the female's abdomen *(bottom photo)*. After he has withdrawn his genitalia, the male sometimes gives her additional, more vigorous rubs with his hind legs. (Photos by D. Perlman.)

Figure 2. The penis of *Notomys mitchelli*, an Australian rodent, is extremely elaborate. The movements of such an organ certainly do not go unnoticed by the female while it is within her. (From Breed 1986; courtesy of W. G. Breed.)

choice (Eberhard 1985). While sperm removal and displacement have been documented in several cases, these are unlikely to be general explanations for the trend of genitalia to evolve and diverge rapidly, because males of many groups with species-specific genitalia do not penetrate deep enough into the female to reach sites where sperm from previous copulations are stored. In the remainder of this article I present some of the evidence supporting the female-choice hypothesis, and show how it calls into question some basic and intuitively "obvious" notions about animal behavior and morphology.

Copulatory courtship and genital stimulation

The obvious function of a male's courtship behavior is to induce the female to mate. Yet males of many species of insects appear to court the female even after they have achieved genital coupling. Their behavior includes typical courtship movements such as waving antennae or colored legs, stroking, tapping, rubbing, or biting the female's body, buzzing the wings in stereotyped patterns, rocking the body back and forth, and singing (Eberhard, unpubl.). A survey of studies of copulation behavior in insects showed that, in just over one-third of 302 species, the male performs behavior apparently designed to stimulate the female (Eberhard, unpubl.). In some species, such as the beetles shown in Figure 1, the male combines movements of body parts such as antennae and hind feet with more or less stereotyped movements of his genitalia. Both genital and non-genital behavior patterns differ in closely related species.

Many male mammals move their genitalia in and out of the female in more or less stereotyped movements prior to insemination (Dewsbury 1972). Some also perform post-ejaculatory intromissions which differ from the earlier ones and which increase the likelihood that the mating will result in the female becoming pregnant (Dewsbury and Sawrey 1984). Male goldeneye ducks perform four different displays after copulation (Dane and van der Kloot 1962).

Other observations also indicate that male genitalia themselves perform copulatory courtship. A study by Lorkovic (1952) showed that the genitalic "claspers" of some male butterflies are rubbed gently back and forth on the sides of the female's abdomen during copulation. Some snails thrust genitalic darts into the female during courtship or copulation (Fretter and Graham 1964), while some moths have elongate, sharp-pointed scales on their penes which are designed to fall off inside the female (Busck 1931), probably delivering stimuli to the female after the male has left. In a variety of groups, ranging from *Drosophila* flies to mice and sheep, copulatory behavior persists even after the male has exhausted his supply of sperm (Dewsbury and Sawrey 1984). In marmosets, male stimulatory effects during copulation have been documented by showing that some female responses disappear when the female's reproductive tract is anesthetized (Dixson 1986). Male cats and some male rodents have backwardly directed spines on their penes which make stimulation of the female inevitable (Fig. 2); ovulation in female cats is known to be induced by mechanical stimulation of the vagina (Greulich 1934; DeWildt et al. 1978). These stimulation devices probably represent mechanical equivalents of the visual displays of erect, brightly colored male penes in some lizards and primates (Bohme 1983; Eckstein and Zuckerman 1956; Hershkovitz 1979).

In some groups, mechanical stimulation and sperm transfer are particularly clear because they are performed separately. Male spiders and millipedes generally transfer sperm to modified structures on their pedipalps or legs, then use these secondary genitalia to introduce sperm into the female. In some species copulation always occurs first with the secondary genitalia empty; the male then withdraws, loads the secondary genitalia with sperm, and copulates again (Austad 1984). A number of species, including beetles, wasps, and rodents, perform a series of preliminary or extra intromissions which apparently do not result in sperm transfer (Cowan 1986; Schincariol and Freitag 1986; Dewsbury, in press). Mallards, one of the few bird groups having intromittent organs, frequently copulate during pair formation, six months before egg-laying, when the male gonads are repressed and sperm are not produced (McKinney et al. 1984).

Copulatory courtship behavior is understandable given the perspective that copulation is only one of a series of events which must occur if a male is to sire offspring. The female must remain still, or at least not actively attempt to terminate copulation prematurely. In rats and fleas, for instance, genitalic stimulation increases a female's tendency to stand still (Rodriguez-Sierra et al. 1975; Humphries 1967). The male's sperm must be transported to the storage site or fertilization site; this process seems to depend to a large extent on female peristalsis or other transport movements rather than on the motility of the sperm (see Overstreet and Katz 1977 on mammals; Davey 1965 on insects). Females of many species have, associated with their sperm-storage organs, glands which must be activated to help keep sperm alive and healthy. In some mammals and arthropods, sperm must be "activated" once inside the female in order to become capable of fertilization (Hamner et al. 1970; Leopold and Degrugillier 1973; Brown 1985). In some species, such as roaches and cats, ovulation and maturation of the eggs are induced by mechanical stimuli associated with copulation (Roth and Stay 1961; Greulich 1934). Brood care in earwigs is thought to be induced by mating (Vancassel cited in Lamb 1976).

Another critical female response often associated with copulation is the lack of further sexual receptivity. Sperm from subsequent matings can offer dangerous competition because it is extremely rare for internal fertilization of eggs to occur immediately following copulation. In some bees and wasps, postcopulatory courtship appears to reduce the frequency of remating by females (van den Assem and Visser 1976; Alcock and Buchmann 1985). Stimuli from both copulation (without spermatophore transfer) and the spermatophore itself reduce further sexual receptivity in the female butterfly

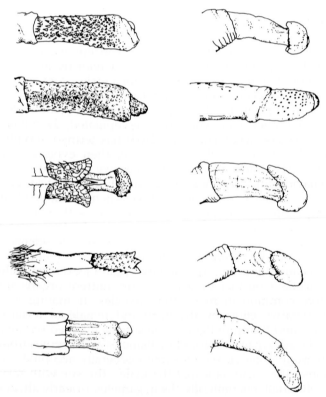

Figure 3. As predicted by the female-choice theory, more elaborate penis morphology occurs in species in which the female is more likely to mate with more than one male, thus being able to choose the father of her offspring. The male genitalia above are from different primate species. Those on the right belong to species in which a single male usually monopolizes the matings of a female during a period of estrous; those on the left belong to species in which receptive females can be mated by more males. (On the left proceeding down: *Galago crassicaudatus, G. garnettii, Arctocebus calabarensis, Euoticus elegantulus, Nycticebus coucang.* **On the right proceeding down:** *Colobus guereza, Callithrix jacchus, Mandrillus sphinx, Erythrocebus patas, Saguinus oedipus.***)(After Dixson 1987; drawn to different scales.)**

Pieris rapae (Obara et al. 1975; Sugawara 1979). Finally, in mice, copulation can inhibit transport of sperm from previous matings (Dewsbury 1985). In sum, a male's reproductive success can be greatly affected by his ability to induce females to perform any of several critical post-coupling activities. Copulatory courtship behavior by males probably serves this end.

Number of mates and coyness

The female-choice hypothesis predicts that genital morphology of males should be under stronger sexual selection in those species in which females mate with more males. Dixson (1987) recently tested this prediction using a sample of 130 primate species. His data show, even when corrected for possible effects of common ancestors, that males of species in which sexually active females are not monopolized by single males have relatively longer genitalia, more highly developed hard spines on the penis, more complex shapes at the tip of the penis, and a more developed baculum (penis bone)(Fig. 3). In addition, males of species in which receptive females are not monopolized display more elaborate copulatory behavior, with prolonged and multiple intromissions. A similar trend occurs in *Heliconius* butterflies; males of species in which females remate more often tend to have more distinctive genitalia (Eberhard 1985).

If reproductive processes in females are triggered by male stimuli after coupling, then sexual selection theory predicts that it will be advantageous for a female to avoid having each copulation result in the fertilization of all her available eggs. This is because females able to favor males proficient at stimulation will have sons that are superior reproducers because the sons will be, on average, proficient stimulators (Fisher 1958). This type of selection can, in theory, give rise to a runaway process in which males develop increasingly elaborate apparatus and females become increasingly discriminating. Along with the probable advantages to females of controlling the timing of fertilization, female choice may help explain the tortuous and complex morphologies of many female reproductive tracts (Fig. 4, 5) and the rarity of designs in which males simply place their sperm at the site of fertilization. Female genitalia may be designed not only to facilitate fertilization, but also to prevent it under certain circumstances.

Perhaps the most dramatic and well-documented case supporting this idea is that of bedbugs and their allies (Carayon (1966). Some male bedbugs have evolved a hypodermic penis which can be inserted at a variety of sites on the female's body; the sperm are injected into her blood, and they migrate to the ovaries where they accumulate in huge masses even after only a single copulation. In some groups the females have responded by evolving a new genital system, complete with an opening on the top of the abdomen, a storage organ, and ducts to the oviduct. This new system would seem unnecessary for sperm transport, and both its developmental origin and mode of action suggest that it serves instead to selectively prevent fertilization. The cells of the new female system are derived from types used to combat infections, and only a very small proportion of the sperm that enter ever reach the ovaries.

Figure 6. The most adaptive direction for sex change can depend on the size of the social group. When the group consists of a single pair, both individuals profit if the larger member of the pair is a female, since she could produce more eggs than a smaller individual. Protandry is most adaptive in this case, and is found in the strictly monogamous anemonefishes of the genus *Amphiprion* (*top*). In larger social groups, the combined egg production of smaller members can easily exceed the egg production of the largest individual, and thus his output is maximized by functioning as a male. Protogyny would be expected here, and is found in the group-living damselfishes of the genus *Dascyllus* (*bottom*), which are closely related to anemonefishes. (Photo at top by H. Fricke; photo at bottom by F. Bam.)

distributed wrasse thus far studied; its population is essentially gonochoristic, with about 50% primary males (Warner 1982). Large territorial males are rather rare and only moderately successful in this species, and nearly all mating takes place in groups.

Certain characteristics of the habitat that allow access to spawning sites by small males apt to engage in "sneaking" should also affect mate monopolization. These characteristics are difficult to measure in a quantitative fashion, but some trends are evident. For example, small parrotfishes that live in beds of sea grass near coral reefs have a higher proportion of small males than species that exist in similar densities on the reefs themselves (Robertson and Warner 1978). In one grass-dwelling species, sex change appears to be entirely absent (Robertson et al. 1982). Sea grasses offer abundant hiding places for small fishes, and dominant males in these habitats suffer interference from smaller males in a high proportion of their matings.

Perhaps the most telling variation within a family occurs in the damselfishes (Pomacentridae), where sex change was only recently discovered (Fricke and Fricke 1977). Small damselfishes called clownfishes or anemonefishes (genus *Amphiprion*) live in or near large stinging anemones in reef areas and thus have extremely limited home ranges. They appear to be unaffected by the stinging cells of the anemone, and may enjoy a certain amount of protection from the close association (Allen 1972). An anemonefish society consists of two mature individuals and a variable number of juveniles. The species are protandrous; the largest individual is a female, the smaller adult a male (Fricke and Fricke 1977; Moyer and Nakazono 1978b). The per capita production of fertilized eggs is higher when the larger individual of a mating pair is the female, and protandry is thus advantageous to both adults (Warner 1978).

Note that the advantage of protandry in this case depends on the fact that the social group is rigidly limited to two adults. If more adults were present, the most adaptive sexual pattern could instead be protogyny. This is because the largest individual, as a male, might be able to fertilize more eggs than it could produce as a female. In accordance with this, protogyny appears in some related damselfishes (genus *Dascyllus*) in which the social groups of adults are larger (Fig. 6; Fricke and Holzberg 1974; Swarz 1980 and pers. com.; Coates 1982).

Social control of sex change

Another way of testing the size-advantage model is through an investigation of the dynamics of sex change within a species. So far, I have stressed the importance of the mating system in determining the advantage of a given sex and size. Within a mating system, it is often relative rather than absolute size that determines reproductive expectations. For example, when dominance depends on size, the probable mating success of a particular male is determined by the sizes of the other males in the local population. It would be most adaptive for individuals to be able to change from female to male when their expectations of successful reproduction as a male increase considerably. Thus the removal of a large, dominant male from a population should result in a change of sex in the next largest individual, but no change should be expected in the rest of the local population.

Such social control of sex change has been noted in several species of protogynous coral-reef fishes. Because haremic species exist in small, localized groups, they have proved to be exceptionally good candidates for studies of this kind. In the cleaner-wrasse *L. dimidiatus*, Robertson (1972) found that if the male is removed from the harem, the largest female rapidly changes sex and takes over the role of harem-master. Within a few hours she adopts male behaviors, including spawning with the females. Within ten days this new male is producing active sperm. By contrast, the other females in the harem remain unchanged.

Social control of sex change has also been found in other haremic species (Moyer and Nakazono 1978a; Hoffman 1980; Coates 1982), as well as in species that live in bigger groups with several large males present (Fishelson 1970; Warner et al. 1975; Shapiro 1979; Warner 1982; Ross et al. 1983). In all cases, it is always the

largest remaining individuals that undergo sex change when the opportunity presents itself. Even when experimental groups consist entirely of small individuals, sex change can still be induced in the largest individuals present, in spite of the fact that they may be far smaller than the size at which sex change normally occurs (Hoffman 1980; Warner 1982; Ross et al. 1983).

The exact behavioral cues used to trigger sex change appear to differ among species. Ross and his co-workers have shown experimentally that the sex-change response in the Hawaiian wrasse *T. duperrey* depends solely on relative size and is independent of the sex and coloration of the other individuals in a group, whereas Shapiro and Lubbock (1980) have suggested that the local sex ratio is the critical factor in the bass *Anthias squamipinnis*. While it is still unclear how sex change is regulated in fishes that live in large groups, the mechanisms appear to operate with some precision. Shapiro (1980) found that the simultaneous removal of up to nine male *Anthias* from a group led to a change of sex in an equivalent number of females.

Social control of sex change occurs in protandrous fishes as well, and in a pattern consistent with the size-advantage model. A resident male anemonefish will change sex if the female is removed (Fricke and Fricke 1977; Moyer and Nakazono 1978b). One of the juveniles—who apparently are otherwise repressed from maturing—then becomes a functional male and the adult couple is reconstituted.

Simultaneous hermaphroditism

In one sense, the adaptive significance of simultaneous hermaphroditism is obvious: by putting most of their energy into egg production and producing just enough sperm to ensure fertilization, a hermaphroditic mating couple can achieve a much higher output of young than a male-female pair (Fig. 7; Leigh 1977; Fischer 1981). The problem, however, rests with the maintenance of simultaneous hermaphroditism in the face of an alternative male strategy. Consider an individual that fertilizes the eggs of a hermaphrodite, but does not reciprocate by producing eggs of its own. Instead, this individual uses the energy thus saved to find and fertilize other hermaphrodites. This strategy would spread rapidly in a purely hermaphroditic population, effectively forcing it to become gonochoristic. It would therefore appear that where simultaneous hermaphroditism is present, there should exist some means of preventing this kind of "cheating" (Leigh 1977; Fischer 1981).

Among the small coral-reef basses (Serranidae) that are known to be simultaneous hermaphrodites, two types of possible anticheating behavior have been observed. The hamlets, small basses common on Caribbean coral reefs (genus *Hypoplectrus*), appear to ensure that investments in eggs are kept nearly even between the members of a spawning pair by what Fischer (1980) has called "egg trading." In this behavior, a pair alternates sex roles over the course of mating (Fig. 8). Each time an individual functions as a female, it extrudes some, but not all, of its eggs. As a male, it fertilizes the eggs of its partner, who also parcels out eggs in several batches. Thus both individuals are forced to demonstrate their commitment to egg production, and neither has the chance for an unreciprocated fertilization of a large batch of eggs.

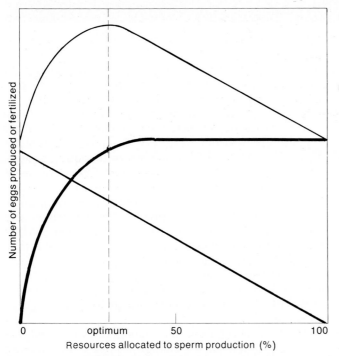

Figure 7. The number of eggs produced by an individual is normally directly related to the amount of energy devoted to their manufacture (*thin line*). In the case of sperm, however, a relatively low output can often produce maximum success (*thick line*) – that is, a small amount of sperm can fertilize all of a partner's eggs, and further investments are superfluous. Natural selection favors the individual with the highest overall reproductive rate combining male and female functions (*black line*), and in this case the optimum result is obtained by putting most of the energy into egg production. A simultaneous hermaphrodite following this strategy has a much higher reproductive success than an individual that is exclusively male or female. (After Fischer 1981.)

Figure 8. Simultaneous hermaphrodites among fishes include the hamlet (genus *Hypoplectrus*), a small bass that may alternate sexual roles as many as four times in the course of a single mating, by turns offering eggs to be fertilized and fertilizing its partner's eggs. Here the fish acting as the male curves its body around the relatively motionless female, cupping the upward-floating eggs as he fertilizes them. The strategy of parceling out eggs in a number of batches means that the egg contributions of the two partners are roughly equal, and reduces the rewards of "cheating" by fertilizing a large batch of eggs and then refusing to offer eggs for fertilization in return. (Photo by S. G. Hoffman.)

Another method of preventing desertion is to reduce the opportunities of your partner to find another mate. Some simultaneously hermaphroditic species of the genus *Serranus* delay their mating until late dusk, just before nightfall. These species do not engage in egg-trading, but presumably the onset of darkness means that time is quite limited before shelter must be taken for the night, and thus further mating is impossible (Pressley 1981).

Although these anticheating behaviors are fascinating in their own right, they give us little insight into the origin of the sexual pattern itself. Simultaneous hermaphroditism has been viewed as an adaptation to extremely low population density: if finding mates is difficult, it helps a great deal to be able to mate with whomever you meet (Tomlinson 1966; Ghiselin 1969). Thus many deep-sea fishes, sparsely distributed in their habitat, are simultaneous hermaphrodites (Mead et al. 1964). Perhaps the small serranids evolved from ancestors who lived at low densities, and developed their anticheating behaviors at a later stage when densities were higher. Alternatively, and perhaps more likely, the existence of late-dusk spawning behavior could have allowed the development of hermaphroditism. Unfortunately, we again run up against the problem of untestability and limited predictive power: several other coral-reef fishes mate late in the day, but they are not simultaneous hermaphrodites.

Broader patterns

The contrast between sex change and simultaneous hermaphroditism is intriguing: sex change, particularly protogyny, appears to be a specific adaptation to certain mating systems that happen to be common on coral reefs. These mating systems may be prevalent in coral habitats because clear water, relatively low mobility, and the absence of paternal care allow a greater degree of dominance by large males. Simultaneous hermaphroditism, on the other hand, is theoretically adaptive in a wider variety of circumstances, but is evolutionarily unstable unless male-type cheating can be prevented. While the wide dispersion of deep-sea fishes automatically works against cheating, I can see no reason why coral reefs are particularly good places for such prevention to come about.

Among hermaphroditic groups other than tropical marine fishes, our knowledge of behavior and ecology is generally insufficient to carry out a similar analysis of the relationship of sexual pattern to mating system. Broad surveys of vertebrates (Warner 1978) or organisms in general (Policansky 1982) have shown large-scale tendencies toward hermaphroditism in some groups, sporadic appearance of the phenomenon in others, and a total lack of sexual flexibility in still others. While major features such as the greater complexity of terrestrial reproduction may help to explain the lack of hermaphroditism in some large groups (Warner 1978), many others must await more thorough investigation of the life histories and behavior of the organisms in question.

On this point, Policansky (1982) has suggested that a major problem of sex-change theory is that among closely related species with similar life histories, some change sex and some do not. In light of this review, I am not yet ready to take such a dim view of the size-advantage model. Sexual expression in fishes is extraordinarily adaptable, and closely related species can have quite different sexual patterns that appear to be predictable from their different mating systems. For the fishes, at least, divergent sexual expressions may be no more surprising than differences in coloration. Detailed considerations of the mating systems and life histories of other sexually labile groups are needed to test the hypothesis further.

References

Allen, G. R., 1972. *The Anemonefishes, Their Classification and Biology.* Neptune City, NJ: T. F. H. Publications.

Charnov, E. L. 1982. *The Theory of Sex Allocation.* Princeton Univ. Press.

Coates, D. 1982. Some observations on the sexuality of humbug damselfish, *Dascyllus aruanus* (Pisces, Pomacentridae) in the field. *Z. für Tierpsychol.* 59:7–18.

Cole, K. S., 1982. Male reproductive behavior and spawning success in a temperate zone goby, *Coryphopterus nicholsi. Can. J. Zool.* 10: 2309–16.

Fischer, E. A. 1980. The relationship between mating system and simultaneous hermaphroditism in the coral reef fish *Hypoplectrus nigricans* (Seranidae). *Anim. Beh.* 28:620–33.

———. 1981. Sexual allocation in a simultaneously hermaphroditic coral reef fish. *Am. Nat.* 117:64–82.

Fishelson, L. 1970. Protogynous sex reversal in the fish *Anthias squamipinnis* (Teleostei, Anthiidae) regulated by presence or absence of male fish. *Nature* 227:90–91.

Fricke, H. W., and S. Fricke. 1977. Monogamy and sex change by aggressive dominance in coral reef fish. *Nature* 266:830–32.

Fricke, H. W., and S. Holzberg. 1974. Social units and hermaphroditism in a pomacentrid fish. *Naturwissensch.* 61:367–68.

Ghiselin, M. T. 1969. The evolution of hermaphroditism among animals. *Quart. Rev. Bio.* 44:189–208.

———. 1974. *The Economy of Nature and the Evolution of Sex.* Univ. of California Press.

Goodman, D. 1982. Optimal life histories, optimal notation, and the value of reproductive value. *Am. Nat.* 119:803–23.

Hoffman, S. G. 1980. Sex-related social, mating, and foraging behavior in some sequentially hermaphroditic reef fishes. Ph.D. diss., Univ. of California, Santa Barbara.

———. 1983. Sex-related foraging behavior in sequentially hermaphroditic hogfishes (*Bodianus* spp.). *Ecology* 64:798–808.

Jones, G. P. 1980. Contribution to the biology of the redbanded perch *Ellerkeldia huntii* (Hector), with a discussion on hermaphroditism. *J. Fish Biol.* 17:197–207.

Larsson, H. O. 1976. Field observations of some labrid fishes (Pisces: Labridae). In *Underwater 75*, vol. 1, ed. John Adolfson, pp. 211–20. Stockholm: SMR.

Lassig, B. R. 1977. Socioecological strategies adapted by obligate coral-dwelling fishes. *Proc. 3rd Int. Symp. Coral Reefs* 1: 565–70.

Leigh, E. G., Jr. 1977. How does selection reconcile individual advantage with the good of the group? PNAS 74:4542–46.

Leigh, E. G., Jr., E. L. Charnov, and R. R. Warner. 1976. Sex ratio, sex change, and natural selection. PNAS 73:3656–60.

Mead, G. W., E. Bertelson, and D. M. Cohen, 1964. Reproduction among deep-sea fishes. *Deep Sea Res.* 11:569–96.

Moyer, J. T., and A. Nakazono. 1978a. Population structure, reproductive behavior and protogynous hermaphroditism in the angelfish *Centropyge interruptus* at Miyake-jima, Japan. *Japan. J. Ichthyol.* 25:25–39.

———. 1978b. Protandrous hermaphroditism in six species of the anemonefish genus *Amphiprion* in Japan. *Japan. J. Ichthyol.* 25: 101–6.

Pressley, P. H. 1981. Pair formation and joint territoriality in a simultaneous hermaphrodite: The coral reef fish *Serranus tigrinus. Z. für Tierpsychol.* 56:33–46.

Policansky, D. 1982. Sex change in plants and animals. *Ann. Rev. Ecol. Syst.* 13:471–95.

Robertson, D. R. 1972. Social control of sex reversal in a coral reef fish. *Science* 1977:1007–9.

Robertson, D. R., R. Reinboth, and R. W. Bruce. 1982. Gonochorism, protogynous sex-change, and spawning in three sparasomatinine parrotfishes from the western Indian Ocean. *Bull. Mar. Sci.* 32: 868–79.

Robertson, D. R., and R. R. Warner. 1978. Sexual patterns in the labroid fishes of the western Caribbean, II: The parrotfishes (Scaridae). *Smithsonian Contributions to Zoology* 255:1–26.

Ross, R. M., G. S. Losey, and M. Diamond. 1983. Sex change in a coral-reef fish: Dependence of stimulation and inhibition on relative size. *Science* 221:574–75.

Shapiro, D. Y. 1979. Social behavior, group structure, and the control of sex reversal in hermaphroditic fish. *Adv. Study Beh.* 10:43–102.

———. 1980. Serial female sex changes after simultaneous removal of males from social groups of a coral reef fish. *Science* 209:1136–37.

Shapiro, D. Y., and R. Lubbock. 1980. Group sex ratio and sex reversal. *J. Theor. Biol.* 82:411–26.

Shen, S.-C., R-P. Lin, and F. C-C. Liu. 1979. Redescription of a protandrous hermaphroditic moray eel (*Rhinomuraena quaesita* Garman). *Bull. Instit. Zool. Acad. Sinica* 18(2):79–87.

Smith, C. L. 1975. The evolution of hermaphroditism in fishes. In *Intersexuality in the Animal Kingdom*, ed. R. Reinboth, pp. 295–310. Springer-Verlag.

Springer, V. G., C. L. Smith, and T. H. Fraser. 1977. *Anisochromis straussi,* new species of protogynous hermaphroditic fish, and synonomy of Anisochromidae, Pseudoplesiopidae, and Pseudochromidae. *Smithsonian Contributions to Zoology* 252:1–15.

Swarz, A. L. 1980. Almost all *Dascyllus reticulatus* are girls! *Bull. Mar. Sci.* 30:328.

Thresher, R. E., and J. T. Moyer. 1983. Male success, courtship complexity, and patterns of sexual selection in three congeneric species of sexually monochromatic and dichromatic damselfishes (Pisces: Pomacentridae). *Anim. Beh.* 31:113–27.

Tomlinson, N. 1966. The advantages of hermaphroditism and parthenogenisis. *J. Theoret. Biol.* 11:54–58.

Trivers, R. L. 1972. Parental investment and sexual selection. In *Sexual Selection and the Descent of Man, 1871–1971*, ed. B. Campbell, pp. 136–79. Chicago: Aldine Publishing Co.

Warner, R. R. 1975. The adaptive significance of sequential hermaphroditism in animals. *Am. Nat.* 109:61–82.

———. 1978. The evolution of hermaphroditism and unisexuality in aquatic and terrestrial vertebrates. In *Contrasts in Behavior*, ed. E. S. Reese and F. J. Lighter, pp. 77–101. Wiley.

———. 1982. Mating systems, sex change, and sexual demography in the rainbow wrasse, *Thalassoma lucasanum. Copeia* 1982:653–61.

Warner, R. R., and S. G. Hoffman, 1980a. Population density and the economics of territorial defense in a coral reef fish. *Ecology* 61: 772–80.

———. 1980b. Local population size as a determinant of a mating system and sexual composition in two tropical reef fishes (*Thalassoma* spp.). *Evolution* 34:508–18.

Warner, R. R., and D. R. Robertson. 1978. Sexual patterns in the labroid fishes of the western Caribbean, I: The wrasses (Labridae). *Smithsonian Contributions to Zoology* 254:1–27.

Warner, R. R., D. R. Robertson, and E. G. Leigh, Jr. 1975. Sex change and sexual selection. *Science* 190:633–38.

Williams, G. C. 1966. *Adaptation and Natural Selection.* Princeton Univ. Press.

Young, P. C., and R. B. Martin. 1982. Evidence for protogynous hermaphroditism in some lethrinid fishes. *J. Fish Biol.* 21:475–84.

The Strategies of Human Mating

A theory of human sexual strategies accounts for the observation that people worldwide are attracted to the same qualities in the opposite sex

David M. Buss

What do men and women want in a mate? Is there anything consistent about human behavior when it comes to the search for a mate? Would a Gujarati of India be attracted to the same traits in a mate as a Zulu of South Africa or a college student in the midwestern United States?

As a psychologist working in the field of human personality and mating preferences, I have come across many attempts to answer such questions and provide a coherent explanation of human mating patterns. Some theories have suggested that people search for mates who resemble archetypical images of the opposite-sex parent (à la Freud and Jung), or mates with characteristics that are either complementary or similar to one's own qualities, or mates with whom to make an equitable exchange of valuable resources.

These theories have played important roles in our understanding of human mating patterns, but few of them have provided specific predictions that can be tested. Fewer still consider the origins and functions of an individual's mating preferences. What possible function is there to mating with an individual who is an archetypical image of one's opposite-sex parent? Most theories also tend to assume that the processes that guide the mating preferences of men and women are identical, and no sex-differentiated predictions can be derived. The context of the mating behavior is also frequently ignored; the same mating tendencies are posited regardless of circumstances.

Despite the complexity of human mating behavior, it is possible to address these issues in a single, coherent theory. David Schmitt of the University of Michigan and I have recently proposed a framework for understanding the logic of human mating patterns from the standpoint of evolutionary theory. Our theory makes several predictions about the behavior of men and women in the context of their respective sexual strategies. In particular, we discuss the changes that occur when men and women shift their goals from short-term mating (casual sex) to long-term mating (a committed relationship).

Some of the studies we discuss are based on surveys of male and female college students in the United States. In these instances, the sexual attitudes of the sample population may not be reflective of the behavior of people in other cultures. In other instances, however, the results represent a much broader spectrum of the human population. In collaboration with 50 other scientists, we surveyed the mating preferences of more than 10,000 men and women in 37 countries over a six-year period spanning 1984 through 1989. Although no survey, short of canvassing the entire human population, can be considered exhaustive, our study crosses a tremendous diversity of geographic, cultural, political, ethnic, religious, racial and economic groups. It is the largest survey ever on mate preferences.

What we found is contrary to much current thinking among social scientists, which holds that the process of choosing a mate is highly culture-bound. Instead, our results are consistent with the notion that human beings, like other animals, exhibit species-typical desires when it comes to the selection of a mate. These patterns can be accounted for by our theory of human sexual strategies.

Competition and Choice

Sexual-strategies theory holds that patterns in mating behavior exist because they are evolutionarily advantageous. We are obviously the descendants of people who were able to mate successfully. Our theory assumes that the sexual strategies of our ancestors evolved because they permitted them to survive and produce offspring. Those people who failed to mate successfully because they did not express these strategies are not our ancestors. One simple example is the urge to mate, which is a universal desire among people in all cultures and which is undeniably evolutionary in origin.

Although the types of behavior we consider are more complicated than simply the urge to mate, a brief overview of the relevant background should be adequate to understand the evolutionary logic of human mating strategies.

David M. Buss is professor of psychology at the University of Michigan. He received his Ph.D. in psychology from the University of California, Berkeley in 1981. He is director of the International Consortium of Personality and Social Psychologists, which conducts cross-cultural research around the world. His most recent book, The Evolution of Desire: Strategies of Human Mating, was published this year by Basic Books. Address: Department of Psychology, University of Michigan, Ann Arbor, MI 48109-1346.

Figure 1. Species-typical mating preferences are expressed by the American businessman Donald Trump and his new wife Marla Maples, here evoking an image of the ideal family for readers of *Vanity Fair*. The traits of a desirable mate appear to be consistent throughout the world: Men prefer to mate with beautiful young women, whereas women prefer to mate with men who have resources and social status. The author argues that these traits offer evolutionarily adaptive advantages to the opposite-sex mate, which account for their ubiquitous desirability.

As with many issues in evolutionary biology, this background begins with the work of Charles Darwin.

Darwin was the first to show that mate preferences could affect human evolution. In his seminal 1871 treatise, *The Descent of Man and Selection in Relation to Sex*, Darwin puzzled over characteristics that seemed to be perplexing when judged merely on the basis of their relative advantage for the animal's survival. How

type of mating	men's reproductive challenges	women's reproductive challenges
short-term	• partner number • identifying women who are sexually accessible • minimizing cost, risk and commitment • identifying women who are fertile	• immediate resource extraction • evaluating short-term mates as possible long-term mates • attaining men with high-quality genes • cultivating potential backup mates
long-term	• paternity confidence • assessing a woman's reproductive value • commitment • identifying women with good parenting skills • attaining women with high-quality genes	• identifying men who are able and willing to invest • physical protection from aggressive men • identifying men who will commit • identifying men with good parenting skills • attaining men with high-quality genes

Figure 2. Mate-selection problems of men and women differ in short-term mating (casual sex) and long-term mating (a committed relationship) because each gender faces a unique set of reproductive challenges. In short-term mating contexts, a man's reproductive success is constrained by the number of fertile women he can inseminate. A man must solve the specialized problems of identifying women who are sexually accessible, identifying women who are fecund, and minimizing commitment and investment in order to effectively pursue short-term matings. In contrast, a woman's short-term mating strategy involves identifying men who would be good long-term mates, identifying men who have "high-quality" genes (are evolutionarily fit), extracting resources from a short-term mate and cultivating potential backup mates. In long-term mating contexts, men must identify women who have high reproductive value, good parenting skills and "high-quality" genes. Men must also assure that they are the father of their mate's offspring. In contrast, women must identify long-term mates who are willing and able to invest resources, can provide physical protection, have good parenting skills and have "high-quality" genes. Because men and women face different reproductive challenges, each gender has evolved different sexual strategies and is attracted to different qualities in the opposite sex.

could the brilliant plumage of a male peacock evolve when it obviously increases the bird's risk of predation? Darwin's answer was sexual selection, the evolution of characteristics that confer a reproductive advantage to an organism (rather than a survival advantage). Darwin further divided sexual selection into two processes: intrasexual competition and preferential mate choice.

Intrasexual competition is the less controversial of the two processes. It involves competition between members of the same sex to gain preferential access to mating partners. Characteristics that lead to success in these same-sex competitions—such as greater strength, size, agility, confidence or cunning—can evolve simply because of the reproductive advantage gained by the victors. Darwin assumed that this is primarily a competitive interaction between males, but recent studies suggest that human females are also very competitive for access to mates.

Preferential mate choice, on the other hand, involves the desire for mating with partners that possess certain characteristics. A consensual desire affects the evolution of characteristics because it gives those possessing the desired characteristics an advantage in obtaining mates over those who do not possess the desired characteristics. Darwin assumed that

preferential mate choice operates primarily through females who prefer particular males. (Indeed, he even called this component of sexual selection *female choice*.)

Darwin's theory of mate-choice selection was controversial in part because Darwin simply assumed that females desire males with certain characteristics. Darwin failed to document how such desires might have arisen and how they might be maintained in a population.

The solution to the problem was not forthcoming until 1972, when Robert Trivers, then at Harvard University, proposed that the relative parental investment of the sexes influences the two processes of sexual selection. Specifically, the sex that invests more in offspring is selected to be more discriminating in choosing a mate, whereas the sex that invests less in offspring is more competitive with members of the same sex for sexual access to the high-investing sex. Parental-investment theory accounts, in part, for both the origin and the evolutionary retention of different sexual strategies in males and females.

Consider the necessary *minimum* parental investment by a woman. After internal fertilization, the gestation period lasts about nine months and is usually followed by lactation, which in tribal societies typically can last several years. In contrast, a man's minimum parental investment can

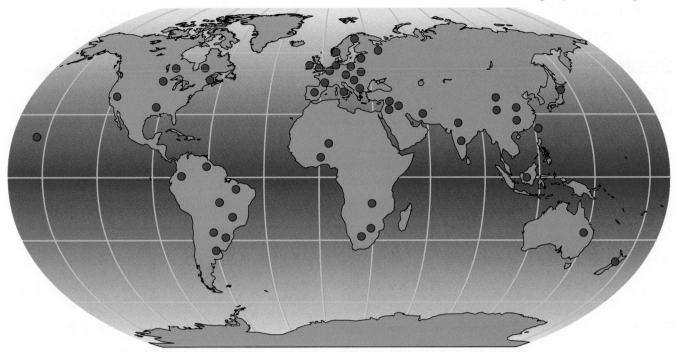

Figure 3. Thirty-seven cultures, distributed as shown above, were examined by the author in his international study of male and female mating preferences. The author and his colleagues surveyed the mating desires of 10,047 people on six continents and five islands. The results provide the largest data base of human mating preferences ever accumulated.

be reduced to the contribution of sperm, an effort requiring as little time as a few minutes. This disparity in parental investment means that the replacement of a child who dies (or is deserted) typically costs more (in time and energy) for women than men. Parental-investment theory predicts that women will be more choosy and selective about their mating partners. Where men can provide resources, women should desire those who are able and willing to commit those resources to her and her children.

Sexual Strategies

Our evolutionary framework is based on three key ingredients. First, human mating is inherently strategic. These strategies exist because they solved specific problems in human evolutionary history. It is important to recognize that the manifestation of these strategies need not be through conscious psychological mechanisms. Indeed, for the most part we are completely unaware of *why* we find certain qualities attractive in a mate. A second component of our theory is that mating strategies are context-dependent. People behave differently depending on whether the situation presents itself as a short-term or long-term mating prospect. Third, men and women have faced different mating problems over the course of human evolution and, as a consequence, have evolved different strategies.

As outlined here, sexual strategies theory consists of nine hypotheses. We can test these hypotheses by making several predictions about the behavior of men and women faced with a particular mating situation. Even though we

make only a few predictions for each hypothesis, it should be clear that many more predictions can be derived to test each hypothesis. We invite the reader to devise his or her own tests of these hypotheses.

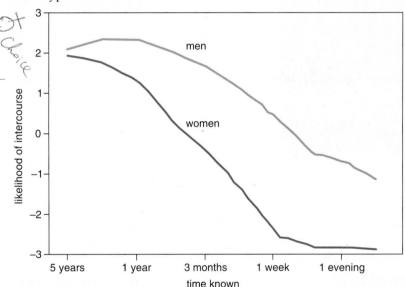

Figure 4. Willingness to have sexual intercourse (measured on a scale from 3, *definitely yes*, to –3, *definitely no*) differs for men and women with respect to the length of time they have been acquainted with their prospective mate. Although men and women are equally likely to engage in sexual intercourse after knowing a mate for five years (both responding with a score of about 2, *probably yes*), women are significantly less inclined to have sex with a prospective mate for all shorter lengths of time. The average man was positive about having intercourse with a woman even after knowing her for only one week, whereas the average women was highly unlikely to have intercourse after such a brief period of time. The data are based on a sample of 148 college students in the midwestern United States. The results support the hypothesis that short-term mating is more important for men than for women.

Hypothesis 1: Short-term mating is more important for men than women.

Figure 5. Stereotypical images of the womanizing male and the marriage-minded female are caricatures of the underlying sexual strategies of men and women. In the television program *Cheers* the character played by the actor Ted Danson exhibited the short-term male sexual strategy of mating with many women. The character played by the actress Shelley Long exhibited the female strategy of seeking a male willing to commit to a long-term relationship. Image courtesy of AP/Wide World Photos.

Hypothesis 2: Men seeking a short-term mate will solve the problem of identifying women who are sexually accessible.

Alain Evrard (Photo Researchers, Inc.)

Hypothesis 1: Short-term mating is more important for men than women. This hypothesis follows from the fact that men can reduce their parental investment to the absolute minimum and still produce offspring. Consequently, short-term mating should be a key component of the sexual strategies of men, and much less so for women. We tested three predictions based on this hypothesis in a sample of 148 college students (75 men and 73 women) in the midwestern United States.

First, we predict that men will express a greater interest in seeking a short-term mate than will women. We asked the students to rate the degree to which they were currently seeking a short-term mate (defined as a one-night stand or a brief affair) and the degree to which they were currently seeking a long-term mate (defined as a marriage partner). They rated their interests on a 7-point scale, where a rating of 1 corresponds to a complete lack of interest and a 7 corresponds to a high level of interest.

We found that although the sexes do not differ in their stated proclivities for seeking a long-term mate (an average rating of about 3.4 for both sexes), men reported a significantly greater interest (an average rating of about 5) in seeking a short-term sexual partner than did women (about 3). The results also showed that at any given time men are more interested in seeking a short-term mate rather than a long-term mate, whereas women are more interested in seeking a long-term mate than a short-term mate.

Second we predict that men will desire a greater number of mates than is desired by women. We asked the same group of college students how many sexual partners they would ideally like to have during a given time interval and during their lifetimes. In this instance men consistently reported that they desired a greater number of sex partners than reported by the

women for every interval of time. For example, the average man desired about eight sex partners during the next two years, whereas the average woman desired to have one sex partner. In the course of a lifetime, the average man reported the desire to have about 18 sex partners, whereas the average woman desired no more than 4 or 5 sex partners.

A third prediction that follows from this hypothesis is that men will be more willing to engage in sexual intercourse a shorter period of time after first meeting a potential sex partner. We asked the sample of 148 college students the following question: "If the conditions were right, would you consider having sexual intercourse with someone you viewed as desirable if you had known that person for *(a time period ranging from one hour to five years)*?" For each of 10 time intervals the students were asked to provide a response ranging from –3 (definitely not) to 3 (definitely yes).

After a period of 5 years, the men and women were equally likely to consent to sexual relations, each giving a score of about 2 (probably yes). For all shorter time intervals, men were consistently more likely to consider sexual intercourse. For example, after knowing a potential sex partner for only one week, the average man was still positive about the possibility of having sex, whereas women said that they were highly unlikely to have sex with someone after knowing him for only one week.

This issue was addressed in a novel way by Russell Clark and Elaine Hatfield of the University of Hawaii. They designed a study in which college students were approached by an attractive member of the opposite sex who posed one of three questions after a brief introduction: "Would you go out on a date with me tonight?" "Would you go back to my apartment with me tonight?" or "Would you have sex with me tonight?"

Of the women who were approached, 50 percent agreed to the date, 6 percent agreed to go to the apartment and none agreed to have sex. Many women found the sexual request from a virtual stranger to be odd or insulting. Of the men approached, 50 percent agreed to the date, 69 percent agreed to go back to the woman's apartment and 75 percent agreed to have sex. In contrast to women, many men found the sexual request flattering. Those few men who declined were apologetic about it, citing a fiancée or an unavoidable obligation that particular evening. Apparently, men are willing to solve the problem of partner number by agreeing to have sex with virtual strangers.

Hypothesis 2: Men seeking a short-term mate will solve the problem of identifying women who are sexually accessible. We can make at least two predictions based on this hypothesis. First, men will value qualities that signal immediate sexual accessibility in a short-term mate highly, and less

Alain Evrard (Photo Researchers, Inc.)

Figure 6. Prostitution is a worldwide phenomenon that is partly a consequence of the short-term mating strategy of males. The relatively rapid exchange of sex and money solves the short-term mating problems of males who can minimize the commitment of resources and quickly identify women who are sexually accessible. Above and on the facing page, interactions on Patpong Street (the prostitution district) in Bangkok play out a scene that is repeated daily in cities around the world.

so in a long-term mate. When we asked men in a college sample of 44 men and 42 women to rate the desirability of promiscuity and sexual experience in a mate, both were significantly more valued in a short-term mate. Although men find promiscuity mildly desirable in a short-term mate, it is clearly undesirable in a long-term mate. It is noteworthy that women find promiscuity extremely undesirable in either context.

We also predict that qualities that signal sexual inaccessibility will be disliked by men seeking short-term mates. We asked men to rate the desirability of mates who have a low sex drive, who are prudish or who lack sexual experience. In each instance men expressed a particular dislike for short-term mates with these qualities. A low sex drive and prudishness are also disliked by men in long-term mates, but less so. In contrast, a lack of sexual experience is slightly valued by men in a long-term mate.

Hypothesis 3: Men seeking a short-term mate will minimize commitment and investment. Here we predict that men will find undesirable any cues

that signal that a short-term mate wants to extract a commitment. We asked the same group of 44 men to rate the variable *wants a commitment* for short-term and long-term mates. Of all the qualities we addressed, this one showed the most striking dependence on context. The attribute of wanting a commitment was strongly desirable in a long-term mate but strongly undesirable in a short-term mate. This distinction was not nearly so strong for women. Although women strongly wanted commitment from a long-term mate, it was only mildly undesirable in a short-term mate.

Hypotheses 4 and 5: Men seeking a short-term mate will solve the problem of identifying fertile women, whereas men seeking a long-term mate will solve the problem of identifying reproductively valuable women. Because these hypotheses are closely linked it is useful to discuss them together. Fertility and reproductive value are related yet distinct concepts. Fertility refers to the probability that a woman is *currently* able to conceive a child. Reproductive value, on the other hand, is defined actuarially in units of expected future

Hypothesis 3: Men seeking a short-term mate will minimize commitment and investment.

Figure 7. Beautiful young women are sexually attractive to men because beauty and youth are closely linked with fertility and reproductive value. In evolutionary history, males who were able to identify and mate with fertile females had the greatest reproductive success. These three young women were photographed at a "fern bar" in Newport Beach, California.

Hypothesis 4: Men seeking a short-term mate will solve the problem of identifying fertile women.

reproduction. In other words, it is the extent to which persons of a given age and sex will contribute, on average, to the ancestry of future generations. For example, a 14-year-old woman has a higher reproductive value than a 24-year-old woman, because her *future* contribution to the gene pool is higher on average. In contrast, the 24-year-old woman is more fertile than the 14-year-old because her *current* probability of reproducing is greater.

Since these qualities cannot be observed directly, men would be expected to be sensitive to cues that might be indicative of a woman's fertility and reproductive value. One might expect that men would prefer younger women as short-term and long-term mates. Again, since age is not something that can be observed directly, men should be sensitive to physical cues that are reliably linked with age. For example, with increasing age, skin tends to wrinkle, hair turns gray and falls out, lips become thinner, ears become larger, facial features become less regular and muscles lose their tone. Men could solve the problem of identifying reproductively valuable women if they attended to physical features linked with age and health, *and* if their standards of attractiveness evolved to correspond to these features.

As an aside, it is worth noting that cultures do differ in their standards of physical beauty, but less so than anthropologists initially assumed. Cultural differences of physical beauty tend to center on whether relative plumpness or thinness is valued. In cultures where food is relatively scarce, plumpness is valued, whereas cultures with greater abundance value thinness. With the exception of plumpness and thinness, however, the physical cues to youth and health are seen as sexually attractive in all known cultures that have been studied. In no culture do people perceive wrinkled skin, open sores and lesions, thin lips, jaundiced eyes, poor muscle tone and irregular facial features to be attractive.

A woman's reproductive success, however, is not similarly dependent on solving the problem of fertility in mates. Because a man's reproductive capacity is less closely linked with age and cannot be assessed as accurately from appearance, youth and physical attractiveness in a mate should be less important to women than it is to men.

Among our sample of American college students we asked men and women to evaluate the relative significance (on a scale from 0, unimportant, to 3, important) of the characteristics *good looking* and *physically attractive* in a short-term and a long-term mate. We found that

Jodi Cobb (© National Geographic Society)

Hypothesis 5: Men seeking a long-term mate will solve the problem of identifying reproductively valuable women.

Figure 8. King Hussein' Ibn Talal' (born in 1935) of Jordan and his wife, Queen Noor (formerly Lisa Halaby of Washington, DC, who was born in 1951), provide an example of the general tendency for men to mate with women who are significantly younger than themselves. Sexual-strategies theory holds that men are attracted to younger women as long-term mates because they have a higher reproductive value than older women. The King and Queen are pictured with some of their children.

men's preference for physical attractiveness in short-term mates approached the upper limit of the rating scale (about 2.71). Interestingly, this preference was stronger in men seeking short-term mates than in men seeking long-term mates (about 2.31). The results are a little surprising to us because we did not predict that men would place a greater significance on the physical attractiveness of a short-term mate compared to a long-term mate.

Women also favored physical attractiveness in a short-term mate (2.43) and a long-term mate (2.10). Here again, physical attractiveness was more important in short-term mating than in long-term mating. In both contexts, however, physical attractiveness was significantly less important to women than it is to men.

We also tested these predictions in our international survey of 37 cultures. My colleagues in each country asked men and women to evaluate the relative importance of the characteristics *good looking* and *physically attractive* in a mate. As in our American college population, men throughout the world placed a high value on physical attractiveness in a partner.

In each of the 37 cultures men valued physical attractiveness and good looks in a mate

more than did their female counterparts. These sex differences are not limited to cultures that are saturated with visual media, Westernized cultures or racial, ethnic, religious or political groups. Worldwide, men place a premium on physical appearance.

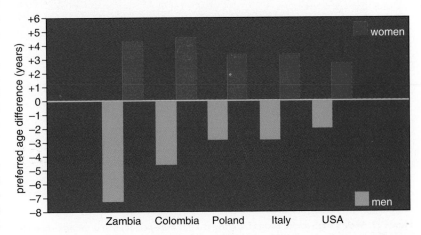

Figure 9. Preferences for an age difference between oneself and one's spouse differ for men and women. Men in each of the 37 cultures examined by the author prefer to mate with younger women, whereas women generally prefer to mate with older men. Here the disparities between the mating preferences of men and women in five countries show some of the cultural variation across the sample.

Hypothesis 6: Men seeking a long-term mate will solve the problem of paternity confidence.

Figure 10. Othello's jealous rage and murder of Desdemona in Shakespeare's tragic play was incited by her presumed sexual infidelity. According to sexual-strategies theory, men have strong emotional responses to a mate's sexual infidelity because the ancestors of modern males who protected themselves against cuckoldry had greater reproductive success. Here the German actor Emil Jannings plays the role of Othello in a movie from the early part of this century.

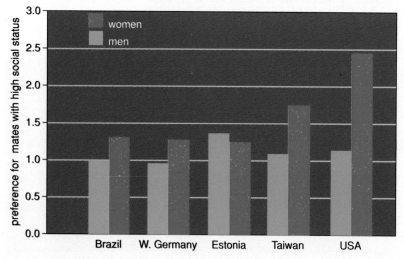

Figure 11. Mate's high social status (ranging from 3, *indispensible*, to 0, *unimportant*) is typically greater for women than for men. Of the 37 cultures examined, only males in Estonia valued the social status of their spouses more than did their female counterparts. (Only a sample of five countries is shown here.) Women generally prefer men with a high social standing because a man's ability to provide resources for her offspring is related to his social status.

A further clue to the significance of reproductive value comes in an international study of divorce. Laura Betzig of the University of Michigan studied the causes of marital dissolution in 89 cultures from around the world. She found that one of the strongest sex-linked causes of divorce was a woman's old age (hence low reproductive value) and the inability to produce children. A woman's old age was significantly more likely to result in divorce than a man's old age.

Hypothesis 6: Men seeking a long-term mate will solve the problem of paternity confidence. Men face an adaptive problem that is not faced by women—the problem of certainty in parenthood. A woman can always be certain that a child is hers, but a man cannot be so sure that his mate's child is his own. Historically, men have sequestered women in various ways through the use of chastity belts, eunuch-guarded harems, surgical procedures and veiling to reduce their sexual attractiveness to other men. Some of these practices continue to this day and have been observed by social scientists in many parts of the world.

Most of these studies have considered three possibilities: (1) the desire for chastity in a mate (cues to *prior* lack of sexual contact with others), (2) the desire for fidelity in mates (cues to no *future* sexual contact with others), and (3) the jealous guarding of mates to prevent sexual contact with other men. We have looked at these issues ourselves in various studies.

In our international study, we examined men's and women's desire for chastity in a potential marriage partner. It proved to be a highly variable trait across cultures. For example, Chinese men and women both feel that it is indispensable in a mate. In the Netherlands and Scandinavia, on the other hand, both sexes see chastity as irrelevant in a mate. Overall, however, in about two-thirds of the international samples, men desire chastity more than women do. Sex differences are especially large among Indonesians, Iranians and Palestinian Arabs. In the remaining one-third of the cultures, no sex differences were found. In no cultures do women desire virginity in a mate more than men. In other words, where there is a difference between the sexes, it is always the case that men place a greater value on chastity.

Although we have yet to examine the desire for mate fidelity in our international sample, in her cross-cultural study Betzig found that the most prevalent cause of divorce was sexual infidelity, a cause that was highly sex-linked. A wife's infidelity was considerably more likely to result in a divorce than a husband's infidelity. Compromising a man's certainty in paternity is apparently seen worldwide as a breach so great that it often causes the irrevocable termination of the long-term marital bond.

We have examined the issue of fidelity among American college students. Indeed,

Photofest

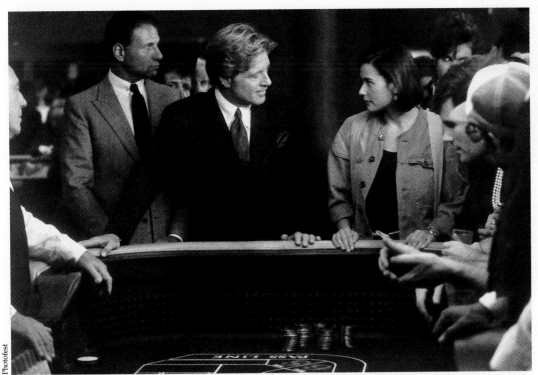

Hypothesis 7: Women seeking a short-term mate will prefer men willing to impart immediate resources.

Figure 12. Female preference for short-term mates who are willing to provide resources was examined in the 1993 film *Indecent Proposal*. Robert Redford played the role of a wealthy older man who offered one million dollars to a younger married woman, played by Demi Moore, in exchange for a short-term mating. Since multiple short-term matings do not directly increase a woman's reproductive success, sexual-strategies theory holds that a woman can increase her reproductive success by acquiring resources.

Schmitt and I found that fidelity is the characteristic most valued by men in a long-term mate. It is also highly valued by women, but it ranks only third or fourth in importance, behind such qualities as honesty. It seems that American men are concerned more about the future fidelity of a mate than with her prior abstinence.

Our studies of jealousy reveal an interesting qualitative distinction between men and women. Randy Larsen, Jennifer Semmelroth, Drew Westen and I conducted a series of interviews in which we asked American college students to imagine two scenarios: (1) their partner having sexual intercourse with someone else, or (2) their partner falling in love and forming a deep emotional attachment to someone else. The majority of the men reported that they would be more upset if their mate had sexual intercourse with another man. In contrast, the majority of the women reported that they would be more upset if their mate formed an emotional attachment to another woman.

We also posed the same two scenarios to another group of 60 men and women, but this time we recorded their physiological responses. We placed electrodes on the corrugator muscle in the brow (which contracts during frowning), on two fingers of the right hand to measure skin conductance (or sweating), and the thumb to measure heart rate.

The results provided a striking confirmation of the verbal responses of our earlier study. Men became more physiologically distressed at the thought of their mate's sexual infidelity than their mate's emotional infidelity. In response to the thought of sexual infidelity, their skin conductances increased by an average of about 1.5 microSiemens, the frowning muscle showed 7.75 microvolt units of contraction and their hearts increased by about five beats per minute. In response to the thought of emotional infideli-

Hypothesis 8: Women will be more selective than men in choosing a short-term mate.

DWF WRITER/PHOTOGRAPHER—creative, pretty, warm-hearted, independent, energetic, NYC—seeks friendship and romance with good-natured, thoughtful, intelligent, successful, fit SW man 45–60. Please no egoists, philistines, or couch potatoes! Box 3389.

L.A. SCIENTIST, M.D., C.E.O., WORKAHOLIC, 51, Joe Mantegna-type, seeks smart, slim, funny, independent SWF to help him slow down, enjoy travel, jazz. Box 3222.

DETROIT. 52 y/o MWF, happy, leggy, athletic. Fine Arts academician. Husband on 12 month sabbatical. Seeks ruggedly handsome younger man for discreet tryst. Box 3998.

FEMALE FRIEND/LOVER sought by tall, slim, handsome, athletic, married WM Ivy Lawyer/Bus. Man, 60s. Box 4235

BABE WITH LOOKS, BOOKS, AND BRAINS, NOW IN LOS ANGELES needs to meet her match for all things men and women do so well together including, but not limited to, great, safe sex. I'm a widowed lawyer, 50+, lively, slim, and trim. You are handsome, successful, well-educated, amusing, trustworthy. Your chronological age does not concern me but your attitude does. I'm looking for a mature man who is young at heart. Letter, phone, photo. No married men, please. Box 1212

Figure 13. Personal advertisements in newspapers for people seeking short-term mates support the hypothesis that women are generally more selective than men. Women tend to define specific qualities they seek in a man, whereas men tend to define their own qualities—attractiveness, high social status, ambition, and professional standing—that they believe will attract women.

Figure 14. Female preferences for long-term mates who are willing to provide resources is parodied in the 1993 film *Addams Family Values*. Joan Cusack *(right)* plays the cultural stereotype of the beautiful "gold-digging" woman, and Christopher Lloyd *(left)* plays the wealthy but unattractive man, Uncle Fester. The character Cousin Itt *(hair in the center)* performs the wedding ceremony.

Hypothesis 9: Women seeking a long-term mate will prefer men who can provide resources for their offspring.

ty, the men's skin conductance showed little change from baseline, their frowning increased by only 1.16 units, and their heart rates did not increase. Women, on the other hand, tended to show the opposite pattern. For example, in response to the thought of emotional infidelity, their frowning increased by 8.12 units, whereas the thought of sexual infidelity elicited a response of only 3.03 units.

Hypothesis 7: Women seeking a short-term mate will prefer men willing to impart immediate resources. Women confront a different set of mating problems than those faced by men. They need not consider the problem of partner number, since mating with 100 men in one year would produce no more offspring than mating with just one. Nor do they have to be concerned about the certainty of genetic parenthood. Women also do not need to identify men with the highest fertility since men in their 50s, 60s and 70s can and do sire children.

In species where males invest parentally in offspring, where resources can be accrued and defended, and where males vary in their ability and willingness to channel these resources, females gain a selective advantage by choosing mates who are willing and able to invest resources. Females so choosing afford their offspring better protection, more food and other material advantages that increase their ability to survive and reproduce. Do human females

exhibit this behavior pattern? If so, we should be able to make a few predictions.

In short-term contexts, women especially value signs that a man will immediately expend resources on them. We asked 50 female subjects to evaluate the desirability of a few characteristics in a short-term and a long-term mate: *spends a lot of money early on, gives gifts early on,* and *has an extravagant lifestyle.* We found that women place greater importance on these qualities in a short-term mate than in a long-term mate, despite the fact that women are generally less exacting in short-term mating contexts.

We would also predict that women will find undesirable any traits that suggest that a man is reluctant to expend resources on her immediately. When we tested this prediction with the same sample population, we found that women especially dislike men who are stingy early on. Although this attribute is undesirable in a long-term mate as well, it is significantly more so in a short-term mate.

Hypothesis 8: Women will be more selective than men in choosing a short-term mate. This hypothesis follows from the fact that women (more than men) use short-term matings to evaluate prospective long-term mates. We can make several predictions based on this hypothesis.

First, women (more than men) will dislike short-term mates who are already in a relationship. We examined the relative undesirability of a prospective mate who was already in a relationship to 42 men and 44 women, using a scale from –3 (extremely undesirable) to 3 (extremely desirable). Although men were only slightly bothered (averaging a score of about –1.04) by this scenario, women were significantly more reluctant to engage in a relationship with such a mate (average score about –1.70).

We would also predict that women (more than men) will dislike short-term mates who are promiscuous. To a woman, promiscuity indicates that a man is seeking short-term relationships and is less likely to commit to a long-term mating. We tested this prediction in the same sample of 42 men and 44 women using the same rating scale as before. Although men found promiscuity to be of neutral value in a short-term mate, women rated the trait as moderately undesirable (an average of about –2.00).

Finally, because one of the hypothesized functions for female short-term mating is protection from aggressive men, women should value attributes such as physical size and strength in short-term mates more than in long-term mates. When we asked men and women to evaluate the notion of a mate being *physically strong,* we found that women preferred physically strong mates in all contexts more than men did, and that women placed a premium on physical strength in a short-term mate. This was true despite the higher standards women generally hold for a long-term mate.

Hypothesis 9: Women seeking a long-term mate will prefer men who can provide resources for her offspring. In a long-term mating context, we would predict that women (more than men) will desire traits such as a potential mate's ambition, earning capacity, professional degrees and wealth.

In one study we asked a group of 58 men and 50 women to rate the desirability (to the average man and woman) of certain characteristics that are indicators of future resource-acquisition potential. These included such qualities as *is likely to succeed in profession, is likely to earn a lot of money,* and *has a reliable future career.* We found that in each case women desired the attribute more in a long-term mate than in a short-term mate. Moreover, women valued each of these characteristics in a long-term mate more than men did.

In our international study, we also examined men's and women's preferences for long-term mates who can acquire resources. In this case we looked at such attributes as *good financial prospects, social status* and *ambition-industriousness*—attributes that typically lead to the acquisition of resources. We found that sex differences in the attitudes of men and women were strikingly consistent around the world. In 36 of the 37 cultures, women placed significantly greater value on financial prospects than did men. Although the sex differences were less profound for the other two qualities, in the overwhelming majority of cultures, women desire *social status* and *ambition-industriousness* in a long-term mate more than their male counterparts do.

Finally, in her international study of divorce, Betzig found that a man's failure to provide proper economic support for his wife and children was a significant sex-linked cause of divorce.

Conclusion

The results of our work and that of others provide strong evidence that the traditional assumptions about mate preferences—that they are arbitrary and culture-bound—are simply wrong. Darwin's initial insights into sexual selection have turned out to be scientifically profound for people, even though he understood neither their functional-adaptive nature nor the importance of relative parental investment for driving the two components of sexual selection.

Men and women have evolved powerful desires for particular characteristics in a mate. These desires are not arbitrary, but are highly patterned and universal. The patterns correspond closely to the specific adaptive problems that men and women have faced during the course of human evolutionary history. These are the problems of paternity certainty, partner number and reproductive capacity for men, and the problems of willingness and ability to invest resources for women.

It turns out that a woman's physical appear-

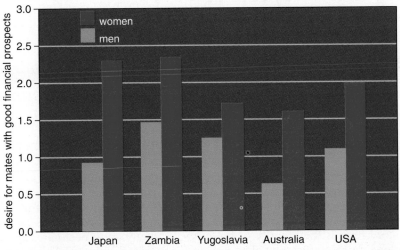

Figure 15. Mate's good financial prospects are consistently more important (measured on a scale from 3, *indispensible,* to 0, *unimportant*) for women than for men throughout the world. In the author's cross-cultural study of 37 countries, women valued the financial prospects of a potential spouse more than their male counterparts did in every culture but one (Spain). A sample of five countries is shown here.

ance is the most powerful predictor of the occupational status of the man she marries. A woman's appearance is more significant than her intelligence, her level of education or even her original socioeconomic status in determining the mate she will marry. Women who possess the qualities men prefer are most able to translate their preferences into actual mating decisions. Similarly, men possessing what women want—the ability to provide resources—are best able to mate according to their preferences.

Some adaptive problems are faced by men and women equally: identifying mates who show a proclivity to cooperate and mates who show evidence of having good parenting skills. Men do not look at women simply as sex objects, nor do women look at men simply as success objects. One of our most robust observations was that both sexes place tremendous importance on mutual love and kindness when seeking a long-term mate.

The similarities among cultures and between sexes implies a degree of psychological unity or species typicality that transcends geographical, racial, political, ethnic and sexual diversity. Future research could fruitfully examine the ecological and historical sources of diversity, while searching for the adaptive functions of the sexual desires that are shared by all members of our species.

Bibliography

Buss, D. 1994. *The Evolution of Desire: Strategies of Human Mating.* New York: Basic Books.

Buss, D. M., et al. 1990. International preferences in selecting mates: A study of 37 cultures. *Journal of Cross-cultural Psychology* 21:5–47.

Buss, D. M., and Schmitt, D. P. 1993. Sexual Strategies Theory: An evolutionary perspective on human mating. *Psychological Review* 100:204–232.

Buss, D. M., Larsen, R., Westen, D., and Semmelroth, J. 1992. Sex differences in jealousy: Evolution, physiology, psychology. *Psychological Science* 3:251–255.

Avian Siblicide

Killing a brother or a sister may be a common adaptive strategy among nestling birds, benefiting both the surviving offspring and the parents

Douglas W. Mock, Hugh Drummond and Christopher H. Stinson

Occasionally, the pen of natural selection writes a murder mystery onto the pages of evolution. But unlike a typical Agatha Christie novel, this story reveals the identity of the murderer in the first scene. The mystery lies not in "whodunit," but in why.

The case at hand involves the murder of nestling birds by their older siblings. Observers in the field have frequently noted brutal assaults by elder nestmates on their siblings, and the subsequent deaths of the younger birds. The method of execution varies among different species, ranging from a simple push out of the nest to a daily barrage of pecks to the head of the younger, smaller chick. Such killings present a challenge to the student of evolutionary biology: Does siblicide promote the fitness of the individuals that practice it, or is such behavior pathological? In other words, are there certain environmental conditions under which killing a close relative is an adaptive behavior? Moreover, are there other behaviors or biological features common to siblicidal birds that distinguish them from nonsiblicidal species?

Avian siblicide holds a special interest for several reasons. First, because nestling birds are relatively easy to observe, a rich descriptive literature exists based on field studies of many species. Second, because birds tend to

Douglas W. Mock is associate professor of zoology at the University of Oklahoma. He was educated at Cornell University and the University of Minnesota, where he received his Ph. D. in ecology and behavioral biology in 1976. Address: Department of Zoology, University of Oklahoma, Norman, OK 73019. Hugh Drummond is a researcher in animal behavior at the Universidad Nacional Autònoma de México. He was educated at Bristol University, the University of Leeds and the University of Tennessee, where he received his Ph. D. in psychology in 1980. Address: Centro de Ecologia, Universidad Nacional Autònoma de México, AP 70-275, 04510 México, D.F. Christopher H. Stinson was educated at Swarthmore College, the College of William and Mary and the University of Washington, where he received his Ph. D. in 1982. Address: 4005 NE 60th Street, Seattle, WA 98115.

be monogamous, siblicide is likely to involve full siblings. (Although recent DNA studies suggest that birds may not be as monogamous as previously thought, most nestmates are still likely to be full siblings.) Third, young birds require a large amount of food during their first few weeks of development, and this results in high levels of competition among nestlings. The competitive squeeze is exacerbated for most species because the parents act as a bottleneck through which all resources arrive. Fourth, some avian parents may not be expending their maximum possible effort toward their current brood's survival (Drent and Daan 1980, Nur 1984, Houston and Davies 1984, Gustafsson and Sutherland 1988, Mock and Lamey in press). Parental restraint may be especially common in long-lived species, in which a given season's reproductive output makes only a modest contribution to the parents' lifetime success (Williams 1966).

Siblicide—or juvenile mortality resulting from the overt aggression of siblings—is not unique to birds. It is also observed, for example, among certain insects and amphibians; in those groups, however, the behavioral pattern is rather different. Most siblicidal insects and amphibians immediately consume their victims as food, whereas in birds (and mammals) siblicide rarely leads to cannibalism. For example, tadpoles of the spadefoot toad acquire massive dentition (the so-called "cannibal morph") with which they consume their broodmates (Bragg 1954), and fig wasps use large, sharp mandibles to kill and devour their brothers (Hamilton 1979). In contrast, among pronghorn antelopes, one of the embryos develops a necrotic tip on its tail with which it skewers the embryo behind it (O'Gara 1969), and piglet littermates use deciduous eyeteeth to battle for the sow's most productive teats (Fraser 1990). Among birds and mammals it seems that the

goal is to secure a greater share of critical parental care.

Although biologists have known of avian siblicide for many years, only recently have quantitative field studies been conducted. The current wave of such work is due largely to the realization that siblicide occurs routinely in some species that breed in dense colonies; such populations provide the large sample sizes needed for formal testing of hypotheses.

Models of Nestling Aggression

Our examination of siblicidal aggression focuses on five species of birds. Two of these, the black eagle (*Aquila verreauxi*) and the osprey (*Pandion haliaetus*), are raptors that belong to the family Accipitridae. A third species, the blue-footed booby (*Sula nebouxii*) is a seabird belonging to the family Sulidae. We also present studies of the great egret (*Casmerodius albus*) and the cattle egret (*Bubulcus ibis*), both of which belong to the family Ardeidae. Each of these species exhibits a distinct behavioral pattern; the range of variation is important to an understanding of siblicide.

The black eagle is one of the first birds in which siblicide was described. This species, also called Verreaux's eagle, lives in the mountainous terrain of southern and northeastern Africa, as well as the western parts of the Middle East. Black eagles generally build their nests on cliff ledges and lay two eggs between April and June. The eaglets hatch about three days apart, and so the older chick is significantly larger than the younger one. The black eagle is of particular interest for the study of

Figure 1 (right). Two cattle egrets peer down at their recently evicted younger sibling. For several days before the eviction, the elder siblings pecked at the head of their smaller nestmate. Here the younger bird holds its bald and bloodied head out of reach. Soon after the photograph was made, the bird was driven to the ground and perished. (Photograph by the authors.)

siblicide because the elder eaglet launches a relentless attack upon its sibling from the moment the younger eaglet hatches. In one well-documented case, the senior eaglet pecked its sibling 1,569 times during the three-day lifespan of the younger nestling (Gargett 1978).

Among ospreys, sibling aggression is neither so severe nor so persistent as it is among black eagles. Ospreys are widely distributed throughout the world, including the coastal and lacustrine regions of North America. The nests are generally built high in trees or on other structures near water. A brood typically consists of three chicks, which usually live in relative harmony. Nevertheless, combative exchanges between siblings do occur in this species;

comparisons between the fighting and the pacifist populations offer insights into the significance of aggression.

The blue-footed booby lives exclusively on oceanic islands along the Pacific coast from Baja California to the northern coast of Peru. Blue-footed boobies are relatively large, ground-nesting birds that typically form dense colonies near a shoreline. Two or three chicks hatch about four days apart, and this results in a considerable size disparity between the siblings. As in many other siblicidal species, the size disparity predicts the direction of the aggression between siblings.

Young nestmates also differ in size in the two egret species we have studied. The larger of these, the great egret, is distributed throughout the middle

latitudes of the world, and also throughout most of the Southern Hemisphere. Great egrets make their nests in trees or reed beds in colonies located near shallow water. The cattle egret also nests in colonies, but not necessarily close to water. Cattle egrets live in the middle latitudes of Asia, Africa and the Americas. As their name suggests, they are almost always found in the company of grazing cattle or other large mammals, riding on their backs and feeding on grasshoppers stirred up by the movement of the animals. Despite their differences in habitat, great egrets and cattle egrets have a number of behaviors in common. Typically, three or four egret nestlings hatch at one- to two-day intervals, and fighting starts almost as

Peter Steyn

Figure 2. Aggression in black eagle nestlings almost always results in the death of the younger sibling. Here a six-day-old black eagle chick tears at a wound it has opened on the back of its day-old sibling.

soon as the second sibling has hatched. Aggressive attacks lead to a "pecking order" that translates into feeding advantages for the elder siblings (Fujioka 1985a, 1985b; Mock 1985; Ploger and Mock 1986). In about a third of the nests, the attacks culminate in siblicide through socially enforced starvation and injury or eviction from the nest.

Obligate and Facultative Siblicide

It is useful to distinguish those species in which one chick almost always kills its sibling from those in which the incidence of siblicide varies with environmental circumstances. Species that practice obligate siblicide typically lay two eggs, and it is usually the older, more powerful chick that kills its nestmate. The black eagle is a good example of an obligate siblicide species. In 200 records from black eagle nests in which both chicks hatched, only one case exists where two chicks fledged (Simmons 1988). Similar patterns of obligate siblicide have been reported for other species that lay two eggs, including certain boobies, pelicans and other eagles (Kepler 1969; Woodward 1972; Stinson 1979; Edwards and Collopy 1983; Cash and Evans 1986; Evans and McMahon 1987; Drummond 1987; Simmons 1988; Anderson 1989, 1990).

A far greater number of birds are facultatively siblicidal. Fighting is frequent among siblings in these species, but it does not always lead to the death of the younger nestling. There are various patterns of facultative siblicide. For example, in species such as the osprey, aggression is entirely absent in some populations, and yet present in others (Stinson 1977; Poole 1979, 1982; Jamieson et al. 1983). In other species aggression occurs at all nests but differs in form and effect. In the case of the blue-footed booby a chick may hit its sibling only a few times per day for several weeks, and then rapidly escalate to a lethal rate of attack (Drummond, Gonzalez and Osorno 1986). Egret broods tend to have frequent sibling fights—there are usually several multiple-blow exchanges per day—but the birds do not always kill each other (Mock 1985, Ploger and Mock 1986).

Traits of Siblicidal Species

Five characteristics are common to virtually all siblicidal birds: resource competition, the provision of food to the nestlings in small units, weaponry, spatial confinement and competitive disparities between siblings. The first

Figure 3. Blue-footed booby nestlings maintain dominance over their younger siblings through a combination of aggression and threats *(upper photograph)*. The assaults do not escalate to the point of eviction unless the food supply is inadequate. An evicted chick has little chance of survival in the face of attacks from neighboring adults *(lower photograph)*. (Photographs courtesy of the authors.)

four traits are considered essential preconditions for the evolution of sibling aggression; the study of their occurrence may shed some light on the origin of siblicidal behavior. The fifth trait—competitive disparities among nestmates resulting from differences in size and age—is also ubiquitous and important, but it is probably not essential for the evolution of siblicide. In fact, competitive disparities may be a consequence rather than a cause of siblicidal behavior; having one bird appreciably stronger than the other reduces the cost of fighting, since asymmetrical fights tend to be brief and it is less likely that both siblings will be hurt during combat (Hahn 1981, Fujioka 1985b, Mock and Ploger 1987).

Of the five traits common to siblicidal species, the competition for resources is probably the most fundamental. Among birds, the competition is primarily for food. Experiments have shown that the provision of additional food often diminishes nestling mortality (Mock, Lamey and Ploger 1987a; Magrath 1989). But "brood reduc-

Figure 6. Food is presented directly to the chick in small units in all known species of siblicidal birds. This direct method of feeding means that a chick may increase its share of food by physically intimidating, and not just by killing, its competing siblings. The young black eagle (*top*) is fed a piece of hyrax meat by the direct-transfer method, even though the bird is well into the fledgling stage. In the blue-footed booby (*lower left*), the parent transfers small pieces of fish from its mouth directly into the mouth of a chick. An osprey chick (*lower right*) receives a piece of meat from its parent while its sibling waits. Osprey chicks take turns feeding, and will fight only if food becomes scarce.

(Lack 1954). Parents must commit themselves to a fixed number of eggs early in the nesting cycle, before the season's bounty or shortcomings can be assessed. Thus, it is often advantageous for parents to produce an additional egg or two, in case later conditions are beneficent, while reserving the small-brood option by making the "bonus" offspring competitively inferior, in case the season's resources are poor. The production of an inferior sibling may be advantageous, since the senior sibling can then eliminate its younger nestmate with greater ease. In fact, experimentally synchronizing the hatchings of cattle egrets results in an increase in fighting, which reduces the reproductive efficiency of the parents (Fujioka 1985b, Mock and Ploger 1987).

Siblicide as an Adaptation

To understand siblicide, we must understand how the killing of a close relative can be favored by natural selection. At first this may seem a simple matter. Eliminating a competitor improves one's own chance of survival, and thereby increases the likelihood that genes promoting such behavior will be represented in the next generation. According to this simple analysis, natural selection should always reward the most selfish act, and siblicide is arguably the epitome of selfishness.

The trouble with this formulation is that it implies that all organisms should be as selfish as possible, which is contrary to observation. (Siblicide is fairly common, but certainly not universal.) A more sophisticated analysis was provided in the 1960s by the British theoretical biologist William D. Hamilton. In Hamilton's view, the fitness of a gene is more than its contribution to the reproduction of the individual. A gene's fitness also depends on the way it influences the reproductive prospects of close genetic relatives.

This expanded definition of evolutionary success, called inclusive fitness, is a property of individual organisms. An organism's inclusive fitness is a measure of its own reproductive success plus the incremental or decremental influences it has on the reproductive success of its kin, multiplied by the degree of relatedness to those kin (Hamilton 1964). Hamilton's theory is generally invoked to explain apparently altruistic behavior, but the theory also specifies the evolutionary limits of selfishness.

An example will help to clarify Hamilton's idea. Suppose a particular gene predisposes its bearer, X, to help a sibling. Since the laws of Mendelian inheritance state that X and its sibling share, on average, half of their genes, X's sibling has a one-half probability of carrying the gene. From the gene's point of view, it is useful for X to promote the reproductive success of a sibling because such an action contributes to the gene's numerical increase. Therefore, helping a sibling should be of selective advantage. It is in this light that we must understand and explain siblicide. Since selection favors genes that promote their own numerical increase, what advantage might there be in destroying a sibling—an organism with a high probability of carrying one's own genes? The solution to the problem lies in the role played by the "marginal" offspring, which may be the victim of siblicide.

In all siblicidal species studied to date there is a striking tendency for the victim to be the youngest member of the brood (Mock and Parker 1986). The youngest sibling is marginal in the sense that its reproductive value can be assessed in terms of what it adds to or subtracts from the success of other family members. Specifically, the marginal individual can embody two kinds of reproductive value. First, if the marginal individual survives in addition to all its siblings, it represents an extra unit of parental success, or extra reproductive value. Such an event is most likely during an especially favorable season, when the needs of the entire brood can be satisfied. Alternatively, the marginal offspring may serve as a replacement for an elder sibling that dies prematurely. In such instances the marginal individual represents a form of insurance against the loss of a senior sibling. The magnitude of this insurance value depends on the probability that the senior sibling will die.

Among species that practice obligate siblicide, the marginal individual offers no extra reproductive value; marginal chicks serve only as insurance against the early loss or infirmity of the senior chick. In these species, if the senior chick is alive but weakened and in-

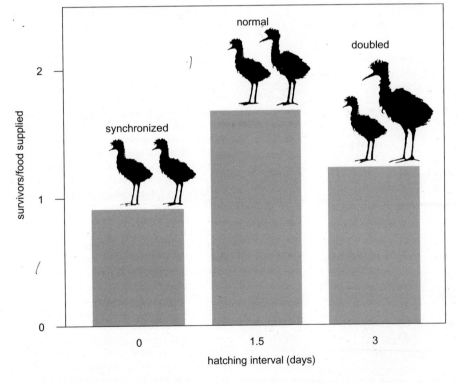

Figure 7. Effect of hatching asynchrony on avian domestic violence was investigated by switching eggs in the nests of cattle egrets. Reproductive efficiency is maximal when chicks hatch at an interval of one and one-half days (as they do under normal conditions). Synchronized hatching (an interval of zero days) increases the amount of fighting between chicks, which results in greater chick mortality. The normal one-and-a-half day interval reduces the amount of fighting since the older chick is able to intimidate the younger chick. Doubling the asynchrony, so that the eggs hatch three days apart, greatly reduces the amount of fighting but exaggerates the competitive asymmetries, so that the youngest nestmates receive little food. The experiments were performed by Douglas Mock and Bonnie Ploger at the University of Oklahoma.

capable of killing the younger chick, the latter may be able to reverse the dominance and kill the senior chick. Such scenarios appear to be played out regularly: In a sample of 22 black eagle nests in which both chicks hatched, the junior chick alone fledged in five of the nests, and the senior chick alone fledged in the remaining 17 cases (Gargett 1977). Similarly, in a sample of 59 nests of the masked booby, the junior chick was the sole fledgling in 13 nests, and the senior chick the sole fledgling in the other 46 nests (Kepler 1969). In

both of these species, the junior chick's chance of being the sole survivor—its insurance reproductive value to the parents—is about 22 percent. Removing the "insurance" eggs results in a reduction in the mean number of fledglings per nest (Cash and Evans 1986). Consequently, the insurance value of the marginal offspring should improve parental fitness if the cost of producing that offspring is reasonable. (In fact, the cost of producing one additional egg seems fairly modest: approximately 2.5 percent of the body weight of the black eagle female.)

Among species that practice facultative siblicide, the marginal offspring may be a source of insurance but may also provide extra reproductive value. The relative contribution of the marginal offspring to the reproductive success of the parents appears to vary considerably within and between species. For example, among great egrets the proportion of nests in which all nestlings survive—the extra reproductive value—varies from 15 to 23 percent, whereas the proportion of the nests in which at least one senior sibling dies and the youngest sibling

lives—the insurance reproductive value—may vary from 0 to 48 percent (Mock and Parker 1986). The blue-footed booby shows great variation in the extra reproductive value provided by the marginal offspring (5 to 67 percent), whereas the insurance reproductive value is generally quite low (5 to 6 percent). In both of these species the magnitudes of the total reproductive values depend on the size of the brood. In general, the marginal offspring provides a greater total reproductive value to the parents when the brood size is smaller.

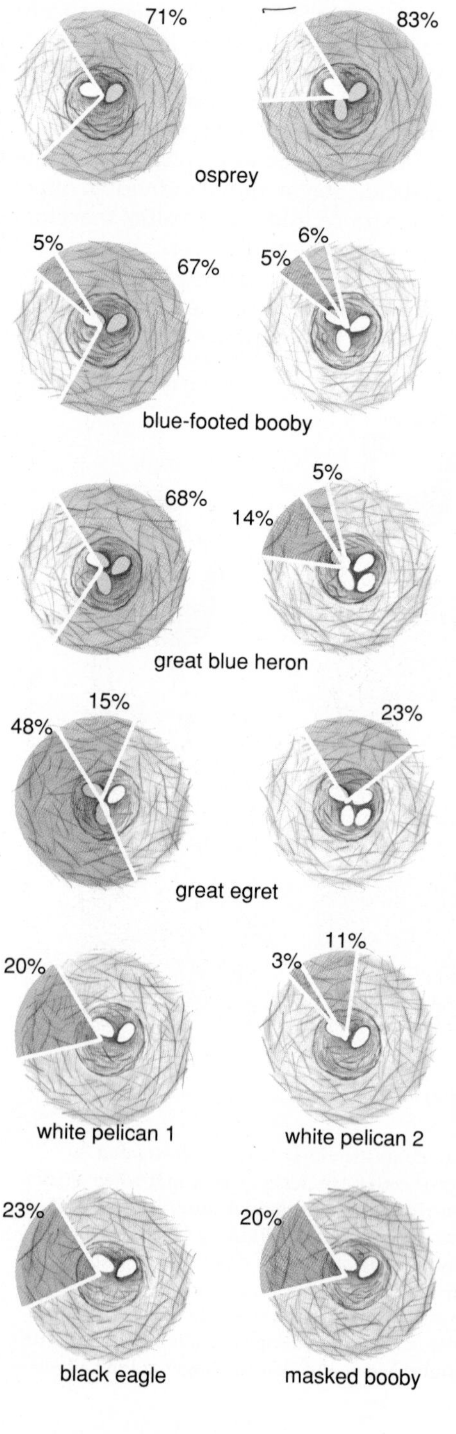

The Timing of the Deed

A senior sibling should kill its younger sibling as soon as two conditions are met: (1) the senior sibling's own viability seems secure; and (2) the resources are inadequate for the survival of both siblings. Killing the junior sibling before these conditions are met would waste the potential fitness the junior sibling could offer in the form of extra reproductive value or insurance reproductive value. Delaying much beyond the point at which the conditions are met also has a cost. First, the food eaten by the victim is a loss of resources, and, second, the cost of execution may increase as the victim gains strength and is more likely to defeat the senior sibling.

In obligate siblicide species, the average food supply is presumably inadequate for supporting two chicks at reasonable levels of parental effort, and as a result the second chick is dispatched as soon as possible after it hatches. For example, the mean longevity of the victim in the case of the masked booby is 3.3 days (Kepler 1969), and only 1.75 days for brown boobies (Cohen et al. in preparation).

Among facultative siblicide species, the mean longevity of the victim is usually greater; in the blue-footed booby it is 18 days (Drummond, Gonzalez and Osorno 1986). Although the senior blue-footed booby chick may peck at the head or wrench the skin of its nestmate, the younger sibling is seldom killed by these direct physical assaults. Instead, death typically results from starvation or violent pecking by adult neighbors when the junior chick is routed from the home nest (Drummond and Garcia Chavelas 1989).

The Causes of Siblicide

The evolutionary difference between the obligate and the facultative forms

Figure 8. Reproductive value of the youngest member of a brood (the usual victim of siblicide) varies across species and brood size. The reproductive value is represented as the proportion of nests (in broods of two, three or four eggs) in which the youngest chick survives. If the youngest chick survives in addition to its elder siblings, it contributes "extra reproductive value" (*blue sections*); when the youngest chick survives as a replacement for an elder sibling that dies early, the junior bird provides "insurance reproductive value" (*red sections*). Among birds that almost always commit siblicide, such as the black eagle and the masked booby, the youngest chick's reproductive value is entirely due to its role as an "insurance policy." In species where siblicide is more occasional, such as the great egret, the youngest chick may provide either form of reproductive value. These estimates of reproductive value are maxima, since they represent survival only part way through the prefledgling period and not recruitment into the breeding population. These data are derived from studies by: Gargett 1977 (black eagle), Stinson 1977 (osprey), Cash and Evans 1986 (white pelican 1), Evans and McMahon 1987 (white pelican 2), Mock and Parker 1986 (great egret and great blue heron), and Drummond (unpublished data on the blue-footed booby). The data on the masked booby are combined from studies by Kepler 1969 and Anderson 1989.

of siblicide may be a function of the risk that a junior chick poses to the welfare of its senior sibling. That risk can be defined both in terms of resource consumption and in terms of the potential for bodily harm. If the resources are adequate only for the survival of a single chick, or if a young chick poses a significant physical threat to an older chick, then the senior sibling might be expected to destroy the younger one. On the other hand, if there is enough food for both chicks, and if the younger sibling can be subjugated so that it does not present a threat, then the survival of the younger sibling is beneficial because it increases the inclusive fitness of the senior sibling. In such circumstances, natural selection should favor a measure of clemency on the part of the senior sibling. Accordingly, we would expect obligate siblicide to evolve in circumstances in which resources are routinely limited and siblings tend to pose a physical threat to one another. In contrast, facultative siblicide should arise in circumstances in which resources are not always limited.

The analysis offered above concerns the inheritance of a long-term predisposition to siblicide. Recent studies suggest that food shortages also act as an immediate stimulus to, or proximal cause of, sibling fighting. A link between the food supply and siblicide was suggested by the finding that brood reductions in the blue-footed booby tend to occur soon after the weight of the senior chick drops about 20 percent below the weight expected at its current age in a good year (Drummond, Gonzalez and Osorno 1986). The relationship between food deprivation and aggression was confirmed by experiments in which the senior chick's neck was taped to prevent it from swallowing food. The experimentally deprived senior chicks pecked their nestmates about three to four times more frequently with the tape in place than without the tape, and they subsequently received a greater share of the food (Drummond and Garcia Chavelas 1989).

In older booby broods, the increase in the amount of aggressive pecking was delayed by about a day after the chick's neck was taped, suggesting that aggression is controlled by a factor that changes progressively over time, such as hunger or growth status. In fact, the increased pecking rate coincided with a 20 percent weight loss by the senior

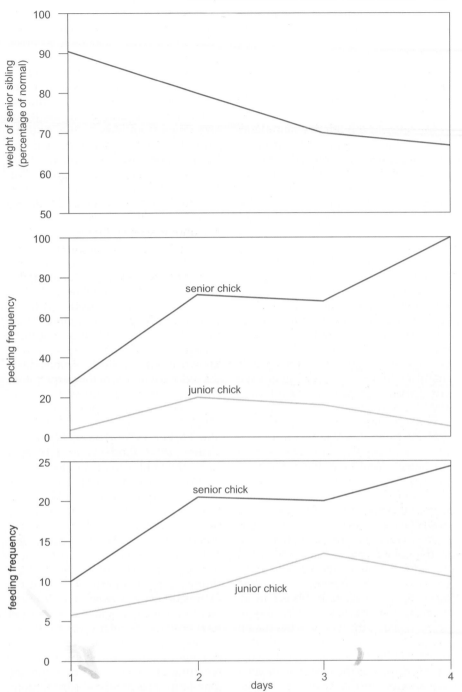

Figure 9. Effect of food deprivation on aggression and food distribution in blue-footed booby nestlings was investigated by taping the senior chick's neck to prevent it from swallowing food. As the weight of the senior chick drops more than 20 percent below normal (*top*), the rate at which it pecks its younger sibling increases more than three-fold (*middle*). The escalating aggression of the elder chick brings it a greater share of food (*bottom*). The experiments were performed by Hugh Drummond and Cecilia Garcia Chavelas at the Universidad Nacional Autònoma de México.

chick. When the tape was removed, the aggressive pecking rate returned toward the baseline level. These results suggest that nestling aggression among certain facultative species is a reversible response that is sensitive to the weight level of the senior chick.

There is also suggestive evidence in other species that practice facultative siblicide that the amount of food available to the nestlings may affect sibling aggression. For example, junior kittiwake siblings are lost from the nest at higher rates following prolonged periods of bad weather, when parental foraging is reduced (Braun and Hunt 1983). Among osprey populations in which there is a high rate of prey delivery to the offspring, the nestlings are amicable and may even take turns feeding (Stinson 1977). In populations where the food delivery rate is lower, the older nestlings frequently attack their younger siblings, although they do not kill them outright (Henny 1988, Poole 1982).

In contrast, the relative abundance of food does not appear to affect the level of aggression in obligate siblicide species. Black eagle nestlings kill their siblings even in the midst of several kilograms of prey, and even while the mother eagle is offering food to the senior sibling. There does not appear to be the same direct relationship between the immediate availability of food and the level of sibling aggression. Since black eagle nestlings require large amounts of food over a period of many weeks, short-term abundance of food may not be an accurate indicator of long-term food levels. As a consequence, aggression and siblicide might be favored in order to obviate any future competition (Anderson 1990, Stinson 1979).

Perhaps the appropriate "sibling aggression policy" is obtained from simple cues available to the chicks from the outset. Assuming that parents deliver food at some optimal rate, then a chick may be able to estimate in advance whether sufficient levels will be available for its own growth. It is interesting to note that the facultatively siblicidal golden eagle (*Aquila chrysaetos*) provides the same amount of food regardless of the number of chicks in the brood (Collopy 1984). If this is typical, then the senior chick may be able to detect whether the food will be enough to support all nestmates. Eagles that practice obligate siblicide generally deliver less food to the nest than facultative species, and consequently no assessment by the chicks is necessary (Bortolotti 1986). The average amount of food provided by the parents may be consistently low enough for natural selection to favor preemptive killing—a system that benefits both the senior chick and its parents. In other words, the insurance policy is canceled.

Even in species that practice facultative siblicide, aggression is sometimes insensitive to food supply. For example, the level of fighting among heron and egret chicks appears to be independent of the amount of food available (Mock, Lamey and Ploger 1987). It may be that the current food level acts as a proximate cue for sibling aggres-

sion only in those species where the current level accurately predicts future food levels. This hypothesis is consistent with the observation that daily food levels are unpredictable and unstable among egrets (Mock, Lamey and Ploger 1987). Interestingly, fighting between egrets ceases when the brood size drops from three to two (thus reducing future food demands) and may be reinstated by restoring the third chick (Mock and Lamey in press). Further studies of other species are necessary to determine whether the degree to which food levels fluctuate is related to aggressive behavior.

Future Directions
The study of siblicide as an adaptive strategy is still in its infancy. Much of the work to date has been devoted to identifying the proximate causes of aggressive behavior and documenting its utility for controlling resources. Less is known about the effects of siblicide on the inclusive fitness of the perpetrators. Although many theoretical models of avian siblicide have been proposed (O'Connor 1978; Stinson 1979; Mock and Parker 1986; Parker, Mock and Lamey 1989, Godfray and Harper 1990), the field data are limited.

Several areas of research need to be explored further. We would like to determine the short-term costs of sibling rivalry, perhaps by comparing the energetics of competitive begging and fighting. Likewise we need to know the long-term costs of temporary food shortages; there is particular interest in the relationship between the development of the chick and the amount of food available. Similarly, what is the relation between the amount of effort parents put into supplying food, the resulting chick survival rate and the long-term costs of reproduction among brood-reducing species? Is there any relation between chick gender, hatching order and siblicide—particularly in siblicidal species that have a large degree of sexual dimorphism? Another area of interest is the role of extra-pair copulations, which reduce the relatedness of nestmates and thereby increase the potential benefits of selfishness; it would be useful to know whether chicks have the ability to discriminate half-siblings from full siblings. Finally, why is it that parents appear not to interfere with the execution process in siblicidal species (O'Connor 1978; Drummond, Gonzalez and Osorno 1986; Mock 1987)? Answers to these

questions can give us a better understanding of how siblicidal behavior may have evolved.

Bibliography
Anderson, D. J. 1989. Adaptive adjustment of hatching asynchrony in two siblicidal booby species. *Behavioral Ecology and Sociobiology* 25:363-368.

Anderson, D. J. 1990. Evolution of obligate siblicide in boobies. I: A test of the insurance egg hypothesis. *American Naturalist* 135:334-350.

Bortolotti, G. R. 1986. Evolution of growth rates in eagles: sibling competition vs. energy considerations. *Ecology* 67:182-194.

Bragg, A. N. 1954. Further study of predation and cannibalism in spadefoot tadpoles. *Herpetologica* 20:17-24.

Braun, B. M., and G. L. Hunt, Jr. 1983. Brood reduction in black-legged kittiwakes. *The Auk* 100:469-476.

Bryant, D. M., and P. Tatner. 1990. Hatching asynchrony, sibling competition and siblicide in nestling birds: studies of swiftlets and bee-eaters. *Animal Behaviour* 39:657-671.

Cash, K., and R. M. Evans. 1986. Brood reduction in the American white pelican, *Pelecanus erythrorhynchos*. *Behavioral Ecology and Sociobiology* 18:413-418.

Collopy, M. 1984. Parental care and feeding ecology of golden eagle nestlings. *The Auk* 101:753-760.

Drent, R. H., and S. Daan. 1980. The prudent parent: energetic adjustments in avian breeding. *Ardea* 68:225-252.

Drummond, H. 1987. Parent-offspring conflict and brood reduction in the Pelecaniformes. *Colonial Waterbirds* 10:1-15.

Drummond, H., E. Gonzalez and J. Osorno. 1986. Parent-offspring cooperation in the blue-footed booby, *Sula nebouxii*. *Behavioral Ecology and Sociobiology* 19:365-392.

Drummond, H., and C. Garcia Chavelas. 1989. Food shortage influences sibling aggression in the blue-footed booby. *Animal Behaviour* 37:806-819.

Edwards, T. C., Jr., and M. W. Collopy. 1983. Obligate and facultative brood reduction in eagles: An examination of factors that influence fratricide. *The Auk* 100:630-635.

Evans, R. M., and B. McMahon. 1987. Within-brood variation in growth and conditions in relation to brood reduction in the American white pelican. *Wilson Bulletin* 99:190-201.

Fraser, D. 1990. Behavioural perspectives on piglet survival. *Journal of Reproduction and Fertility, Supplement* 40:355-370.

Fujioka, M. 1985a. Sibling competition and siblicide in asynchronously-hatching broods of the cattle egret, *Bubulcus ibis*. *Animal Behaviour* 33:1228-1242.

Fujioka, M. 1985b. Food delivery and sibling competition in experimentally even-aged broods of the cattle egret. *Behavioral Ecology and Sociobiology* 17:67-74.

Gargett, V. 1977. A 13-year population study of the black eagles in the Matopos, Rhodesia, 1964-1976. *Ostrich* 48:17-27.

Gargett, V. 1978. Sibling aggression in the black eagle in the Matopos, Rhodesia. *Ostrich* 49:57-63.

Godfray, H. C. J., and A. B. Harper. 1990. The evolution of brood reduction by siblicide in birds. *Journal of Theoretical Biology* 145:163-175.

Gustafsson, L., and W. J. Sutherland. 1988. The costs of reproduction in the collared flycatcher *Ficedula albicollis*. *Nature* 335:813-815.

Hahn, D. C. 1981. Asynchronous hatching in the laughing gull: Cutting losses and reducing rivalry. *Animal Behaviour* 29:421-427.

Hamilton, W. D. 1964. The genetical evolution of social behaviour. *Journal of Theoretical Biology* 7:1-52.

Henny, C. J. 1988. Reproduction of the osprey. In *Handbook of North American Birds*, ed. R. E. Palmer. Yale University Press.

Houston, A. I., and N. B. Davies. 1985. The evolution of cooperation and life history in the dunnock *Prunella modularis*. In R. Sibly and R. Smith (eds.) *Behavioral Ecology: The Ecological Consequences of Adaptive Behaviour.* Blackwell: Oxford. pp. 471-487.

Jamieson, I. G., N. R. Seymour, R. P. Bancroft and R. Sullivan. 1983. Sibling aggression in nestling ospreys in Nova Scotia. *Canadian Journal of Zoology* 61:466-469.

Kepler, C. B. 1969. Breeding biology of the blue-faced booby on Green Island, Kure Atoll. *Publications of the Nuttal Ornithology Club* 8.

Lack, D. 1954. *The Natural Regulation of Animal Numbers*. Clarendon Press: Oxford.

Magrath, R. 1989. Hatch asynchrony and reproductive success in the blackbird. *Nature* 339:536-538.

Mock, D. W. 1984. Siblicidal aggression and resource monopolization in birds. *Science* 225:731-733.

Mock, D. W. 1985. Siblicidal brood reduction: The prey-size hypothesis. *American Naturalist* 125:327-343.

Mock, D. W. 1987. Siblicide, parent-offspring conflict, and unequal parental investment by egrets and herons. *Behavioral Ecology and Sociobiology* 20:247-256.

Mock, D. W., and T.C. Lamey. In press. The role of brood size in regulating egret sibling aggression. *American Naturalist*.

Mock, D. W., and G.A. Parker. 1986. Advantages and disadvantages of ardeid brood reduction. *Evolution* 40:459-470.

Mock, D. W., and B.J. Ploger. 1987. Parental manipulation of optimal hatch asynchrony in cattle egrets: An experimental study. *Animal Behaviour* 35:150160.

Mock, D. W., T. C. Lamey and B. J. Ploger. 1987. Proximate and ultimate roles of food amount in regulating egret sibling aggression. *Ecology* 68:1760-1772.

Mock, D. W., T. C. Lamey, C.F. Williams and A. Pelletier. 1987. Flexibility in the development of heron sibling aggression: An intraspecific test of the prey-size hypothesis. *Animal Behaviour* 35:1386-1393.

Nur, N. 1984. Feeding frequencies of nestling blue tits (*Parus coeruleus*): Costs, benefits, and a model of optimal feeding frequency. *Oecologia* 65:125-137.

O'Connor, R. J. 1978. Brood reduction in birds: Selection for infanticide, fratricide, and suicide? *Animal Behaviour* 26:79-96.

O'Gara, B. W. 1969. Unique aspects of reproduction in the female pronghorn, *Antilocapra americana*. *American Journal of Anatomy* 125:217-232.

Parker, G. A., D. W. Mock and T. C. Lamey. 1989. How selfish should stronger sibs be? *American Naturalist* 133:846-868.

Ploger, B. J., and D. W. Mock. 1986. Role of sibling aggression in distribution of food to nestling cattle egrets, *Bubulcus ibis*. *The Auk* 103:768-776.

Poole, A. 1979. Sibling aggression among nestling ospreys in Florida Bay. *The Auk* 96:415-417.

Poole, A. 1982. Brood reduction in temperate and sub-tropical ospreys. *Oecologia* 53:111-119.

Simmons, R. 1988. Offspring quality and the evolution of Cainism. *Ibis* 130:339-357.

Stinson, C. H. 1977. Growth and behaviour of young ospreys, *Pandion haliaetus*. *Oikos* 28:299-303.

Stinson, C. H. 1979. On the selective advantage of fratricide in raptors. *Evolution* 33:1219-1225.

Williams, G. C. 1966. Natural selection, the costs of reproduction, and a refinement of Lack's principle. *American Naturalist* 100:687-690.

Woodward, P. W. 1972. The natural history of Kure Atoll, northwestern Hawaiian Islands. *Atoll Research Bulletin* 164.

PART V

The Adaptive Value of Social Behavior

Social behavior is of particular interest to humans, perhaps because we are among the most social of organisms ourselves, but also because the evolution of sociality poses some especially challenging puzzles. How can it be adaptive for some animals to share food rather than monopolizing any food bonanza they happen to find? Why do members of some species form family groups while others do not? Why do some family members in some species fail to reproduce, and instead spend most or all of their lives laboring for others? How could such self-sacrificing behavior spread when one would expect natural selection to eliminate fitness-reducing behavioral traits from the populations in which they occur?

We begin with Bernd Heinrich and John Marzluff's look at why ravens share food. Heinrich and Marzluff explain why it is odd that ravens will call others to a spot where they have found a large edible carcass. Why don't the finders just remain silent and keep the location of the food a secret? Heinrich and Marzluff discuss alternative explanations for how some individuals might actually gain by attracting others to carrion. Readers can outline the authors' use of the hypothetico-deductive method (see Woodward and Goodstein's article in Part I), which enabled them to discriminate among these explanations and to conclude that the "selfish" interests of individuals can be served by helping others—under special circumstances.

In addition, ravens sometimes roost communally in the evening, forming another kind of society away from a food source. Heinrich and Marzluff argue that these groups exist because of the advantages that juvenile ravens gain from sharing information about the location of food, and they provide convincing evidence that this is indeed the reason why the young birds are social. They note in passing, however, that the "information center hypothesis" is not the only possible explanation for the formation of nocturnal bird roosts. What alternatives can you develop, and what information would you need to have in order to reject or accept these other hypotheses on the evolution of simple avian societies?

Next, Michelle Pellissier Scott describes another species in which groups form at carcasses. Burying beetles seek out small mammal carcasses, which they bury as food for their offspring. These insects may have the appearance of small automatons, but they are, like many other insects, capable of highly sophisticated social decisions. Under some circumstances, beetles that have discovered a dead mouse will share it with others, but under other conditions, a carcass-finder will attempt to

drive away competitors of the same sex. Try to restate Scott's explanation for these different responses in terms of the reproductive costs and benefits of cooperation for individuals that have found a carcass.

As is true of ravens, the social status of an individual beetle affects its behavior and the gains it can derive from associating with others. How have these birds and insects converged with respect to the behavior of subordinate individuals? More generally, what are the fundamental similarities between Scott's explanation for communal breeding in burying beetles and Heinrich and Marzluff's conclusions about why ravens sometimes share information about food?

In the species examined in the first two articles in Part V, cooperators are not closely related. In some species, however, family groups form (see the article by Getz and Carter in Part IV as well), creating conditions that may be especially favorable for the evolution of certain kinds of self-sacrificing behavior, or altruism. Stephen Emlen, Peter Wrege, and Natalie Demong examine such a family-forming species, the white-fronted bee-eater. This bird is one of a fairly large number of species in which family groups consist of breeding adults and nonbreeding helpers at the nest. These helpers provide food for chicks that are not their own. Unlike the cooperative ravens and burying beetles, helper white-fronted bee-eaters work to increase the number of offspring produced by others while reducing their own direct reproductive success through their altruism.

Altruism poses a special, but not insuperable, problem for evolutionary biologists. As Emlen and his coauthors explain, it is possible for nonbreeding helpers to propagate their genes, provided the genetic cost of not reproducing during a breeding season is outweighed by the increased reproductive success of the relatives they assist. The authors explain the concept of inclusive fitness, and show that white-fronted bee-eaters are extremely skillful decision makers in terms of adopting the behavioral option that will maximize their inclusive fitness. Their analysis demonstrates that it can be adaptive for young birds to be altruists that help their parents.

The special power of the inclusive fitness approach is apparent when one realizes that it can be used to understand conflict, as well as cooperation, between offspring and their parents. After reading the explanation for family disharmony in bee-eater societies, readers may wish to identify the predictions that Emlen and company tested in order to check their hypothesis.

If you now return to Stephen Jay Gould's article in Part III, you can re-read his critique of "self-styled Darwinian evolutionists" who make claims about the optimal nature of behavioral traits while neglecting the forces of history that impose imperfection on all evolved attributes. Would Gould include Emlen and his co-workers (and all the others whose articles appear in Parts IV and V) in this group of misguided scientists? You will note that these researchers make little or no mention of historical constraints on evolutionary change in their articles. Are adaptationists missing the boat when they test hypotheses based on the assumption that animal behavior helps individuals maximize their inclusive fitness? Or can studies of history and of fitness consequences remain apart because they represent different levels of analysis, as Holekamp and Sherman argue (in Part II)?

Rodney Honeycutt's article revisits the issue of altruism and its evolution in a bizarre little subterranean mammal in which helpers at the burrow lead a celibate life—a far more extreme kind of altruism than the facultative and generally temporary helping at the nest of white-fronted bee-eaters. Prior to 1981, the only animals known to have permanent nonbreeding helpers—or workers, as they are more commonly known—were insects: the ants, the termites, some bees and some wasps, a few beetles, several aphids, and a handful of thrips. In these insects, workers may be physiologically sterile, unable to reproduce on their own, and they spend their lives helping their mother produce additional siblings, some of which will reproduce and pass on the genes they share in common with their sterile helpers.

In 1981, the naked mole-rat, a buck-toothed East African rodent, burst on the scene as the first vertebrate known to have a nonbreeding caste of workers. Naked mole-rats are wonderfully odd (and some would say downright ugly) animals that live in colonies of dozens dominated by a single reproducing female and one to three reproducing males. The other members of the colony forage for food and excavate, maintain, and defend the colony's vast subterranean burrow system, but they do not breed. Like ants and termites, as well as other less dramatically altruistic species such as bee-eaters, naked mole-rat workers aid close relatives, with whom they share a relatively high proportion of their genetic material.

But the discovery that helpers, whether permanently sterile or not, assist other family members cannot by itself explain the evolution of helping at the nest or burrow. After all, in all diploid species, full siblings are on average as closely related to one another as parents are to their offspring, and yet relatively few species of animals have nonbreeding helpers. The evolution of reproductive self-sacrifice must require special ecological and demographic factors, as Honeycutt argues. What are these factors, and might some of the same ecological pressures that favored helping in naked mole-rats and bee-eaters also have contributed to the evolution of nonbreeding workers in honeybees? In many human families, some offspring help their parents raise other offspring, their younger brothers and sisters. Is such behavior altruistic? In what ways are humans "eusocial," and in what ways are they not?

We conclude with two articles on honeybees, one by Mark Winston and Keith Slessor and the other by Thomas Seeley. The single most striking feature of a honeybee colony from an evolutionary perspective is that only one female, the queen, reproduces regularly, while all the other female colony members generally fail to reproduce, and instead work to maintain the conditions needed for the survival of the queen and her brood. How do the workers know what to do and when to do it? Both articles answer this question by describing the remarkable communication system that enables tens of thousands of individuals to do all the things the colony needs to survive and multiply.

On the one hand, Seeley makes the case that workers exchange a great deal of information without centralized control from the queen. Winston and Slessor, on the other hand, suggest that the queen bee does indeed have much to say about what her workers do. Queen bees are not physically imposing or aggressive, like naked mole-rat queens, but their mandibular glands produce massive amounts of a potent chemical blend. Some workers receive this pheromone orally from the queen and then distribute it to nestmates, which pass it on to still other nestmates, all of which are reproductively suppressed as a result. Winston and Slessor used traditional experimental procedures to establish precisely what chemicals—and in what proportions—are passed around by the workers. Their painstaking chemical sleuthing paid off when they finally isolated the "essence of royalty," a mixture of three decenoic acids and two aromatic compounds.

The queen's mandibular pheromone and the other mechanisms of communication within the hive produce a remarkable unity of purpose in the colony. The cohesive nature of the workers' activities in helping their queen makes it possible for Seeley to argue that the colony can be considered a superorganism, essentially a single "individual" composed of thousands of units, all of which labor for the same end: the production of relatives capable of reproducing. If all goes well, a colony will succeed in producing a daughter queen who will inherit about half the colony workforce when the old queen and the other half of the workers swarm away to find a new nest site. According to Seeley, the shared goals of workers and queens make it advantageous for workers to accept the chemical signals they receive from their mother, blocking their own reproduction in order to advance their mother's reproduction and their own genetic interests. The sophisticated nature of information flow in the hive, which among other things, leads foragers to work harder when there are more open honeycomb cells, lends credence to Seeley's view.

Workers are, however, no more closely related genetically to their mother than human offspring are to their parents. Perhaps conflicts occur in honeybee colonies, just as they do in families of other organisms. Indeed, as Winston and Slessor suggest, queens may be attempting to "force" workers to follow orders by giving them chemical propaganda, which the workers consume not to accept these orders, but to digest and destroy the signal, the better to do what is best for their own genes, which are not all identical to those of their queen mother. According to this view, the degree of cooperation in a colony is far less complete than it appears on the surface, with the queen trying to control a population that may be attempting to defy her.

How would you test the validity of these two competing views of honeybee sociality? If workers are truly cooperative members of a superorganism, would you expect them ever to reproduce? If workers fight against the reproductive suppressing effects of the queen's pheromone, would you expect them to accept the material from other workers? How should selfish workers behave toward a sister-worker that attempts to lay eggs of her own? Because queens mate with many males, workers can be either full or half-sisters. Can you use this fact to design a test of the alternative superorganism versus selfish individual views of honeybee societies?

The behaviors of ravens, burying beetles, bee-eaters, naked mole-rats, and honeybees illustrate why social interactions offer such a range of interesting phenomena for evolutionary biologists to explain. The last word has yet to be written on sociality, but the articles in this section demonstrate how the scientific process has been used to identify major questions about social evolution, and how scientists have begun the satisfying task of solving these puzzles.

Why Ravens Share

Young ravens eat regularly, even when food is rare, because they direct one another to food bonanzas and fend off adults by feeding in large crowds

Bernd Heinrich and John Marzluff

In a forest in northern New England, a moose dies in a spruce thicket. Coyotes soon find the dead moose and feed on it at night. The next day, a hungry young common raven discovers this bonanza of food. But the raven does not feed: It circles above the carcass, then flies off. A few days afterward, daybreak reveals a raucously calling string of about 40 ravens, flying in for a feast. Within a week, more than 100 ravens have joined in consuming more than 90 percent of the 1,000-pound carcass.

We have observed this scenario, or scenarios much like it, more than 100 times while studying ravens in the Maine forests. Our findings indicate that most of the birds that come to feed on such a treasure—a lifeline during a harsh winter—learned of its location from the raven that made the original discovery. Such communication might be expected within a closely knit group of related individuals. But a feasting flock of ravens hardly fits this definition: It consists of birds that usually defend exclusive domains, and that wander widely before settling down and eventually mating. Among such animals, ecological and evolutionary theory suggests that a large carcass, such as a moose, should be defended, not shared. After all, it provides a source of food that might last an entire winter for a bird lucky enough to find

Bernd Heinrich is professor of biology at the University of Vermont. He earned his Ph.D. in 1970 from the University of California at Los Angeles. He continues to work on Maine ravens. John Marzluff is a research scientist at Sustainable Ecosystems Institute in Boise, Idaho. He earned his Ph.D. in 1987 from Northern Arizona University. He is currently working on western corvids. Address for Heinrich: Department of Biology, Marsh Life Science Building, University of Vermont, Burlington, VT 05405–0086.

and defend it. Sharing a carcass represents altruism—a selfless act—because a bird that shares a heap of food might starve later in winter.

Nevertheless, when a young raven finds a large supply of food, it brings in other ravens from as far away as 30 miles. We wondered how the birds communicate the location of food, and whether sharing proves truly selfless, or advantageous. We suspected, and found, that juvenile ravens possess immediately selfish reasons for this apparently altruistic act. In fact, food sharing turns out to be a successful strategy for maximizing survival in an environment where food is sparsely and unevenly distributed in space and time, and where young birds must cooperate in order to defend and feed on a carcass at the same time.

Origins of Altruism

According to Darwin's original theory of evolution, altruism cannot evolve because it requires an animal to sacrifice its reproductive fitness to generate benefits for other animals. But nature provides many examples in which an animal behaves in a way that benefits other animals. In some cases, an animal might even sacrifice its own life to help others. These examples generate a theoretical problem, because selfless behavior cannot be transmitted genetically.

In general, self-sacrificing behavior, or helping, buys delayed or hidden benefits. In other words, selfishness lies behind seemingly selfless behavior. An organism, for example, may give up resources, such as a big pile of food, if that favor will be repaid in the future, or if food is being shared with a relative who can pass on the sharer's genes to future generations. A termite soldier, for instance, may sacrifice its life in defending its colony against an ant attack. Al-

though such behavior appears altruistic from a soldier's perspective, it proves selfish from a genetic perspective—provided that a soldier's behavior, on average, enhances the survival of its colony, whose members are closely related, and the propagation of its genes by others.

During winter in New England, ravens might share food because they have difficulty finding carcasses on a regular basis. Finding carcasses is difficult because they are rare, can be hidden in thick brush and often become covered with snow. The short daylight hours of winter make searching even more difficult. Ravens must feed on a regular basis to survive. In addition, a raven cannot penetrate the hide of deer or most other large animals that it eats from, so a raven usually feeds on a carcass that has already been torn apart by a mammalian carnivore or another scavenger, which means that some of the food has already been consumed. When a raven does find a carcass, it may provide enough of a meal for a short bout of sharing, but the carcass will not last long, and the raven soon returns to its main task: finding the next carcass.

Ravens might improve their chances of finding spatially and temporally unreliable food by forming alliances composed of many individuals that search independently for carcasses and then get together when a carcass is found. An individual raven that shares an ample-size carcass might be giving up little because, after the ravens eat their fill, the remaining meat might be consumed by waiting carnivores anyway. By sharing finds, ravens reduce the uncertainty of finding another meal. In fact, the larger and rarer carcasses are, the smaller is the cost of sharing and the larger is the benefit of foraging cooperatively. A carcass hidden under snow or brush may

Figure 1. Pairs of adult ravens in the forests of northern New England establish territories and defend their discoveries of food during the harsh winter. But their control may be short-lived: If a juvenile raven comes upon a defended carcass, it communicates the location to many other young ravens, which soon arrive at the carcass and feed as a group. The adult pair's behavior is well explained by evolutionary theory, but the sharing of food by juveniles poses a challenge to the theory. The authors suspected that the seemingly altruistic behavior of juveniles may actually be rooted in self-interest. (Photographs courtesy of Bernd Heinrich.)

escape the attention of one raven, but having more pairs of looking eyes increases the likelihood that all birds will be fed, and on a continuous basis.

The food-sharing scenario, however, suffers from a potential flaw. If an individual raven refuses to share, that bird gains an immediate advantage. If all ravens share their food, on the other hand, the entire population benefits because they all eat regularly. In other words, the system of food sharing among ravens relies on trust—each bird trusting that all other birds will share. Might ravens be the first truly communistic social organization, based on trust and rationality, where individuals give according to their ability and receive according to their needs?

Scout or Squad

Determining whether ravens really share food and why requires knowledge of how they find food and who eats it. Do single individuals or crowds of ravens usually find carcasses? Answering that question took more than 1,000 hours of experimentation: Putting out a carcass (cattle, deer, moose, sheep and so on), obtained from a farmer or game warden, and then watching it from a snow-covered spruce-fir blind until one or more ravens happened to find the food. A single flying raven or two ravens flying together discovered each of the 25 carcasses put out in this experiment.

A crowd of ravens arrived only after one or two ravens discovered the bait. In addition, the crowd usually came at dawn, and most of birds in the crowd flew in from the same direction. These observations suggest that a crowd of ravens feeding on a carcass does not develop from a flock happening to find it, and a feeding crowd must assemble as a group the night before flying to a carcass. It appears, therefore, that a lone discover or a pair of discoverers brings in the crowds of dozens of ravens from some assembly location. At a carcass, ravens often "yell," or make a special type of vocalization. Playing recorded yells attracts nearby ravens. Clearly, ravens at a carcass do not try to hide the discovery.

Still, we wondered which birds make up a crowd at a carcass. To find out, we captured 463 ravens and marked them with colored and numbered wing tags and radio transmitters. That procedure did not disperse the birds, because the largest number of marked birds could always be found immediately after being released. In addition, we demonstrated that the tags stayed in place for at least two years on birds that lived in a large aviary. Markers on wild birds allowed us to identify ravens feeding at a carcass. On every carcass that we put out, new birds came and "old" birds left on an hourly basis. No two groups of eating ravens consisted of the same individuals, because the composition of a group changed constantly.

Watching these very shy birds from blinds constructed from spruce and balsam-fir boughs in the forest near our baits, we were surprised by the continual turnover of different birds from one day to the next. On some very large baits, for instance, some ravens returned daily for up to six days, but most stayed for only a couple of days, and many stayed for only an hour or less. Up to 500 ravens partic-

Figure 2. Raven yells, attracting its cohorts after discovering a food bonanza. The yell (*see Figure 7*) is different from calls made by infant and fledgling ravens.

Figure 3. Winter forces ravens to survive on a variable supply of food. Carcasses are scarce and can be difficult to locate. The short days of winter reduce the available searching time, and a carcass can be hidden in brush or snow. Furthermore, feeding on a frozen, picked-over carcass does not provide quick satiation. Feeding in crowds may give subadults the ability to retain control of a carcass long enough to satisfy their hunger.

ipated in consuming one "super bonanza," two skinned cows, but usually no more than 50 birds fed at once.

Working with Patricia Parker of Ohio State University and Thomas A. Waite then of Simon Fraser University, we found by DNA fingerprinting that the ravens feeding in a group are not family members that stayed together on a home turf. In fact, ravens that fed together were no more closely related than ravens that fed in separate groups. We concluded that "vagrants"—ravens that have no home domain or that wander over a very large range—make up a crowd of feeding ravens. In fact, one bird that we marked was sighted 200 miles away. Combining those results suggests that sharing food provides little or no opportunity for a raven to promote the propagation of its own genes by sacrificing food to feed a relative. In addition, vagrancy reduces the odds that a raven will receive a future favor by sharing food with an unrelated bird, because the chances of meeting that same bird again are slim.

Individual Eaters

By marking ravens, we discovered two distinct groups: residents and wanderers. The residents, mostly adults, defend any food bonanza located in their domain. The wanderers, mostly juveniles, get little or no access to defended food unless they come in sufficient numbers to neutralize the adults' defenses. In other words, juveniles must recruit larger numbers of birds to get access to a carcass.

Several lines of evidence support the idea that wandering subadults create ad hoc gangs, which allows them to feed at a carcass, even when it is defended by territorial adults. A pair of resident adults may dominate a carcass for weeks and then be swamped by a crowd of young ravens in just one or two days. Second, radio-tagging studies showed that, although residents roost alone, vagrant ravens sleep communally, in a group consisting of from 20 to 60 birds (about the same number that appears at a carcass at one time), and at a site near a carcass. Third, if a juvenile arrives at a carcass, the resident adults make dominance displays and attack the juvenile, which leads to submissive postures and vocalizations by the juvenile. Finally, a juvenile's calls attract other vagrant juveniles.

We tested our gang hypothesis in an

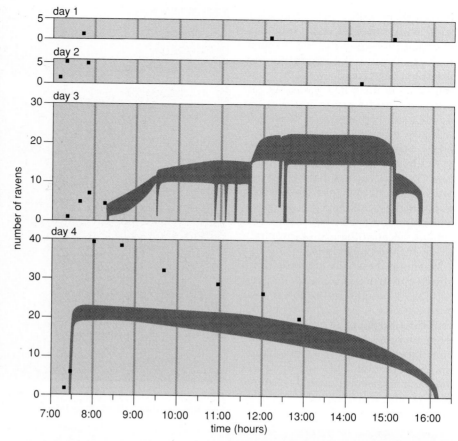

Figure 4. Ravens arrive in a group to feed on a carcass. A sheep carcass, which the authors put out as bait, was discovered by a raven on day 1. (Individual appearances are denoted by *black squares*.) That bird apparently returned several times but did not eat. By day 2, ravens were feeding, but in small numbers. By day 3, a group of ravens (*gray* areas, where thickness indicates variability in the number of ravens in a group) shared the sheep. On day 4, about 20 ravens gathered in the morning and consumed the remainder of the carcass.

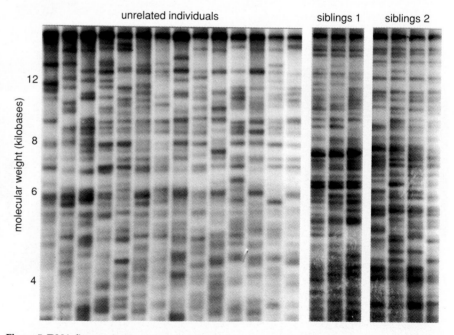

Figure 5. DNA fingerprinting shows that unrelated ravens make up feeding groups. Two groups of siblings possess similar DNA (*right*), but ravens in one feeding group show wide variations in their DNA profiles (*left*), meaning that they are not related. So ravens do not share food in hopes that a feeding relative will pass on shared genes.

Figure 6. Adult ravens *(left)* may control a carcass found in their territory for a week or so, until a juvenile raven flies over and sees the carcass. When the juvenile's signals attract other juveniles, a struggle for control ensues *(right)*. A small group of juveniles cannot overpower a pair of adults. So a juvenile that discovers a carcass must recruit more juveniles, or none of the young birds will eat.

outdoor aviary, which enclosed 7,100 cubic meters. That huge aviary, located at our study site in Maine, consisted of a main area that housed about 20 subadults and side-arms leading to two ancillary aviaries, one of which housed a pair of resident adults that were captured from the wild. From a central observation hut, we could watch the birds through one-way glass, and we controlled access between aviaries by opening and closing gates with guy wires. Our studies in the aviary showed that, as in the wild, adults viciously attacked hungry subadults that attempted to reach food, and it required a gang of at least nine subadults to feed relatively unmolested.

Eating from a frozen, picked-over carcass, however, precludes quick satiation, because a raven cannot rush in and grab a billfull of meat. Instead, a raven can only chip off tiny pieces of frozen meat. Moreover, bones and skin obscure most of the meat. A raven might feed for several hours before becoming satiated. The limited number of choice feeding sites on a carcass might suggest that each feeding bird eats less as the overall number of birds increases. Nevertheless, the average amount of feeding increased with larger group sizes, at least for medium-sized carcasses and larger ones, including deer and cows. In the presence of adults, juveniles benefit even more from a crowd, because the adults attack the dominant subadults, and that distraction allows subordinate

subadults to eat. So feeding in crowds provides most subadults with an immediate advantage: eating more, especially in the presence of territorial adults. As already indicated, sharing also provides a long-term advantage: reducing the patchiness of an otherwise temporally unreliable food resource.

What's in a Yell?

Delving deeper into the mechanism behind food sharing by ravens steered us toward sounds. Ravens possess a large and varied vocal repertoire, which might include recruitment calls. Many of the calls advertise territories or attract mates, but we concentrated on calls that ravens give when crowds feed. The first call, a plaintive "yell," quickly recruits nearby ravens when a recording of it is played in the wild. Given that a yell attracts other ravens and that givers of the call gain feeding privileges (access to defended meat), yelling should lead to a demonstrated advantage. But what causes the yelling behavior?

Manipulating the food available to ravens in our aviary showed that yelling increases as a function of hunger when a subadult sees food. A yell resembles other food-related calls, including one made by young birds when their parents come near them and another call that a female makes when she begs for her mate to feed her while she sits on a nest. In young birds that are out of the nest, a juvenile's yell tells a parent where to find the juvenile,

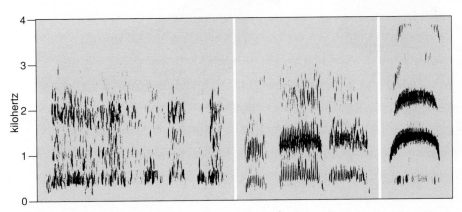

Figure 7. Calls made by juvenile ravens can lead to eating. Before fledging, ravens beg their parents for food with a raspy call, depicted in a sonogram *(left)*. Soon after fledging, a raven's beg possesses more distinct tones *(middle)*. A couple of months after fledging, a raven's beg develops into a so-called yell *(right)* that attracts other juveniles to a discovered food bonanza.

Figure 8. Kettle of young ravens broadcasts an impending move to a new roost. Most days, ravens return to their nearby roost after feeding at a carcass and settle in for the night. After consuming all of a carcass, however, ravens return to their roost and circle over it, soaring as high as 2,000 feet. Then the ravens fly to a new roost, presumably near a new carcass.

as well as which one is hungry. In addition, juvenile yells tell other members of the same brood (after they have left the nest) where their parents and food can be found. The juvenile yell persists for months or years after leaving the nest, presumably because it continues to provide food to a caller. In subadults, however, the call attracts other vagrants, not parents.

One might expect that the loudest yells would come from a lone vagrant located near adults with food. In our aviary, however, lone subadults did not yell when they saw meat near the adults. It turns out that social status affects yelling. Even among subadult vagrants, the most dominant bird does most of the yelling, and it suppresses yelling in others. If we re-

Figure 9. Postures and feather configurations portray a bird's status. A raven at an uncontested food source (*a*) holds its head up and keeps its feathers smooth across its head. A raven first approaching food (*b*) lowers its head. A vagrant at an adult-protected carcass (*c*) keeps its head up and the feathers on its head fluff out. When juveniles swamp a carcass, a resident adult performs a dominance display (*d*), which includes erect posture, raised bill, raised earlike feathers and fluffed-out throat and leg feathers.

Figure 10. Sharing by ravens depends largely on communication. Although resident adults try to defend a carcass *(a)*, a lone juvenile (identified by a *yellow* wing tag) will eventually discover it *(b)*. That juvenile relays the location of the carcass to a roost of other ravens *(c)*, and the group flies to the carcass and overthrows the resident adults *(d)*. That cooperation helps young ravens eat consistently, even when food runs scarce.

moved the most dominant raven, then the second-most dominant one did the yelling. Removing the second in line stimulated yelling by the third, and so on. So hunger stimulates yelling, and socially superior ravens suppress it. A lone vagrant does not yell near a territorial male for fear of being beaten. Dominant ravens in a crowd of vagrants do yell, which tells other vagrants that they might join the

feeding with little interference from defending territorial adults.

Subadult ravens make another call that also attracts other birds. In our aviary, when a subadult approached adults with food, the adults attacked, and the subadult made a begging call that attracted the rest of the subadults. Playing recorded begs also attracted subadults. Begging calls also reduce

adult aggression, probably because an adult wants to limit the calling that recruits other subadults.

Although both yelling and begging attract other ravens, calls near a carcass attract only nearby individuals. When a carcass is first discovered and few vagrants happen to be in the area, yelling and begging would not attract the crowds that gather. In fact, when we broadcast yells in a forest, we did not attract any ravens, presumably because their low density reduces the chance of being within a couple of miles of a bird.

Roost Reporting

Nocturnal roosts provide the primary crowd-forming capability of ravens. Subadult ravens often sleep in communal roosts that form at dusk in pine groves that lie within a few miles of a food bonanza. In general, most or all of the ravens from a roost leave *en masse* at dawn and fly directly to the food. Throughout the day, ravens feed at a carcass and also disperse, possibly looking for other carcasses, but only feeding on one. In the evening, the ravens return to a roost, coming in from many directions. According to our radio-tagging studies, ravens often join different roosts on different nights. After depleting a carcass, the ravens either disperse or move on to another source of food.

In 1971 the Israeli biologist Amotz Zahavi proposed that communal bird roosts serve as information centers, and that hypothesis spawned a long debate in the ornithological literature. Our results prove that they do. When we released naive ravens (which had been held in captivity for at least several weeks) in the evening and near a roost, they immediately joined the strangers. The following morning, the naive ravens appeared at a carcass that the roost birds were eating. By contrast, other naive ravens that were released in the evening and at equal distance from a carcass but not in the vicinity of a roost did not appear at the carcass. These results suggest that wandering ravens need merely locate a roost to be led directly to food.

Carcass size probably regulates roost size. If a deer carcass, for example, offers room for 15 feeding birds, then additional birds would gain little access without fighting constantly, and so they move on. On the other hand, if a moose carcass provides places for 40 ravens, then the feeding crowd grows, as does the roost. In Maine, roosts usually consisted of less than 50 birds, about the limit that we observed feeding at any time at moose or other large carcasses that we provided. In the western United States, where ravens eat insects and grain in open rangeland, relatively permanent raven roosts of more than 1,000 individuals exist.

Although most ravens feed at a carcass for only a short time, often not returning again until days or even weeks later, the number of birds feeding on a carcass remains high, suggesting that many birds know of a carcass's location. In addition, ravens probably know the location of more than one carcass at a time. So if members of a roost know about several different potential feeding sites, why do all the ravens in a roost go to the same carcass?

Soaring

Several observations suggest that roost relocation depends on consensus. We routinely climbed to the tops of tall spruce trees for the panoramic view necessary to watch roost formation in late afternoon, from two hours before dusk until dark. During 328 nights of roost watching, we observed 72 cases in which from 3 to 103 ravens circled over a roost, and then the whole roost rose out of the trees and all of the birds disappeared into the distance. In addition, we also saw a single radio-tagged bird discover a fresh carcass, and then return the following dawn with 20 or more followers. (Such observations are rare because as long as ravens know of a feeding site they have no reason to go to a new site even when a roost member discovers one.)

Spectacular soaring displays usually accompany roost moves. When a carcass has been nearly cleaned up, the ravens return to their old roost in the late afternoon or evening, but instead of flying directly into the trees to roost, as they usually do, the ravens ascend high into the air, sometimes to 2,000 feet or more, and fly in large "kettles." A kettle consists of birds flying noisily, diving and tumbling. The displays last from 15 minutes to more than two hours, and neighboring ravens keep joining in, making the soaring crowd grow. Finally, the growing aggregation stops circling, and then the birds fly in long lines, all traveling in the same direction as they disappear over the horizon. The next day, the ravens do not return to their old carcass, because the new roost probably lies near another feeding site. We have also observed this sequence from the other direction. After putting out bait, we may see only a pair of resident adults for many days. One evening, though, soaring ravens will settle nearby and begin feeding the next day. The resident adults will be swamped, no longer able to defend the meat against vagrants.

Almost all of the hundreds of carcasses that we have provided in Maine over the last 11 years were ultimately shared by crowds of ravens. Birds from the surrounding hundreds of square miles eventually participate in eating a large carcass. Working with Delia Kaye, Kristin Schaumburg and Ted Knight, for example, we radio-tagged 10 birds and then attempted to find them on a daily and nightly basis for two months. Most of those birds ranged over more than 1,000 square miles. In another experiment, we spread 10 carcasses over a linear distance of 30 miles, and nine of the 10 baits were eaten in turn by the marked birds and others.

Our data show that carcass-sharing behavior by ravens did not evolve because of altruism acting through intelligence and foresight, or generosity. Instead, ravens share because their system serves the common good by harnessing self-interest, not suppressing it. That combination of self-interest and common good gives the common raven—in the forests of New England and presumably elsewhere—a large edge over all other species in harvesting a rich resource of food, which is not available on a steady basis to any other bird. Surprisingly, harnessing the most selfish of motivations in an extremely aggressive species creates amazing cooperation for the common good.

Bibliography

Heinrich, B. 1988. Winter foraging at carcasses by three sympatric corvids with emphasis on recruitment by the raven, *Corvus corax. Behavioral Ecology and Sociobiology* 23:141–156.

Heinrich, B. 1989. *Ravens in Winter.* New York: Summit Books.

Heinrich, B. 1993. A birdbrain nevermore. *Natural History* 102:51–56.

Heinrich, B. 1994. Does the early bird get (and show) the meat? *The Auk* 111:764–769.

Heinrich, B., D. Kaye, T. Knight and K. Schaumburg. 1994. Dispersal and association among a "flock" of common ravens, *Corvus corax. The Condor* 96:545–551.

Heinrich, B., and J. M. Marzluff. 1991. Do common ravens yell because they want to attract others? *Behavioral Ecology and Sociobiology* 28:13–21.

Heinrich, B., J. M. Marzluff and C. S. Marzluff. 1993. Ravens are attracted to the appeasement calls of discoverers when they are attacked at defended food. *The Auk* 110:247–254.

Marzluff, J. M., B. Heinrich and C. S. Marzluff. (in press). Raven roosts are mobile information centers. *Animal Behavior.*

Parker, P. G., T. A. Waite, B. Heinrich and J. M. Marzluff. 1994. Do common ravens share ephemeral food resources with kin? DNA fingerprinting evidence. *Animal Behavior* 48:1085–1093.

Communal Breeding in Burying Beetles

Groups of males and females may engage in reproductive cooperation, sharing resources and caring for each other's offspring

Michelle Pellissier Scott

When a young chipmunk dies in the woods, a wide variety of organisms—from mammalian scavengers to insects, fungi and microbes—compete for this sudden food bonanza. In New England, burying beetles rank as one of the most successful competitors. A chipmunk is a relatively large carcass for these 1.5 centimeter–long insects, and such a food source may be discovered and buried by various combinations of adult burying beetles. As they bury the carcass, they shave it and roll it into a ball, readying it to become food for a single large brood of young, or grubs. Once finished, some of the adults may remain in the brood chamber to tend to the carcass and to feed and defend the larvae until their development is complete.

Among biologists, such an example of apparent cooperation excites special interest, because this behavior appears at first glance to oppose an individual beetle's best interest. For instance, it seems that a dominant beetle could monopolize a carcass, thereby leaving more offspring of its own, rather than sharing it with another beetle of the same sex. Likewise, a subordinate beetle could look for its own carcass, rather

Michelle Pellissier Scott is an associate professor of zoology at the University of New Hampshire. She earned a Ph.D. from Harvard University in 1984 and was a science fellow at the Bunting Institute of Radcliffe College and a research associate at Boston University before going to New Hampshire. She has been investigating the reproductive ecology of burying beetles since she and a colleague discovered one trying to bury a mummified mouse in the walls of an old farmhouse. Her research has focused on the evolution of male parental care and more recently on cooperative associations of females. Address: Department of Zoology, University of New Hampshire, Durham, NH 03824. Internet: mps@christa.unh.edu.

than devote valuable time to helping to rear offspring, most of which are not hers. Societies in which individuals undertake cooperative activities, such as communal breeding, usually also include conflict. Individuals are selected to cooperate only when they gain more than they lose from participating. The balance between cooperation and conflict—between feeding or killing the offspring of another—is a delicate one, and it depends on many factors, including the relatedness of the participants, the alternative options open to each, the relative efficiency of cooperation and the ease with which one individual can dominate others.

Reproductive cooperation and conflict lie at the heart of the evolution of sociality. Communal breeding, in which adults share a nest or a mate and may provide care to another's young, has been identified as one path leading to complex social systems in both insects and vertebrates. Communal breeding exists in birds, some canids, social insects and only rarely in other groups of animals. Often the participants are related, which gives them some indirect benefit from helping to rear the offspring of a relative.

I have studied the ecology and evolution of reproductive behavior of burying beetles in southern New Hampshire since 1984. They are very informative animals to study, because they are abundant and easy to rear in the laboratory, and because possible causal factors for the evolution of their cooperation can be manipulated experimentally.

Behavior and Ecology
Southern New Hampshire hosts four species of burying beetles. Although they partition their habitat temporally by reproducing at somewhat different

times during the summer, they overlap in the size of carcass that they are able to bury. The smaller species may rear a few larvae on very small carrion, such as a mouse, which is not buried by a larger species. The medium-size *Nicrophorus tomentosus* shows the greatest readiness to breed communally. Cooperating groups of *N. tomentosus* can sometimes bury and prepare larger carcasses that would be difficult for a single pair.

Burying beetles possess highly sensitive chemical receptors in their antennae, which allow them quickly to discover small vertebrate carcasses. Usually a single male and female prepare and bury a carcass. If more than one beetle of the same sex discovers a carcass, they compete fiercely, and usually the larger beetle wins; losers are driven off to nearby vegetation. However, larger carcasses, such as a vole or turkey chick, may be buried by a group of males and females. Arriving beetles first assess the suitability and size of the carcass by walking around its circumference and tasting and lifting it. They may move it to a better spot for burial.

Although beetles can get the carcass out of sight in a few hours, it takes several days to fully bury and prepare a carcass. Females start to lay eggs in the soil nearby 18 to 24 hours after they begin work. These eggs hatch three days later, and the larvae make their way to the top of the carcass, where their parents have cut a hole in its skin. Parents facilitate larval feeding with their proteolytic oral secretions and through direct regurgitation. They also keep the inside and outside of the carcass clean of fungi. Larvae grow very quickly and complete development in seven to nine days, at which time the carcass is generally completely consumed. The du-

Mark Moffett, Minden Pictures

Figure 1. Burying beetles use a small animal carcass as a food source for a brood of beetles. This male and female burying beetle (*Nicrophorus tomentosus*) have discovered a dead mouse, which will be used to feed their offspring. Several factors—including the species of burying beetle, the sex and number of beetles that first discover the carcass and the size of the carcass—determine how the carcass will be used. It may be guarded by a pair of beetles or shared among a group of beetles that breed communally.

ration of maternal and paternal care varies among species, and females usually remain longer than males.

The readiness to breed communally depends, at least in part, on the size of a carcass. No species of burying beetle breeds communally on a small carcass, such as a white-footed mouse. In *N. tomentosus*, greater carcass size leads to an increased rate of communal breeding: 47 percent on medium (about 45 grams) carcasses, such as a voles, and 75 percent on large (about 70 grams) carcasses, such as juvenile chipmunks. Communal breeding may be more common on larger than on smaller carcasses because the size of the carcass largely determines the number of larvae that can be raised, and females are somewhat limited in the number of eggs that they can lay that will hatch

more or less synchronously. Although a female can fully utilize a small carcass by herself, a larger carcass can support many more young than she can produce, and it would also be more difficult for her alone to prepare. Thus a female probably loses less in sharing a large carcass with another female, and both females gain additional assistance.

Although more than one male frequently remains after eggs are laid, females are more likely than males to bury a carcass together. The female that remains in the brood chamber the longest is usually one of the largest. Secondary females may remain after eggs hatch, but secondary males often leave soon after eggs are laid. Rarely do more than a few adults remain in the burial chamber after the larvae appear. Females may be present simultaneously in the burial chamber before eggs hatch, and they do not show any behavioral signs of competition, but observations on brood chambers after larvae are present indicate that two females are seldom present simultaneously to feed and care for larvae. Cooperation, especially in the early stages, may make it possible to raise a larger brood in total, but it is expected to be in the best interests of each female to monopolize reproduction as much as possible.

To understand the origins of different reproductive strategies in these beetles, I had to assess the relative success of each alternative. The first step in that process was identifying which female in a communal nest was the mother of each young beetle. To determine that,

Scott Williams of Boston University and I used molecular techniques to match parents with their offspring. In laboratory broods with only two males and two females present, which simplified the task, we found that 70 percent of the females shared a carcass and reared a mixed brood. Likewise, 70 percent of the males also shared reproduction—each male inseminated one or both females. The division of maternity of these mixed broods was surprisingly varied, ranging from equitable (nearly 50/50) to extremely inequitable. In the 10 experimental broods, the larger of the two females invariably produced more offspring and remained with the brood longer. The larger male sired more young in 9 of 10 cases, and the male with more young always remained longer.

Why Breed Communally?

Resident beetles must defend their carcass against competitors, especially flies and other burying beetles. Flies have access to a carcass before it is discovered by beetles. And other burying beetles can discover a carcass (even after it has been buried four to five centimeters under ground), evict the residents, kill their brood and produce their own young. So it pays to protect a carcass.

One might expect that smaller beetles would be more likely to cooperate in larger groups as a means of protecting a carcass from other beetles. For instance, does the smaller *N. tomentosus* breed communally to defend a carcass against the larger burying beetle, *N. orbicollis*, which also breeds in late summer? Despite the appeal of this idea, my experiments showed that two *N. tomentosus* can protect a carcass nearly as well as four of them can. For example, a foursome of *N. tomentosus* can fend off a female *N. orbicollis*, but a male *N. orbicollis* often defeats two or four *N. tomentosus* residents. When

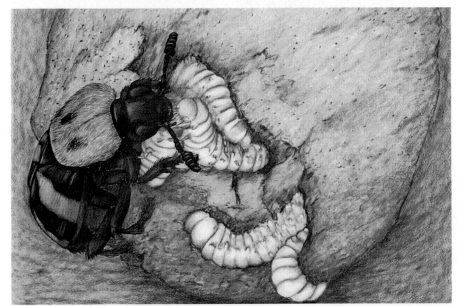

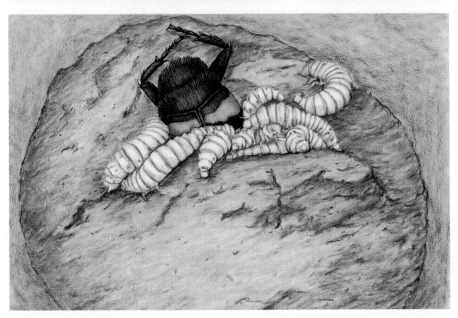

Figure 2. After inspecting a carcass, a pair of beetles shave it and roll it into a ball as they bury it. Beetles spread secretions from their hind gut over the carcass as a preservative, and they keep its surface clean with their mandibles. Young hatch and make their way to the top of the brood ball (*top*). Although larvae of this species are able to survive without parental regurgitations, they are fed by males and females (*middle*). Parents must continue to keep the carcass clean and free of fungi and to treat the inside with proteolytic oral fluids as the larvae consume it (*bottom*).

tested against other *N. tomentosus*, a pair could defend the brood chamber as well as a foursome. So *N. tomentosus* does not appear to breed communally simply because of its small size.

It turns out that the most important advantage of communal breeding comes during the first few days after discovering a carcass, when the extra beetles help destroy fly eggs and maggots, which could consume the treasure. Flies pose a particularly troublesome problem for *N. tomentosus*, which is the only species of burying beetle in New Hampshire that is active during the day. Any carcass that these beetles find has probably also been discovered by flies, thereby reducing its value. A pair of beetles rears fewer young on a medium or large fly-infested carcass than do pairs on clean carcasses or foursomes on fly-infested carcasses.

On average the dominant female benefits most from breeding communally. Parentage analysis on laboratory broods reared on medium carcasses indicates that the dominant female produces 80 percent of the total brood. One might ask, why would a subordinate female—who may be the mother of only 20 percent of the brood—remain and help? Although it may seem like too much work for too little profit, keep in mind that she faces an even less appealing alternative—finding her own carcass and breeding alone. Finding a carcass and breeding probably happens only once or twice in a beetle's short life.

Flies cause a dominant female to gain more than she loses from breeding communally. A communal brood produced in competition with flies may be 32 percent larger on average, compared with the brood of a dominant female reproducing alone. Some of the fiercest competition with flies is in August, when the other species of burying beetles in New Hampshire lose most of the carcasses that they bury to flies and abandon them. *N. tomentosus* rarely loses a carcass to flies.

As one might expect, the duration of parental care does not depend on the presence or absence of competition from flies. The threat from flies ends after the first few days, when the fly eggs and maggots have been destroyed. In an experiment that compared when adult beetles left a carcass, only the secondary male (the first to leave) stayed a little longer on fly-infested carcasses than on clean ones. Given that help

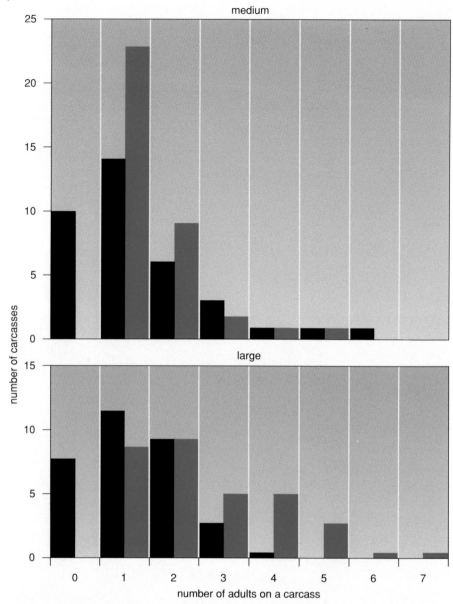

Figure 3. Carcass size affects how many beetles participate in burial. The data in these graphs come from 36 medium *(top)* and 32 large *(bottom)* carcasses, on which I counted the number of males *(black)* and females *(green)* tending the broods. Medium-size carcasses were usually buried by one or two beetles. As the 1 bin shows, I found one female (and some number of males) on 23 carcasses, and 14 carcasses included one adult male (and some number of females). In addition, the 0 bin *(far left)* shows that 10 medium carcasses had zero adult males on them, but no carcass had zero females. Large carcasses were frequently buried by larger associations of adults.

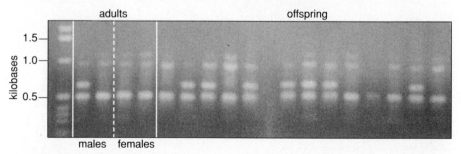

Figure 4. DNA fingerprinting identifies an offspring's parents. Reading from right to left this gel shows segments of DNA—amplified by the polymerase chain reaction—for 13 larvae and their possible parents: two females and two males. In this example, only the first male's DNA reveals a band at 0.7 kilobases, meaning that the seven larvae with the same band must be his offspring. Paternity of the other larvae was resolved with additional markers.

proves crucial in the early stages of preparing and ridding a carcass of fly competitors, a dominant female is especially tolerant at that time of both additional males and females in her burial chamber. That time also serves as the critical period for reproductive competition, when males compete to inseminate females, and females deposit their complement of eggs in the nearby soil.

Skewed Reproduction

Despite the potential benefits of communal breeding, a dominant female must still do what she can to limit the number of offspring of other females and to enhance the survival of her offspring. How reproduction is partitioned among adults has recently come to be considered a key characteristic for describing animal societies. Cooperative breeding in birds and mammals and eusociality in complex insect societies form a continuum in which the distribution of lifetime reproductive success among group members provides an important descriptor. Along that continuum, reproduction may be shared equitably, or it may be skewed in favor of some and at the expense of others.

These communally breeding burying beetles share many characteristics with "primitively" social insects, including paper wasps, which form relatively small groups and found colonies with a single female or with a few females that may or may not be related. These insects may establish reproductive dominance through a number of mechanisms, including producing more eggs and then protecting them. Competitors can also be excluded from egg-laying sites, and they can be prevented from engaging in the behavior that promotes egg laying. Social insects with large and complex societies, such as honey bees, are more likely to achieve and maintain reproductive dominance through pheromonal and behavioral suppression of ovarian development.

In order to get a large sample of broods from which to measure the partitioning of reproduction and to investigate how a dominant female might skew reproduction in her favor, I fed one of the female burying beetles in each brood chamber a colored dye that passed quickly to her eggs. When I dug up the brood chamber and surrounding soil and located the eggs, I was able to estimate the number of offspring from each of two females.

Mark Moffett, Minden Pictures

Figure 5. A buried carcass may be discovered by other burying beetles. This male (*N. orbicollis*) has driven off the resident parents (*N. tomentosus*) and is killing their larvae. Later, he will attract a mate, and they will produce their own brood on the remains of this carcass.

Four days after the carcass was buried, just before eggs were expected to hatch, the distribution of eggs laid by two females on a medium carcass was equitable, or at least not statistically different from random, in 42 percent of the broods. The rest of the broods, however, displayed either significant skew or complete monopoly, usually by the larger female. (On large carcasses, reproduction was almost never statistically different from equitable.) Nonetheless, the total number of eggs in these shared broods was significantly less than the sum expected from single females, suggesting that females either suppressed each other's rate of egg laying or that they destroyed each other's eggs.

When females are searching for carcasses, their mature ovaries contain partially developed eggs. After locating a carcass, a female's behavior of as-

sessment and burial cause her eggs to rapidly finish development. Although the dominant female does not prevent the subordinate's access to the carcass to slow egg development, cooperative burial does significantly stimulate ovarian development for the dominant beetle and suppresses it for the subordinate one. These physiological effects may be mediated hormonally through the behavior of the dominant and subordinate females. The difference in ovarian development, however, is short lived, and both females lay eggs within 24 hours.

Over the next two days, the total number of eggs present in the soil decreases dramatically. Apparently, marauding females can recognize kin and destroy the eggs of competitors. Both the larger and smaller female destroy each other's eggs in experimental broods, but the larger usually protects her eggs significantly better, even while destroying the smaller beetle's eggs. When eggs laid by two communally breeding females were identified and counted two days after the dye feeding, the reproductive skew resembled that seen when both females were still present on the fourth day, just before eggs hatched. When the larger, and presumably dominant, female was removed on the second day, the smaller female significantly skewed the brood in her favor by the fourth day. This strongly suggests that females can recognize and destroy another beetle's eggs, and that egg destruction is the most important mechanism for skewing reproduction. Once eggs hatch, adults apparently cannot recognize kin in order to differentially kill unrelated larvae. In fact, adults feed larvae indiscriminately.

Shaping the Skew

Several factors may be important in determining whether reproduction is shared equitably or skewed in favor of a dominant beetle. In many animal societies, one individual may help to rear kin and even forgo reproduction. This seemingly altruistic behavior provides an evolutionary advantage in lifetime fitness through the indirect gain of rearing kin, with whom some genes are shared. Nevertheless, even unrelated communally breeding females should not necessarily exhibit equitable reproduction. The proportion of the brood that a female captures should depend on the probability and relative success of independent breeding and the relative fighting ability of the communally breeding females.

Although I have never been able to measure the probability of finding a carcass and breeding in a natural population, the intense competition over carcasses suggests it is a rare event. I have estimated that an average beetle might breed only once, possibly twice, in its lifetime. So the probability is low that independent breeding can serve as an alternative to joining a communal association. Moreover, the lower this probability is, the more a dominant female can take advantage of a subordinate one.

Relative size correlates strongly with relative fighting ability in burying beetles, making it a good predictor of which female will be dominant. Nevertheless, relative size does not predict whether a subordinate female will get close to half or nearly none of a brood. My laboratory experiments showed that the larger female was reproductively dominant about 83 percent of the time, and was the mother of 68 percent and 59 percent of the brood on medium and large carcasses, respectively. But the size ratio between the two females did not predict whether the brood was shared more or less equitably, whether it was strongly skewed or whether one female was excluded completely. Once relative status is established there is probably little chance of reversal. So the subordinate is not expected to fight for a bigger share of the brood, unless the probability that she could win exceeds the probability that she could breed independently.

Although relative size is an important determinant in the establishment of dominance, other factors play a role as well. As mentioned above, the discovery of a carcass and its assessment and preparation cause rapid hormonal changes that trigger speedy egg maturation. These endocrine changes probably produce behavioral effects, too. If a female begins to work on a carcass and is joined in two to three hours by a larger female, the smaller female has a greater chance of being reproductively dominant than if the two had discovered the carcass at the same time. In the wild, then, chance events—such as the order in which beetles discover a carcass—must be factored in with ecological parameters—including carcass availability and the degree of competition from other species—to evaluate the outcome of competition between female burying beetles.

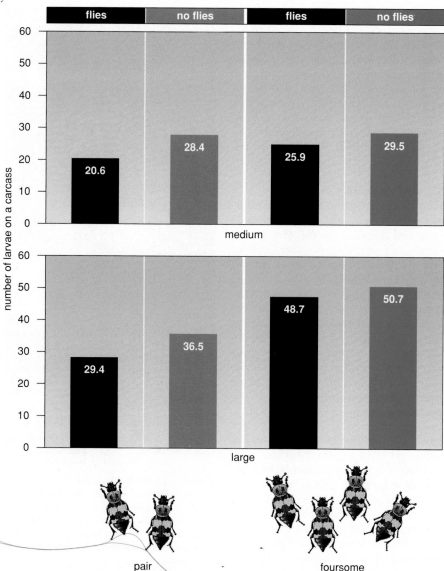

Figure 6. Cooperative associations provide an advantage over pairs when flies have laid eggs on a carcass. On both medium-size *(top)* and large *(bottom)* carcasses, pairs rearing broods on fly-infested carcasses had fewer young on average than did pairs on clean carcasses or foursomes on flyblown carcasses. On medium carcasses, the presence of flies had a significant effect on average brood size for both pairs and foursomes, but on large carcasses, foursomes reared more offspring than did pairs regardless of competition from flies.

Figure 7. Eggs can be matched to the female that laid them by first feeding females mealworms injected with a fat-soluble dye. This dye passes immediately to the eggs, seen here dyed pink and blue. (Photograph courtesy of the author.)

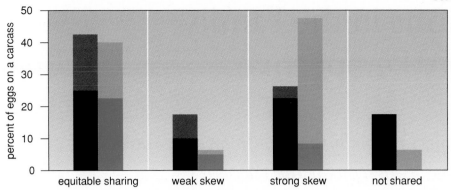

Figure 8. Dominant females can skew broods in their favor. When two females are present in a brood (*gray*), **both of them usually have about the same number of eggs—equitable sharing—before hatching. In some cases, however, a smaller female** (*light gray*) **may not do as well; the larger, and presumably dominant, female** (*dark gray*) **can gain a slight edge (weak skew), a significant edge (strong skew) or even monopolize reproduction (not shared). In another set of experiments** (*green*), **the author removed the dominant female soon after eggs were laid. In some cases, the remaining female** (*light green*) **skewed reproduction in her favor by destroying many of the dominant female's eggs** (*dark green*).

number of larvae that can be raised. Nevertheless, only a fine line separates cooperation and conflict, and reproductive competition produces extremely variable results.

Acknowledgments
The author wishes to thank the National Science Foundation, which has supported this research, and Steve Trumbo, Kern Reeve and Steve Rebach for their helpful comments.

As these results show, the evolution of cooperation among nonrelatives depends on a net gain for each of the participants. A subordinate female burying beetle, for example, joins another as long as she can produce some offspring. Rather than laying eggs and deserting, a subordinate female beetle remains to protect her own eggs and to destroy those of other females. After eggs hatch, the subordinate female continues to feed and guard all larvae indiscriminately, as long as her assistance increases the probability that the brood will survive. A dominant female faces a different choice: She can evict subordinates or allow them to remain and limit their share of the brood to the extent possible. She has the least to lose on a large carcass, because she may not be able to fully utilize it alone, and assistance from other adults increases the

Bibliography
Eggert, A. K., and J. K. Müller. 1992. Joint breeding in female burying beetles. *Behavioural Ecology and Sociobiology* 31:237–242.

Scott, M. P. 1994. Competition with flies promotes communal breeding in the burying beetle, *Nicrophorus tomentosus*. *Behavioural Ecology and Sociobiology* 34:367–373.

Scott, M. P., and J. F. A. Traniello. 1990. Behavioural and ecological correlates of male and female parental care and reproductive success in burying beetles (*Nicrophorus* spp.) *Animal Behaviour* 39:274–283.

Scott, M. P., and S. M. Williams. 1993. Comparative reproductive success of communally-breeding burying beetles as assessed by PCR with randomly amplified polymorphic DNA. *Proceedings of the National Academy of Science* 90:2242–2245.

Trumbo, S. T. 1992. Monogamy to communal breeding: exploitation of a broad resource base by burying beetles (*Nicrophorus*). *Ecological Entomology* 17:289–298.

Making Decisions in the Family: An Evolutionary Perspective

The complex social interactions in a family of white-fronted bee-eaters are governed by some simple rules of reproductive success

Stephen T. Emlen, Peter H. Wrege and Natalie J. Demong

The family has been the fundamental social unit throughout much of human evolutionary history. For countless generations, most people were born, matured and died as members of extended families. However, human beings are not the only animals that form such social structures. Some of the most outstanding examples can be found among birds, of whom nearly 300 species form social bonds that are unquestionably recognizable as family units. In most cases, the family appears to play a crucial role in the socialization and survival of the individual.

The significance of the family to the development of the individual is not lost on biologists, who are inclined to ask whether certain social interactions between family members might be better understood in an evolutionary framework. Given the intensity of the interactions within a family, it is natural to expect that natural selection has shaped many of the behaviors that emerge. Could the same forces that act on birds act also on the human species? Such questions are controversial but compelling.

The evolutionary framework that is used to understand most social interaction is the theory of kin selection, for-

Stephen Emlen is a professor of animal behavior at Cornell University. His research focuses on cooperation and conflict in animal societies and on animal mating systems. He has also worked on the orientation, navigation and acoustic communication of birds. Peter Wrege is a research associate at Cornell where he received his Ph.D. in 1980 for his studies on the social foraging strategies of the white ibis. Natalie J. Demong is a freelance writer and photographer who specializes in avian field studies. Emlen's address: Department of Neurobiology and Behavior, Mudd Hall, Cornell University, Ithaca, NY 14853-2702. Internet: ste1@cornell.edu

malized by William D. Hamilton in 1964. Hamilton emphasized that individuals can contribute genetically to future generations in two ways: directly, through the production of their own offspring, and indirectly, through their positive effects on the reproductive success of their relatives. This is because a relative's offspring also carry genes that are identical to one's own by virtue of common descent. The closer the genetic relationship, the greater the proportion of shared genes. The sum of an individual's direct and indirect contributions to the future gene pool is his or her inclusive fitness.

Because of this genetic relatedness, the social dynamics of family life is expected to differ in significant ways from the dynamics of other types of group living. The degree of kinship is predicted to influence the types of behavior exhibited among individuals. All else being equal, closely related individuals are expected to engage in fewer actions that have detrimental reproductive consequences for one another, and more actions with beneficial reproductive consequences. Although we expect significant amounts of cooperation within families, we must also recognize that not all familial interactions will be harmonious. Kinship may temper selfish behavior, but it does not eliminate it. Individuals will often differ in their degrees of relatedness to one another, in their opportunities to benefit from others, and in their abilities to wield leverage over others. These variables should predict the contexts of within-family conflicts, the identity of the participants and even the probable outcomes.

Human beings are notoriously difficult subjects for such studies because so much of our behavior is sculpted by cultural forces. In contrast, family-

dwelling birds provide excellent opportunities for testing evolutionary predictions about social interactions among relatives. They have a large repertoire of complex social behavior, yet they have few culturally transmitted behaviors that might confound the analysis. They are a natural system in which to search for fundamental biological rules of social interaction.

It is in this light that we spent eight years studying the white-fronted bee-eaters at Lake Nakuru National Park in Kenya. Our original motivation was to study the altruistic behavior of these birds, in particular their tendency to help others at the nest. We came to realize, however, that the birds simultaneously engaged in a number of selfish behaviors as well. Indeed, the birds displayed a wide range of subtle tactics, some mutually beneficial but others clearly exploitative.

An Extended Family

In biological terms, a family exists when offspring continue to interact with their parents into adulthood. This distinguishes families from temporary child-rearing associations in which young members disperse from their parents when they reach sexual maturity. We can further narrow the definition by stipulating that the parents must maintain a preferential social and sexual bond with each other. The white-fronted bee-eaters of Kenya fulfill these qualifications.

Indeed, the heart of the bee-eater society is the extended family, a multigenerational group consisting of 3 to 17 individuals. A typical family contains two or three mated pairs plus a small assortment of single birds (the unpaired and the widowed). A young bee-eater matures in a group of close relatives,

Figure 1. White-fronted bee-eaters of Kenya provide a culture-free animal model for studying the complex dynamics within a group of closely related individuals—a family unit. Colorful tags on each member of a family allow the authors to document the interactions between specific birds. Such studies reveal that bee-eaters make sophisticated decisions based on the status and genetic relatedness of the individual with which they interact. (Photograph courtesy of Marie Read.)

and most continue to interact with parents, siblings, grandparents, uncles, aunts, nephews and nieces into adulthood. Families can even include step-relatives (stepparents and half-siblings) when individuals remate after the death or divorce of a partner. As a result, bee-eater families often have very complex genealogies.

About 15 to 25 families (100 to 200 birds) roost and nest together in a colony. The nests are excavated in sandy cliff faces where the birds dig meter-long tunnels that end in enlarged nesting chambers. Late in the afternoon all bee-eaters congregate at their colony to socialize and roost.

During pair formation, one member leaves its own family and moves to that of the other. This dispersal rule reduces the likelihood of within-family pairings; indeed, we have never witnessed an in-

cestuous pairing among bee-eaters. As a consequence, the resident member of the pair continues to live in a network of close genetic kin. As in most species of birds, it is the bee-eater females that usually disperse. A paired female becomes socially integrated into her mate's family, but the genetic kinship links are lacking. Unrelated females are the functional equivalents of "in-laws."

Once paired, bee-eaters are socially monogamous, exhibiting high mate fidelity over years. Divorce rates are low, with the effect that most individuals remain paired to the same partner for life. Both sexes share equally and heavily in all aspects of parental care.

In many respects, the social structure of bee-eaters has similarities to the supposed organization of ancestral human beings, who are thought to have formed long-term pair bonds, who lived in vil-

lages consisting of several extended-family groups, and whose families included both related and unrelated (in-law) members.

Helping Whom?

The most dramatic aspect of bee-eater reproductive behavior is the phenomenon of helping at the nest. Helpers play a major role in almost every aspect of nesting except copulation. Even before breeding begins, helpers aid in digging the nest chamber, a task that may take 10 to 14 days. Helpers also bring food to breeding females during the week in which they are energetically burdened by egg production. After the eggs have been laid, helpers of both sexes undergo physiological changes, enabling them to incubate the clutch. Helpers will defend the young birds for weeks after they are hatched and for several weeks after

Figure 2. Colony of white-fronted bee-eaters in the face of a sandy cliff may contain as many as 25 families, or 200 birds. The birds excavate a nesting chamber at the end of a meter-long tunnel in the wall. The colony serves as a year-round site for nesting and roosting. (Photograph courtesy of Natalie Demong.)

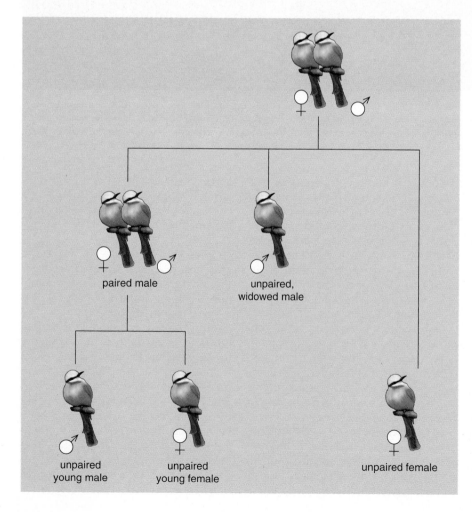

paired male

unpaired, widowed male

unpaired young male

unpaired young female

unpaired female

they are fledged. Helping usually ceases only when the young are completely self-sufficient.

By far the most significant component of helping is providing food for the young. Because the abundance of the bee-eaters' staple food (flying insects) varies unpredictably, the extraparental helpers can have a major effect on a pair's breeding success. In our study, one-half of all nestlings died of starvation before leaving the nest. However, the presence of even a single helper reduced the starvation losses to the point of doubling the fledgling success of an unaided pair!

Because bee-eaters tend to provide aid only to family members (genetic relatives), helpers play a major role in the reproductive success of their nondescendent kin. This means that helpers are indirectly increasing their own inclusive fitness. Interestingly, bee-eater helpers gain no measurable direct benefits from helping. In many other cooperative breeders (species with helpers at the nest or den) the experience of helping often translates into increased personal reproductive success later in life. This is true if the act of helping increases the likelihood that a helper will become a breeder in the future or if helping provides a better breeding slot. It is also true if the experience of helping makes one a better parent in the future. None of these personal benefits accrued to the bee-eater helpers at Nakuru; their helping behavior appears to be maintained entirely through kin selection.

If the major benefit the helpers accrue is through kin selection, bee-eaters should be sensitive to their degree of kinship to different family members. This is indeed the case. When a bee-eater faced the choice of helping one of several relatives, the helper chose to aid the most closely related breeding pair in over 90 percent of the cases (108 of 115).

Kin-selection theory also helps to explain why nearly half (44 percent) of all bee-eaters neither breed nor provide

Figure 3. Extended family of white-fronted bee-eaters may contain three or more generations of birds. Males remain in their natal family after taking a mate and are surrounded by close genetic relatives throughout most of their lives. When females pair they leave their natal group to live with their mate's family and consequently are not closely related to the birds in their new home. The difference of living with or without genetic relatives is associated with striking differences in the social interactions of paired males and females.

help in any given year. There is little profit in helping distant kin. Indeed, most nonhelpers are individuals with no close relatives in their social group. The largest subset of nonhelpers are the females who separated from their own families at the time of pairing. Helping does not increase the inclusive fitness of such females until they have raised fully grown (and breeding) offspring of their own. At this point, they again become helpers, selectively aiding their breeding sons to produce grand-offspring.

On the other hand, the benefits of helping close kin also explain instances in which birds whose own nesting attempts fail, change roles and addresses and become helpers at nests of other breeders in the family. Through such redirected helping, they can recoup much of their lost inclusive fitness. This "insurance" option is typically available only to the males, since they are more likely to be surrounded by close genetic relatives. As predicted, the vast majority (90 percent) of redirected helping involves males. Although females typically relocate to the new nesting chamber with their mate, they rarely participate in rearing unrelated young. The contrasting behaviors of the male and the female are especially striking in the light of all the stimuli—eggs, incubating adults, begging nestlings and attending adults feeding the nestlings—that

would seemingly induce the female to help at the nest.

Coercion by Parents

Since helpers have a large positive effect on nesting success, their services are a valuable resource in a bee-eater family. As a result, we would expect some competition among breeders for a helper's services, and even occasional conflicts between breeders and potential helpers over whether the latter should help. In some instances, helping at the nest might be forcefully "encouraged."

Bee-eaters do, in fact, engage in seemingly coercive behaviors that result in the disruption of nesting attempts of subordinate birds and their subsequent recruitment as helpers at the nest of the disrupter. Older birds will repeatedly interfere with the courtship feeding of a newly formed pair and block the pair from gaining access to its nesting chamber. Both actions increase the probability that the harassed pair will fail to initiate breeding and that the kin-related subordinate bird will help at the nest of the older bird.

The surprise is that the harassing birds are close genetic relatives of the pair they disrupt. Indeed, parents (mostly fathers) are the most frequent harassers; they disrupt the breeding attempts of their own sons. Over half (54 percent) of one-year-old sons whose

parents are breeding fail to breed themselves apparently because they are successfully recruited. This proportion drops as the sons become older and gain in dominance status. By the time sons are three years old, they are practically immune to coercion attempts.

The existence and the resolution of this conflict become understandable when we consider the relatively large net fitness benefit to the breeder and the small net cost to the potential helper when the latter is a son. For one thing, a son is equally related to his own offspring and his parents' offspring (which are his full siblings, provided that no cuckoldry or parasitic egg dumping has occurred). Since an unaided breeder (such as a subordinate son) produces only slightly more young on his own than he does if he contributes as a helper at another's nest, the genetic cost of the tradeoff is minimal to him. Sons apparently do not resist, because the fitness benefits of the two options are nearly equal for them. In contrast, the parents gain considerably more genetic fitness for themselves by using their son to help them increase the production of their own offspring (each of whom shares one-half of a parent's genes by descent) than they would if their son bred and produced grand-offspring (each of whom shares only one-quarter of a grandparent's genes by descent). In

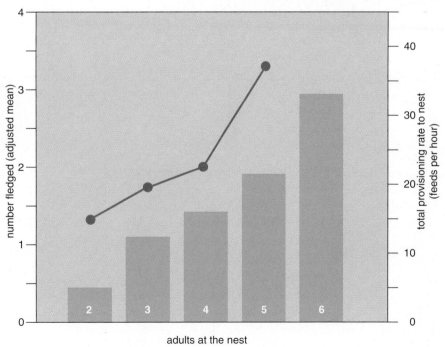

Figure 4. Feeding of juveniles by adult members of a family is crucial to the survival of the younger birds. The number of adults at the nest (*right*) affects the rate at which juveniles are fed (*orange line*), which is closely associated with the number of birds that survive to fledgling status (*green bars*). Here adults bring food to nestlings waiting within the tunnels (*left*). (Photograph courtesy of Natalie Demong.)

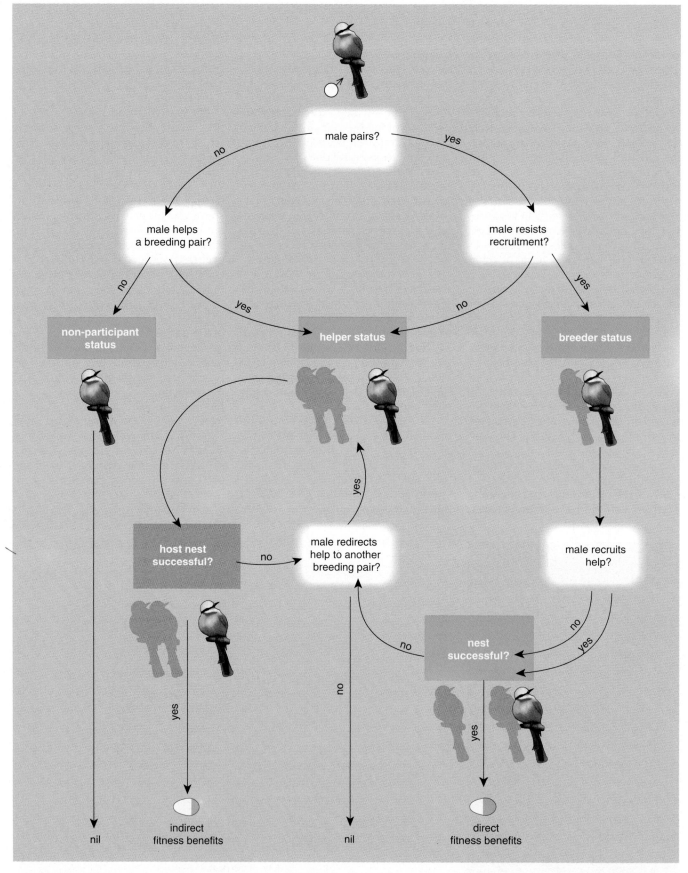

Figure 5. Male bee-eater faces a number of decisions every breeding season that influence his direct and indirect reproductive fitness. A male gets *direct* **fitness benefits (corresponding to 0.5 units for each offspring produced,** *green half egg***) by acquiring a mate and tending a successful nest. A male gets** *indirect* **fitness benefits (varying from near zero to 0.5 for each offspring produced,** *green quarter egg***) by helping a close relative raise young, rather than breeding on his own. At each decision point the male generally chooses the option that maximizes his inclusive fitness.**

this light, the harassment of the son by the parents makes evolutionary sense.

Other members of the family find themselves in a very different situation. Although a breeder will always gain by recruiting a helper, the cost to the helper increases dramatically when he or she is more distantly related to the harasser. Potential helpers who are distantly related to the harasser should, and do, show much greater resistance to recruitment attempts. An older dominant bird can exert leverage over a younger subordinate, but only to a point. It is not surprising then that harassers preferentially select the youngest, most closely related male family members as their targets.

The Female's Options

Since female bee-eaters break the social bonds with their natal families when they pair, their choice of reproductive options differs from those of male bee-eaters. For one thing, they largely forfeit the ability to obtain indirect benefits by helping. Unlike her mate, a female's inclusive fitness (after pairing) depends almost entirely on her success in breeding.

Since a female bee-eater lives with her mate's family, her breeding success is strongly affected by the composition and social dynamics of his family. The likelihood that the new pair will have helpers of its own, or will be able to breed unharassed by others, depends on the male's social and genealogical position within his family. We would expect females to incorporate social components of male quality in their mating choices. Females should pay attention to the prospective mate's social dominance and to the nature of his kin, who may be potential helpers or harassers.

These predictions have been confirmed. Widowed or divorced, older males with offspring of their own were nearly twice as likely to become paired as were young males with older close relatives. The older males were more likely to provide the pairing female with helpers at her initial nesting, whereas younger males were more likely to have their initial nesting disrupted.

Unpaired females who postpone the decision to take a mate retain the option of gaining indirect benefits from helping members of their natal family. Females with close breeding kin should be more likely to remain single. Again, this prediction was borne out: Females with both parents breeding were nearly twice

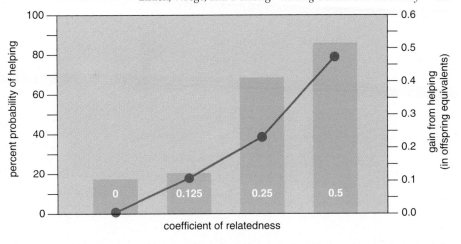

Figure 6. Helper's gain in indirect reproductive benefits *(orange line)* is proportional to the degree of genetic relatedness between the helper and the juvenile being helped. Not surprisingly, the degree of genetic relatedness is a strong predictor of the probability *(green bars)* that one bee-eater will help another.

as likely to remain single as were females with only distantly related breeders in their family.

Females appear to be making a very sophisticated assessment of their options. They act as if they compare the expected benefits of helping versus breeding. We compared the females' actual decisions to those predicted on the basis of the expected benefits given their circumstances (the identity of their breeding natal relatives and the status of their chosen mate within his family). We found that more than 90 percent of the females (67 of 74 cases) behaved as our model predicted. They paired when a potential mate was in a social position

that provided a net increase in their expected inclusive fitness benefits, but they remained in their natal families when their benefits were greater as unpaired helpers. For many females it is better to delay breeding for a season than to accept a mate of poor social standing.

After pairing, a female bee-eater is faced with another series of reproductive choices. If she succeeds in mating with a male in good standing, her problems are solved. But what options remain if her nesting attempts end in failure? Returning to her natal family to help at the nest seems to be an obvious choice, but we have seen this be-

Figure 7. Aggressive interactions take place when dominant bee-eaters attempt to recruit subordinate relatives to help raise the aggressor's offspring, rather than permit the relatives to breed on their own. Most coercive interactions take place between a father and a young son. The existence and the outcomes of such conflicts can be predicted on the basis of the reproductive benefits to the individuals. (Photograph courtesy of Natalie Demong.)

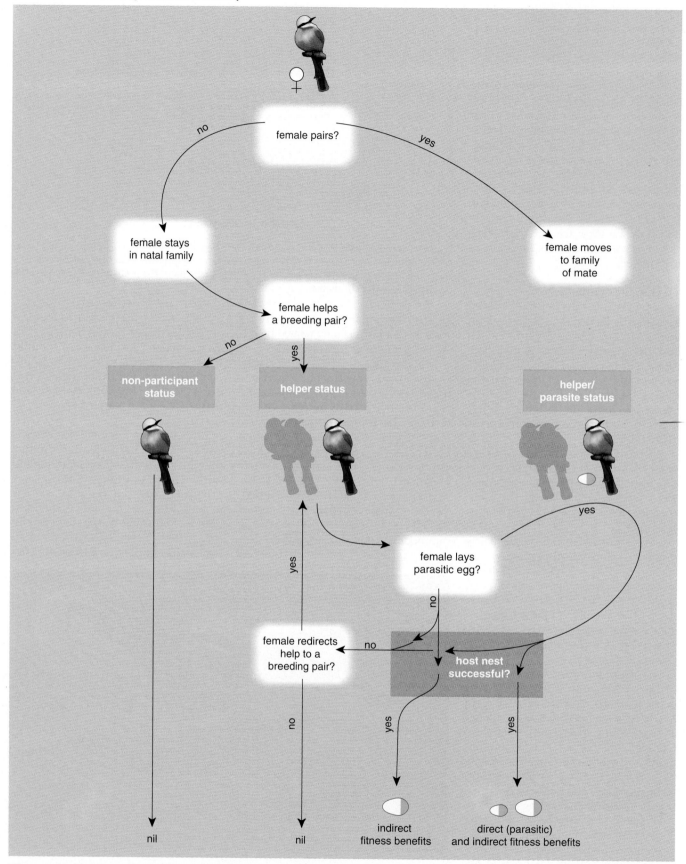

Figure 8. Female bee-eaters that choose not to mate during a breeding season stay in the natal family group. If they choose not to help during a breeding season, their net fitness benefit is nil. If they choose to help at a relative's nest, they receive *indirect* fitness benefits. Occasionally an unpaired female may copulate with a neighboring paired male and then return to the natal nest. If such a female lays a fertile egg in the nest of the relative she is helping she will receive both *direct (small, pink half egg)* and *indirect* fitness benefits (*large, pink quarter egg*).

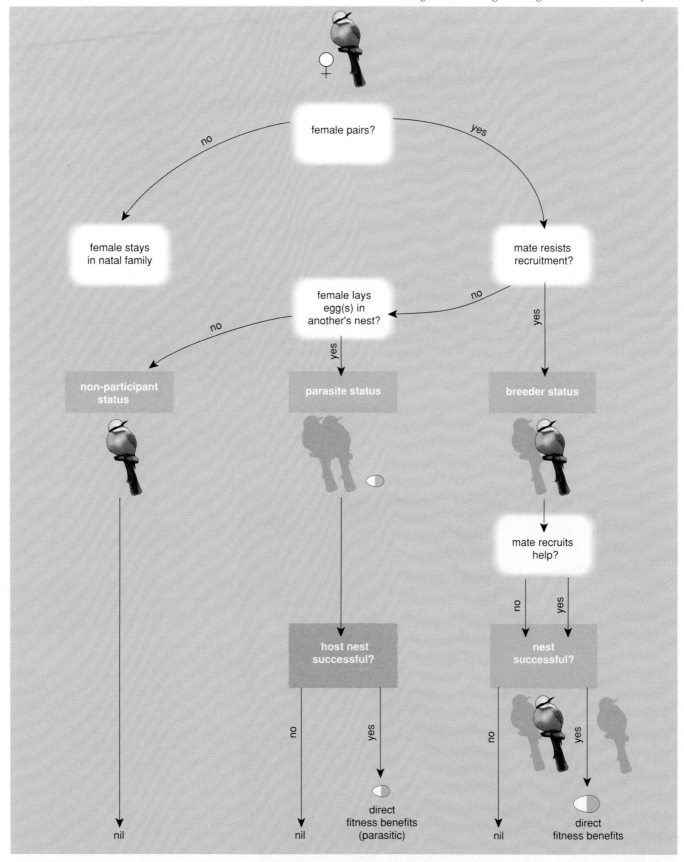

Figure 9. Female bee-eaters that take a mate leave the natal group to live with the male's family group. If the female's mate resists recruitment by his relatives, the pair can establish their own nest and receive *direct* fitness benefits (*large egg*) by having offspring. If the female's mate does not resist recruitment, a female has the option of parasitically laying her egg(s) in another's nest and still gaining direct fitness benefits (*small egg*). Parasitism is a relatively frequent tactic, with about 16 percent of nests containing a parasitic egg. If the female's mate is recruited, the female usually remains in her mate's family group but does not help raise the offspring of her mate's relatives.

factors in a female's decision whether to mate or stay in her natal group	
concerns within natal family	• kinship relationships to breeders • natal group size
concerns within mate's family	• mate's kinship relationships • mate's family group size • mate's age • mate's relative dominance poisition

Figure 10. Female bee-eaters appear to make a very sophisticated assessment of their potential fitness benefits when they decide whether to take a mate or stay at home during a breeding season. A female must weigh the indirect fitness benefits she might gain by helping at the nest of close relatives in her family group compared to the potential for direct fitness benefits she could acquire by taking a mate and producing offspring of her own. She must assess whether the potential mate can recruit help from his family on the basis of his genetic relationships and his dominance status, or whether he is likely to be recruited to help at the nest of other members of his family. On the basis of these factors the authors were able to correctly predict the females decision in 67 of 74 cases.

havior only a handful of times in eight years of studying these birds. We can only speculate that returning home entails some hidden costs. One possibility is that a prolonged separation from one's mate increases the risk of dissolving the pair bond.

It turns out that a female who fails at nesting has another option. If nest failure takes place while she is still at the egg-laying stage, she can deposit her remaining egg(s) in the nest of another bee-eater. The large number of active nests in a colony provides ample opportunity for such parasitic behavior. Indeed, parasitism was common in our study populations: About 16 percent of bee-eater nests were parasitized, and 7 percent of all eggs were laid by foreign females. Despite its frequency, parasitism is a low-yield tactic: Parasitic females usually lay only one egg, and many of these do not survive.

The reproductive costs and benefits of parasitism have resulted in behavioral adaptations by parasites and counter-adaptations by hosts. Breeders and helpers take turns guarding their nests against all trespassers, and breeding females actively remove foreign eggs found in their nest chambers until they have laid their own first egg. Parasites must locate a potential host at the appropriate stage in the nesting cycle and gain access to the chamber when there is a lapse in the host's defenses. Eggs laid too early will be removed, whereas eggs laid too late will fail to hatch before incubation ends.

There is an interesting twist to the story of parasitism among bee-eaters: Not all parasitic females are unrelated to their hosts. About one-third of the parasites are unpaired daughters who were assisting at the nest of their parents. The parasitic daughter actively defends the nest against nonfamily members, but she slips an egg of her own into her mother's (or stepmother's) clutch. In one instance, a daughter removed one of her mother's eggs before laying her own in its place. These intrafamilial parasites remain active as helpers at the nest, sharing in incubation and providing food.

How are these females fertilized? We have watched a few such daughters closely: They actively trespassed onto the territory of a neighboring family, where they solicited a copulation with a paired male! Thus their eggs are not the result of an incestuous mating. Rather, parasitism by a daughter appears to be a tactic involving a specific series of complex behaviors.

Intrafamilial parasitism offers a single female the option of achieving *direct* fitness benefits in addition to the *indirect* benefits gained by helping. However, the daughter's gain comes at the expense of the parent. It is not clear whether the parents tolerate their daughter's egg dumping to retain her as a helper or whether the daughter is

Figure 11. Breeding female actively removes eggs laid by parasitic females before she initiates her own clutch. A successful parasite must overcome the host's defenses *and* lay her eggs within the two- or three-day period that the host lays her own eggs. (Photograph courtesy of Marie Read.)

surreptitiously taking advantage of her parents. In either case, the existence of this form of parasitism underscores the flexibility of the bee-eaters' reproductive options and the subtle conflicts that take place in this species.

Conclusion

The tactics that individuals use in their interactions with one another have only recently become the subject of evolutionary analysis. This is because the expression of social tactics is very plastic: Most organisms can adopt a variety of roles according to the situation and the identity of the other participants. Early workers found it difficult to reconcile this plasticity with the view that specific genes literally determine specific behaviors. It is now recognized that natural selection can operate on the *decision-making process* itself.

As long as there is heritable variation in the decision rules that the birds use, natural selection will favor variants that result in the expression of situation-dependent behaviors that maximize the inclusive fitness of the actor. One of the pioneers of this approach, Robin Dunbar of the University College of London speaks heuristically of organisms as "fitness maximizers." They make decisions based on their ability to assess the costs and benefits of the options available to them.

We have observed that bee-eaters behave as if they assess the relative costs and benefits of pursuing different options in very complex social situations. Gender, dominance and kinship all influence the fitness tradeoffs of the various tactical alternatives available to bee-eaters. Knowing these variables allows us to predict with considerable accuracy whether an individual will attempt to breed, whether it will help at a nest and whom it will help. We can also ascertain whether a bird will be harassed and whether harassment will be successful. Differences in the behavior of genetic and nongenetic members in an extended family group would remain mysterious if it were not for the explanatory power of inclusive-fitness theory.

Gender, dominance and kinship should be important predictors of family dynamics in any species that exhibits long-term pair bonding, sex-biased dispersal and interactions where one family member can influence the reproductive success of another. Cases of breeding harassment and even reproductive suppression are common features of many species that live in family-based societies. Analyzing the fitness consequences of such behavior from the perspectives of the various participants provides an evolutionary framework for understanding such social dynamics.

Can we learn anything about the dynamics in a human family from the behavior of the bee-eaters? More than any other species, the behavior of human beings is shaped by culture. The rewards and punishments that accompany human social actions are largely determined by society. The currency human beings use in assessing the costs and benefits of a particular tactic is no longer solely based on reproductive fitness. But this does not mean that we do not possess a set of behavioral predispositions based on flexible decision rules that were adaptive in our evolutionary past. Such tendencies would have been molded during our long history of living in extended family groups. It is these underpinnings that surface more clearly in animal studies of family-dwelling species.

A small but growing number of psychologists and anthropologists are incorporating an evolutionary perspective into their studies of human families. Investigation of the roles of nonparental family members in childrearing has focused on the role of siblings (especially the mother's brother) and grandparents as human analogues of helpers at the nest. Martin Daly and Margo Wilson of McMaster University have studied the effects of relatedness (parent versus stepparent) on child abuse. Robert Trivers, now of Rutgers University, has looked at the theoretical basis for parent-offspring conflict. Trivers and his colleague Dan Willard have proposed an evolutionary hypothesis to explain why some parents invest unequally in their sons and daughters.

We believe that an evolutionary framework has great potential for increasing our understanding of the social dynamics of family-based societies. By focusing on the fitness consequences of different actions to different individuals, it provides a functional explanation for why particular behavioral predispositions may have evolved. It also provides a theoretical basis for predicting the social roles that different individuals will adopt under differing circumstances. We fully expect that the same general variables found to be important predictors of bee-eater behavior—gender, dominance and kinship—will be important predictors of cooperation, conflict and the resolution of conflict, in most other social species, including human beings. We expect that the incorporation of this Darwinian approach into the social sciences will provide a valuable additional perspective to our understanding of human family interactions.

Bibliography

Betzig, L. L., M. Borgerhoff Mulder and P. Turke. 1988. *Human Reproductive Behaviour: A Darwinian Perspective.* Cambridge: Cambridge University Press.

Daly, M., and M. Wilson. 1985. Child abuse and other risks of not living with both parents. *Ethology and Sociobiology* 6:197–210.

Daly, M., and M. Wilson. 1987. Evolutionary psychology and family violence. In *Sociobiology and Psychology,* ed. C. Crawford, M. Smith and D. Krebs. Hillsdale, New Jersey: Lawrence Erlbaum Associates.

Dunbar, R. 1989. *Reproductive Decisions: An Economic Analysis of Gelada Baboon Social Strategies.* Princeton: Princeton University Press.

Emlen, S. T. 1991. The evolution of cooperative breeding in birds and mammals. In *Behavioural Ecology: An Evolutionary Approach,* ed. J. Krebs and N. Davies, pp. 301–337. Blackwell Scientific Publishers.

Emlen, S. T. 1994. Benefits, constraints and the evolution of the family. *Trends in Ecology and Evolution* 9:282–285.

Emlen, S. T., and P. H. Wrege. 1986. Forced copulations and intra-specific parasitism: Two costs of social living in the white-fronted bee-eater. *Ethology* 71:2–29.

Emlen, S. T., and P. H. Wrege. 1988. The role of kinship in helping decisions among white-fronted bee-eaters. *Behavioral Ecology and Sociobiology* 23:305–315.

Emlen, S. T., and P. H. Wrege. 1989. A test of alternate hypotheses for helping behavior in white-fronted bee-eaters. *Behavioral Ecology and Sociobiology* 25:303–319.

Emlen, S. T., and P. H. Wrege. 1991. Breeding biology of white-fronted bee-eaters at Nakuru: The influence of helpers on breeding success. *Journal of Animal Ecology* 60:309–326.

Emlen, S. T., and P. H. Wrege. 1992. Parent-offspring conflict and the recruitment of helpers among bee-eaters. *Nature* 356:331–333.

Emlen, S. T., and P. H. Wrege. 1994. Gender, status and family fortunes in the white-fronted bee-eater. *Nature* 367:129–132.

Hamilton, W. D. 1964. The genetical evolution of social behaviour. *Journal of Theoretical Biology* 7:1–52.

Hegner, R. E., S. T. Emlen and N. J. Demong. 1982. Spatial organization of the white-fronted bee-eater. *Nature* 296:702–703.

Smith, E. A., and B. Winterhalder. 1991. *Ecology, Evolution and Human Behavior.* New York: Aldine de Gruyter.

Trivers, R. L. 1974. Parent-offspring conflict. *American Zoologist* 14:249–264.

Trivers, R. L., and D. E. Willard. 1973. Natural selection of parental ability to vary the sex ratio of children. *Science* 179:90–92.

Wrege, P. H., and S. T. Emlen (1994). Family structure influences mate choice in white-fronted bee-eaters. *Behavioral Ecology and Sociobiology* 35:185–191.

Naked Mole-Rats

Like bees and termites, they cooperate in defense, food gathering and even breeding. How could altruistic behavior evolve in a mammalian species?

Rodney L. Honeycutt

Biological evolution is generally seen as a competition, a contest among individuals struggling to survive and reproduce. At first glance, it appears that natural selection strongly favors those who act in self-interest. But in human society, and among other animal species, there are many kinds of behavior that do not fit the competitive model. Individuals often cooperate, forming associations for their mutual benefit and protection; sometimes they even appear to sacrifice their own opportunities to survive and reproduce for the good of others. In fact, apparent acts of altruism are common in many animal species.

It is easy to admire altruism, charity and philanthropy, but it is hard to understand how self-sacrificing behavior could evolve. The evolutionary process is based on differences in individual fitness—that is, in reproductive success. If each organism strives to increase its own fitness, how could natural selection ever favor selfless devotion to the welfare of others? This question has perplexed evolutionary biologists ever since Charles Darwin put forth the concepts of natural selection and individual fitness. An altruistic act—one that benefits the recipient at the expense of

Rodney L. Honeycutt is an associate professor in the Department of Wildlife and Fisheries Sciences and a member of the Faculty of Genetics at Texas A&M University. He began his research on the genetics and systematics of African mole-rats in 1983 during his tenure as assistant professor of biology at Harvard University and assistant curator of mammals at Harvard's Museum of Comparative Zoology. He is interested in the evolution and systematics of mammals and has taught courses in mammalian biology for the past seven years. His research has taken him to regions of Africa, South America, Central America and Australia. Address: Department of Wildlife and Fisheries Sciences, Texas A&M University, 210 Nagle Hall, College Station, TX 77843.

the individual performing the act—represents one of the central paradoxes of the theory of evolution.

In seeking to explain this paradox, biologists have focused their attention on the social insects—ants, bees, wasps and termites. These species exhibit an extreme form of what has been called reproductive altruism, whereby individuals forgo reproduction entirely and actually help other individuals reproduce, forming entire castes of sterile workers. Since reproductive success is the ultimate goal of each player in the game of natural selection, reproductive altruism is a remarkable type of self-sacrifice.

Helping behavior is common in vertebrate societies as well, and some species cooperate in breeding. But until recently there did not appear to be a close vertebrate analogue to the extreme form of altruism observed in social insects. Such a society may now have been found in the arid Horn of Africa, where biologists have been studying underground colonies of a singularly unattractive but highly social rodent.

The naked mole-rat, *Heterocephalus glaber*, appears to be a eusocial, or truly social, mammal. It fits the classical definition of eusociality developed by Charles Michener (1969) and E. O. Wilson (1971), who extensively studied the social insects. In the burrow colonies of naked mole-rats there are overlapping adult generations, and as in insect societies brood care and other duties are performed cooperatively by workers or helpers that are more or less nonreproductive. A naked mole-rat colony is ruled, as is a beehive, by a queen who breeds with a few select males. Furthermore, the other tasks necessary to underground life—food gathering, transporting of nest material, tunnel expansion and cleaning and defense against predators—appear to

be divided among nonreproductive individuals based on size, much as labor in insect societies is performed by the sterile worker castes.

The naked mole-rat is not the only vertebrate that can be described as eusocial, but no other vertebrate society mimics the behavior of the eusocial insects so closely. The fact that highly social behavior could evolve in a rodent population suggests that it is time to reexamine some old theories about how eusocial behavior could come into being—theories that were based on the characteristics of certain insects and their societies. In the past decade, since Jennifer U. M. Jarvis first revealed the unusual social structure of a naked mole-rat colony, a number of biologists have been at work considering how a eusocial rodent could evolve. I shall discuss the state of that work briefly here, examining what is known about the naked mole-rat's ecology, behavior and evolution and about altruistic animal societies.

Introducing the Naked Mole-Rat

The naked mole-rat is a member of the family Bathyergidae, the African mole-rats—so named because they resemble rats but live like moles. Many rodents burrow and spend at least part of their life underground; all 12 species of Bathyergidae live exclusively underground, and they share a set of features that reflect their subterranean lifestyle and that demonstrate evolutionary convergence, the independent development of similar characteristics. Like the more familiar garden mole, a mole-rat has a stout, cylindrical body, a robust skull, eyes that are small or absent, reduced external ears, short limbs, powerful incisors and sometimes claws for digging, and a somewhat unusual physiology adapted to the difficulties of life underground, including a burrow

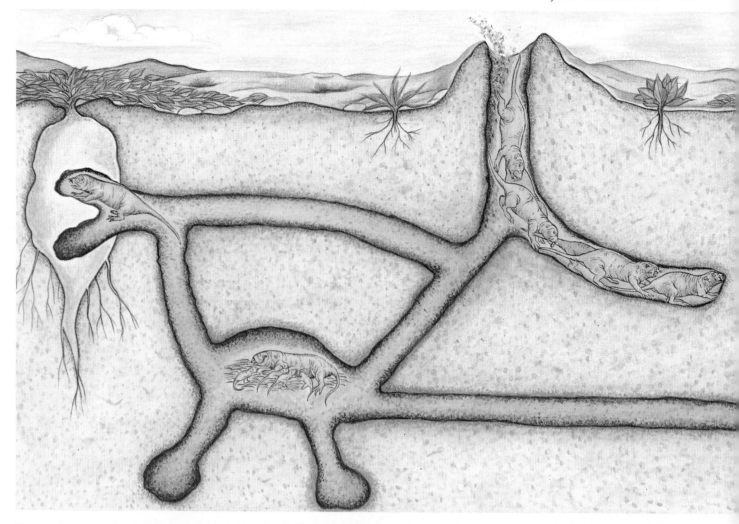

Figure 1. Burrow system built by naked mole-rats beneath the East African desert illustrates the complex social organization that makes the subterranean species unusual. Reproduction in a naked mole-rat colony, which usually has 70 to 80 members, is controlled by a queen, the only breeding female, shown here nursing newborns in a nest chamber. Digging tunnels to forage for food is one of the functions of

atmosphere high in carbon dioxide. All Bathyergidae species are herbivorous, and all but one sport fur coats.

Field biologists who encountered naked mole-rats in the 19th century thought that these small rodents—only three to six inches long at maturity, with weights averaging 20 to 30 grams— were the young of a haired adult. But subsequent expeditions showed that adult members of the species are hairless except for a sparse covering of tactile hairs. Oldfield Thomas, noting wide variations in the morphological characteristics of the naked mole-rats, identified what he thought were several species. *H. glaber* is currently considered a single species, within which there is great variation in adult body size.

Naked mole-rats inhabit the hot, dry regions of Ethiopia, Somalia and Kenya. Like most of the Bathyergidae species, they build elaborate tunnel systems. The tunnels form a sealed, compartmentalized system interconnecting nest sites, toilets, food stores, retreat routes and an elaborate tunnel system allowing underground foraging for tubers *(Figure 1)*. Like the morphology of the animals, the tunnel system is an example of convergent evolution, being similar to those of the other mole-rats in its compartmentalization, atmosphere and more or less constant temperature and humidity. Naked mole-rats subsist primarily on geophytic plants (perennials that overwinter in the form of bulbs or tubers), which are randomly and patchily distributed. The mole-rats forage broadly by expanding their burrows, but their distribution is limited by food supply and soil types. Like most rodents that live underground, they are not able to disperse over long distances.

The tunnel systems of naked mole-rats can be quite large, containing as many as two miles of burrows. The average colony is thought to have 70 to 80 members. In order to study the social organization of the naked mole-rats, biologists have had to devise ways to capture whole colonies and recreate their burrow systems in the laboratory. This is not an easy task, but it is possible because the rodents have a habit of investigating opened sections of their burrow systems and then blocking them. One can create an opening, then capture the naked mole-rats as they come to seal it. Cutting off their retreat requires quick work with a spade, hoe or knife, and the procedure must be repeated in various parts of the tunnel system in order to retrieve an entire colony. A carefully reconstructed colony can survive quite well in captivity, and naked mole-rats are beginning to become an attraction at zoos.

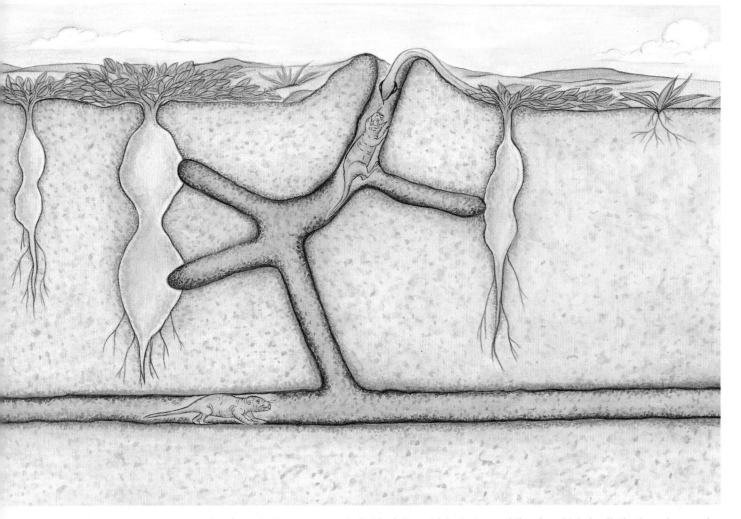

nonreproductive workers, which often form digging teams; one individual digs with its incisors while others kick the dirt backward to a mole-rat that kicks it out of the tunnel. The molehills or "volcanoes" formed in this way are plugged to create a closed environment and deter predators such as the rufous-beaked snake. Tubers and bulbs are the naked mole-rats' food source.

Most African mole-rats excavate by digging with their large incisors, removing the dirt from the burrow with their hind feet. The digging behavior of naked mole-rats, which are most active during periods when the soil in their arid habitat is moist, appears to be unlike that of the other mole-rats in two respects. First, instead of plugging the surface opening to a tunnel during excavation, the naked mole-rats "volcano," kicking soil through an open hole to form a tiny volcano-shaped mound. When excavation is complete, the tunnel is plugged to form a relatively airtight, watertight and predator-proof seal *(Figure 4)*. Second, naked mole-rats have been observed digging cooperatively in a wonderfully efficient arrangement that resembles a bucket brigade. One animal digs while a chain of animals behind move the dirt backward to an

Figure 2. Wrinkled, squinty-eyed and nearly hairless, the first naked mole-rats found by biologists were thought to be the young of a haired adult. The rodents are just three to six inches long at maturity, although there is great variation in body size within each colony. Other morphological features reflect the fact that the naked mole-rats live entirely underground: small eyes, two pairs of large incisors for digging, and reduced external ears. (Except where noted, photographs courtesy of the author.)

Figure 3. Habitat of the naked mole-rats is hot, dry and dotted with patches of vegetation. Visible in the foreground of this photograph, taken in Kenya, are the molehills formed by the rodents.

Figure 4. "Volcanoes" formed when naked mole-rats kick sand out of a tunnel, then plug the opening, make the animals' burrows easy to find. Naked mole-rats are most vulnerable to predators while forming volcanoes; the activity often attracts the attention of snakes.

animal at the end, which kicks the dirt from the burrow. One 87-member colony was seen to remove about 500 kilograms of soil per month by this process. Another colony of similar size moved an estimated 13.5 kilograms in an hour—about 380 times the mean body weight of a naked mole-rat. A team kicking dirt through a surface opening is vulnerable to attack from snakes; the mounds also make *H. glaber*'s colonies easy for scientists to find.

Naked mole-rats are long-lived animals and prolific breeders. Several individuals caught in the wild are surviving after 16 years in captivity; two of these are females that still breed. In captive colonies females have produced litters as large as 27, and in wild populations litter sizes can be as high as 12. The naked mole-rat breeds year-round, giving birth about every 70 to 80 days. This fecundity is unusual among the Bathyergidae. The other highly social species of African mole-rat, *Cryptomys damarensis*, is also a year-round breeder but produces smaller litters, with an average size of five.

The major threat to the longevity of a naked mole-rat, and probably to all of the mole-rats, is predation. On at least two occasions I have encountered the rufous-beaked snake in a mole-rat burrow; one snake had three mole-rats in its stomach. Similar field observations have been made by other investigators. Encounters between mole-rats and snakes in the laboratory suggest that avoidance may not be the mole-rat's only strategy against predators; individuals have also been seen attacking the predator in their defense of the colony.

The naked mole-rat's closest relatives are the 11 other species in the Bathyergidae, which are all of exclusively African origin and distribution *(Figure 5)*. It has been difficult to determine which of the 32 other rodent families shares a common ancestry with the Bathyergidae, but a consensus arising from recent studies places the family in the rodent suborder Hystricognathi, which includes caviomorph rodents from the New World—porcupines, guinea pigs and chinchillas—and porcupines and cane rats from the Old World. The naked mole-rat is the most divergent species within the Bathyergidae, its evolutionary branch splitting off at the base of the family's phylogenetic tree *(Figure 6)*.

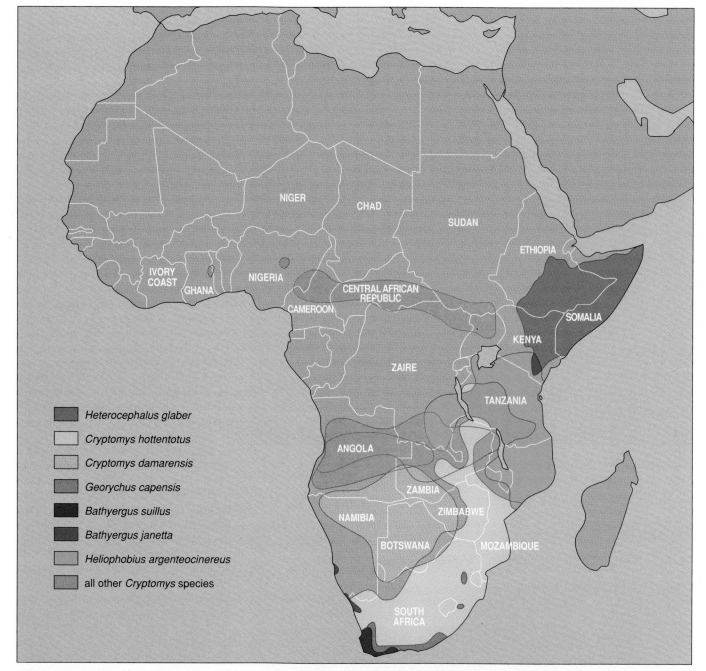

Figure 5. Geographic range of the naked mole-rat, *Heterocephalus glaber,* **is limited to the hot, dry region called the Horn of Africa—parts of Ethiopia, Kenya and Somalia. On the map are shown the areas inhabited by other species of African mole-rats. All species in the family Bathyergidae live entirely underground. Most are solitary or colonial; the other species with a highly developed social structure,** *Cryptomys damarensis,* **is found in Southern Africa.**

How Do Altruistic Societies Evolve?

Darwin called the development of sterile castes in insect societies a "special difficulty" that initially threatened to be fatal to his theory of natural selection. His solution to the problem was surprisingly close to current hypotheses based on genetic relatedness, even though he did not have a knowledge of genetics. Darwin suggested that traits, such as helping, that were observed in sterile form could survive if individuals that expressed the traits contributed to the reproductive success of those individuals that had the trait but did not express it.

Today the notion of *inclusive fitness* forms the foundation for theories about how reproductive altruism might evolve. The idea arose in 1964 from William Hamilton's remarkable genetic studies of the Hymenoptera, the insect order that includes the social ants, bees and wasps. Hamilton showed that if the genetic ties within a generation are closer than the ties between generations, each member of the generation might be motivated to invest in a parent's reproductive success rather than his or her own. Inclusive fitness is a combination of one's own reproductive success and that of close relatives.

In the Hymenoptera, Hamilton found an asymmetric genetic system that could contribute to the development of reproductive altruism by giving individuals chances to maximize their inclusive fitness without reproducing. Hymenopteran males arise from unfertilized eggs and thus have only one set of chromosomes (from the mother); females have one set from each parent. The males are called haploid, the females diploid, and this system of sex determination is referred to as *haplodiploidy (Figure 9)*. The daughters of a monogamous mother share identical genes from their father and half their mother's genes; they thus have three-quarters of their genes in common. A female who is more closely related to her sister than to her mother or her offspring can propagate her own genes most effectively by helping create more sisters. Sterile workers in hymenopteran insect colonies are all female.

Hamilton's work prompted a flurry of interest in genetic asymmetry, but he and others recognized that it was not a general explanation for how eusocial societies might evolve. There are many limitations; for instance, multiple matings by females reduce the closeness of relationships between sisters, and it is hard to explain the incentives for females to tend juvenile males, which are not as closely related as are sisters. Furthermore, although eusociality has evolved more times in the Hymenoptera than in any other order, it has also evolved in parts of the animal world in which both sexes are diploid— namely Isoptera, which includes the social termites, and Rodentia, the order that includes the naked mole-rat. Finally, there are many arthropod species that are haplodiploid and have not developed highly social behavior.

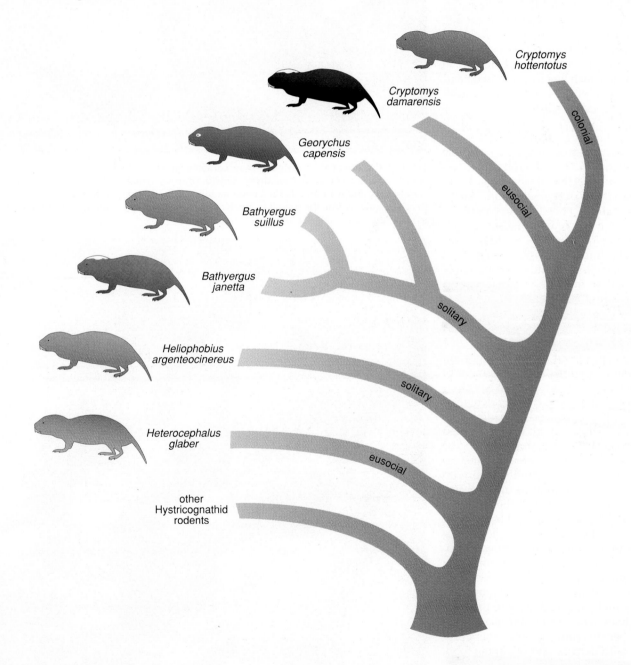

Figure 6. Phylogenetic tree for the family Bathyergidae, the African mole-rats, shows that the two eusocial, or truly social, species are quite divergent. Among other rodents, the suborder Hystricognathi, which includes porcupines, guinea pigs and chinchillas, appears to have the closest genetic link with the African mole-rats. Although there is much similarity among the Bathyergidae species in their physiological characteristics and their subterranean lifestyle, the phylogenetic distance between the eusocial species of mole-rats suggests that complex social behavior evolved separately in the two cases.

There is another way that close kinship might develop among the members of a generation, and it is considered a possible explanation for the evolution of the termite and naked mole-rat societies. Several generations of inbreeding could result in a higher degree of relatedness among siblings than between parents and offspring *(Figure 10)*. When male and female mates are unrelated, but each is the product of intense inbreeding, their offspring can be genetically identical and might be expected to stay and assist their parents for the same reasons set forth in the haplodiploid model. The inbreeding model was developed by Stephen Bartz in 1979 to explain the development of eusocial behavior in termites, which live in a contained and protected nest site conducive to multigenerational breeding.

Genetics alone cannot provide a comprehensive explanation for the evolution of eusociality. Other possible explanations, especially relevant to termites and vertebrate helpers, lie in combinations of ecological and behavioral factors. These factors perhaps provided preconditions or starting points for the eventual evolution of a eusocial lineage or species. The best way to understand the development of eusociality may be to consider the costs and benefits associated with remaining in the natal group and helping, as compared to the costs and benefits of dispersing and breeding.

Probably one of the most important preconditions for the development of eusociality is parental care in a protected nest, where offspring are defended against predators and provided with food. If there is a high cost associated with dispersal—in terms of restricted access to food, lack of breeding success or increased vulnerability to predators—then there may be an incentive for juveniles to remain in the protected nest and become helpers. Helpers that remain in the nest for multiple generations may forgo reproduction indefinitely as a consequence of maternal manipulation.

The short-term benefits of group living seem to accrue mainly to those individuals who are reproducing, since they benefit from the help others provide with defense and obtaining food. In fact, there is a correlation between the breeder's reproductive fitness and the number of helpers in cooperatively breeding vertebrate species. Thus the long-term effect of helping may be an

Figure 7. Catching naked mole-rats requires some understanding of their behavior. Mole-rat catchers create an opening from the surface to a burrow, which is normally kept sealed by the animals, and wait quietly for a mole-rat to investigate. A spade, hoe, pick or knife blade is driven quickly into the tunnel to block the mole-rat's escape. (Photograph courtesy of Stan Braude, University of Missouri at St. Louis.)

Figure 8. Captive naked mole-rats, carrying identifying tattoos, adapt well to being placed together in bins, apparently because the highly social animals tend to huddle together for warmth in their burrows in the wild. These rodents are part of Jennifer U. M. Jarvis's collection at the University of Cape Town.

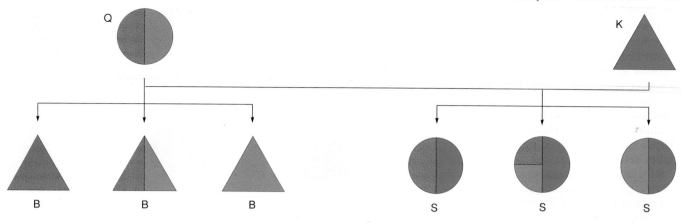

degrees of relatedness in haplodiploid species

	daughter	son	mother	father	sister	brother
female	$\frac{1}{2}$	$\frac{1}{2}$	$\frac{1}{2}$	$\frac{1}{2}$	$\frac{3}{4}$	$\frac{1}{4}$
male	1	0	1	0	$\frac{1}{2}$	$\frac{1}{2}$

Figure 9. Haplodiploidy, an asymmetric genetic system, is thought to contribute to the development of reproductive altruism in ants, bees and wasps—species with intricate social systems that include sterile castes of workers. In a haplodiploid species, males *(triangles)* arise from unfertilized eggs and have only one set of chromosomes, whereas females *(circles)* have one set of chromosomes from each parent. The relatedness between sisters—the fraction of their genes that are shared—is thus greater than the relatedness of mother and daughter *(bottom panel)*. William D. Hamilton hypothesized that females seeking to increase their inclusive fitness—a combination of their own reproductive success and that of close relatives—might in a haplodiploid species become helpers, advancing the continuation of their own genetic heritage by helping with the reproduction of sisters rather than their own offspring. Although haplodiploidy is not considered a full explanation of how eusocial behavior would evolve in ants, bees and wasps, it is notable that most species in which reproductive altruism has evolved are haplodiploid, and that the sterile workers among the haplodiploid insects are all female. In this illustration, the parents are labeled *Q* and *K* and the offspring *S* and *B*, following the scheme in Figure 10; for simplicity, the effects of any recombination of genes are not depicted.

increase in inclusive fitness for the helpers. This may prove to be a very important consideration in species where the probability of a dispersing individual procuring a nest site and eventually breeding is extremely low.

Naked Mole-Rat Society
In some ways the social organization observed in naked mole-rat colonies is more akin to the societies of the social insects than to the social organization of any other vertebrate species. In other respects, mole-rats are unique and may always remain a bit of a mystery.

Some similarities between naked mole-rat societies and the insect societies are striking. A naked mole-rat colony, like a beehive, wasp's nest or termite mound, is ruled by its queen or reproducing female. Other adult female mole-rats neither ovulate nor breed. The queen is the largest member of the colony, and she maintains her breeding status through a mixture of behavioral and, presumably, chemical control. She is aggressive and domineering; queenly behavior in a naked mole-rat includes facing a subordinate and shoving it along a burrow for a distance. Queens have been long-lived in captivity, and when they die or are removed from a colony one sees violent fighting among the larger remaining females, leading to a takeover by a new queen.

Most adult males produce sperm, but only one to three of the larger males in a colony breed with the queen, who initiates courtship. There is little aggression between breeding males, even upon removal of the queen. The queen and breeding males do not participate in the defense or maintenance of the colony; instead, they concern themselves with the handling, grooming and care of newborns.

Eusocial insect societies have a rigid caste system, defined on the basis of distinctions in behavior, morphology and physiology. Mole-rat societies, on the other hand, demonstrate behavioral asymmetries related primarily to reproductive status (reproduction being limited to the queen and a few males), body size and perhaps age. Smaller nonbreeding members, both male and female, seem to participate more in gathering food, transporting nest material and clearing tunnels. Larger nonbreeders are more active in defending the colony and perhaps in removing dirt from the tunnels. Jarvis has suggested that differences in growth rates may influence the length of time that an individual performs a task, regardless of its age.

Naked mole-rats, being diploid in both sexes, do not have an asymmetric genetic system such as haplodiploidy. As Bartz has proposed for termites, inbreeding in naked mole-rats may create a genetic asymmetry that mimics the result of haplodiploidy. There is genetic evidence suggesting that naked mole-

Figure 1. Signals are stimuli that convey information and have been molded by natural selection to do so; cues are stimuli that contain information but have not been shaped by natural selection specifically to convey information. Information can also pass between the members of a colony indirectly, through any component of their shared environment. Shown at the left are bees following another bee performing waggle dances, which are elaborate signals that indicate with precision the distance and direction of rich patches of flowers. In the top photograph, a food-storer bee (*left*) is unloading nectar from a forager. The delay a forager experiences before she can pass off her nectar is a cue that indicates the colony's nutritional status. In the photograph above, bees are fanning their wings in order to expel warm, moist air from their hive. The effect of this fanning—a cooler, drier atmosphere inside the hive—conveys information to other bees about the colony's need for ventilation. (Photos courtesy of P. K. Visscher, *left and top*, and S. Camazine, *above*.)

which they evolved (Fig. 1). In a colony of honey bees two levels of biological organization—organism and superorganism—coexist with equal prominence. The dual nature of such societies provides us with a special window on the evolution of biological organization, through which we can see how natural selection has taken thousands of organisms that were built for solitary life and merged them into a superorganism.

Is it a superorganism?

The term "superorganism" was coined by William Morton Wheeler (1928) to denote insect societies that possess features of organization analogous to the physiological processes of individual organisms. These include advanced social insects like army ants, leaf-cutter ants, fungus-growing termites, stingless bees, and honey bees. Although sociologists dealing with insects have used the superorganism concept more as a heuristic device than as a category of societal complexity (Lüscher 1962; Southwick 1983), recent insights into the logic of natural selection support the use of this term in a manner close to Wheeler's original intent (Hull 1980; Dawkins 1982; Wilson and Sober 1989). It seems correct to classify a group of *organisms* as a superorganism when the organisms form a cooperative unit to propagate their genes, just as we classify a group of *cells* as an organism when the cells form a cooperative unit to propagate their genes. By this definition, most groups of organisms are not perfect superorganisms because there is usually intense intragroup conflict when members compete for reproductive success (Trivers 1985). Indeed, in many species of social insects the female members of a colony (queens and workers) fight over who will lay eggs (West-Eberhard 1981; Bourke 1988). In the most advanced species of social insects, however, there appears to be little if any conflict within colonies, so that these colonies do represent superorganisms.

How complete is the cooperation in a honey bee colony, and thus to what extent is a colony of honey bees truly a superorganism? The best way to answer these questions is to determine the degree of congruence in the genetic interests of a colony's members. Consider the typical situation of a colony comprising one queen and some 20,000 workers, all daughters of the queen. At first glance, it might seem that there will be tremendous divergence of genetic interests within the colony. As a result of sexual reproduction, the queen's genotype does not match that of her workers; furthermore, although the workers are all offspring of the queen, because of segregation and recombination of the queen's genes during meiosis and because the queen has mated with ten or more males (Page 1986), the workers possess substantially different genotypes.

A closer look, however, reveals several features of the biology of honey bees that indicate a close alignment of genetic interests among the members of a colony, despite these genetic differences (Ratnieks 1988). Although worker bees possess ovaries and will lay eggs to produce sons if they lose their queen (Page and Erickson 1988), in the presence of the queen, workers engage in essentially no direct, personal reproduction. Workers cannot mate, so their only possible avenue of direct reproduction is through haploid sons from unfer-

tilized eggs. A recent study in which the extent of worker reproduction in colonies with queens was measured using genetic markers to distinguish drones from queen-laid and worker-laid eggs, reported that only one in one thousand drones in a colony is the offspring of workers (Visscher, in press). This means that as long as the queen is present there is a reproductive bottleneck in which every individual's gene propagation occurs virtually exclusively through a common pathway—the reproductive offspring (queens and drones) of the mother queen. This situation promotes strong cooperation among the queen and all workers; ultimately each worker focuses her efforts on the welfare and reproductive success of one individual, the queen.

This reproductive bottleneck does not, however, indicate that a perfect alignment of the genetic interests of a colony's members has evolved. The workers in colonies with queens may still disagree over which eggs should be reared into queens when it is time to produce new queens. This potential conflict of interest traces to the multiple mating of honey bee queens, which produces a set of patrilines within each colony. Because workers share three times as many genes with full-sister queens (same patriline) as with half-sister queens (different patriline), they are expected to prefer that queens produced in a colony be their full sisters. Over the last few years several investigators have searched for intracolony competition during queen rearing, and a growing body of evidence indicates that some patrilines within a colony do achieve a small bias in their favor (Noonan 1986; Visscher 1986; Page et al. 1989). However, all studies that have reported preferential rearing of more closely related queens involved somewhat artificial test conditions, such as transfers of larval queens between colonies or use of colonies containing only two or three instead of the normal number (ten or more) of patrilines. It may be that even the slight bias in queen rearing observed in these studies is greater than what occurs under natural conditions (Hogendoorn and Velthuis 1988).

Given the bottleneck for gene propagation and the strong indication that workers have nearly equal genetic stakes in a colony's production of reproductives (due to meiosis in the queen, together with little patriline bias in queen rearing), we can conclude that the genetic interests of the workers in a colony led by a queen are nearly, though not perfectly, congruent. Furthermore, we know that the mother queen and the workers have evolved similar interests in matters such as who lays the eggs that produce the colony's drones, the ratio of the colony's investment in queens and drones, and the timing of replacement of the queen (Seeley 1985; Ratnieks 1988). Thus it appears that there is minimal conflict within honey bee colonies as long as the mother queen is present. Therefore, we may conclude that honey bee colonies containing queens are nearly true superorganisms.

This conclusion, based on analyses of the genetic interests of a colony's members, is reinforced by the picture of pervasive cooperation which has emerged from analyses of colony functioning. In choosing a nest site, building a nest, collecting food, regulating the nest temperature, and deterring predators, a honey bee colony containing a queen resembles a smoothly running machine in which each part always contributes to the efficient operation of the whole (Seeley 1985; Winston 1987). As we will see, in a normal honey bee colony, food, information, and aid appear to pass freely among the members in ways that apparently promote the economic success of the whole colony.

It should be very revealing, and at most only slightly misleading, to view a honey bee colony as an integrated biological machine that promotes the success of the colony's genes. Given this perspective, the outstanding biological question becomes: How did evolution take a large number of organisms built for solitary life and forge them into a single vehicle of gene survival? The answer to this question has two parts. One concerns the ultimate forces of natural selection, which caused the evolution of unified colonies; the other involves the proximate mechanisms by which colonies function as integrated wholes. This article focuses on the second half of the answer. The key to understanding this aspect of the puzzle involves understanding the flow of information within colonies. Coordination in any complex system depends upon each part having access to appropriate information at the right time and place (Wiener 1961). Coherence implies communication.

Architecture of information flow

Coordination of the activities in a honey bee colony arises without any centralized decision making. There is no evidence of an information and control hierarchy, with some individuals taking in information about the colony, deciding what needs to be done, and issuing commands to other individuals who then perform necessary tasks. As the biblical King Solomon observed, there is "neither guide, overseer, nor ruler." In particular, it is clear that the queen does not supervise the activities of her workers. She does emit a chemical signal, the queen-substance pheromone, which plays a role in regulating the colony's production of additional queens (Free 1987), but this signal cannot provide comprehensive supervision of the activities of the tens of thousands of workers in a colony.

A colony's coherence depends instead upon the ability of its members to circulate throughout the hive, gather information about the colony's needs, and adjust the supply of their labor to the demands they sense. This idea was suggested in the early 1950s by Martin Lindauer (1952), who painstakingly followed individual workers within colonies living in glass-walled observation hives. He learned that the bees devote about 30% of their time to walking about the nest, and that this patrolling is punctuated by bouts of activity in a wide variety of tasks (Fig. 2). A typical 30-minute segment from Lindauer's records reveals the following behavior for a seven-day-old bee: patrolling, shaping comb, patrolling, feeding young brood, cleaning cells, patrolling, shaping comb, eating pollen, resting, patrolling, shaping comb. The task performed at any given moment presumably depends upon the specific labor need sensed by the bee.

Why are honey bee colonies organized in this way? Decentralized control is possibly superior to centralized control for bees. Systems with decentralized control generally have faster responses to local stresses than

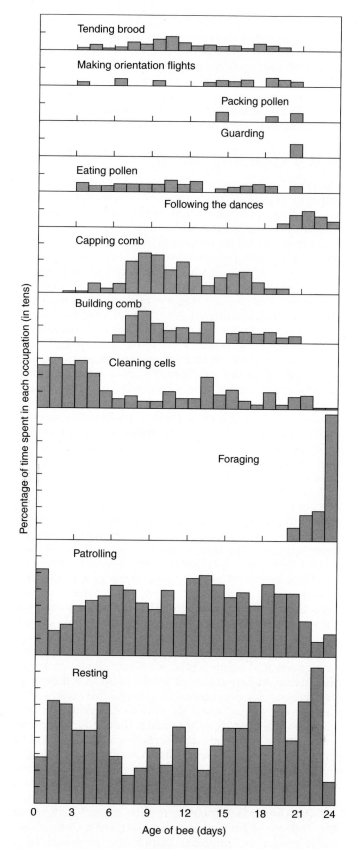

Figure 2. In the course of her life, a worker honey bee performs a variety of tasks. As the distribution above indicates, she can perform several different tasks on any given day and at any given age. She thereby behaves flexibly, responding to different needs encountered in the hive. The large amount of time spent patrolling is evidently related to the gathering of information—through cues, signals, and the shared environment—about the colony's labor needs. (After Lindauer 1952.)

those with centralized control (Miller 1978), and this may be extremely valuable to colonies of honey bees. A colony may be able to respond to a predator's attack at the nest entrance or to a temperature rise in the central broodnest much more quickly if the workers at the trouble sites can perform corrective actions immediately than if information has to be sent to a supervisor, who would then issue instructions.

A perhaps more likely explanation of decentralized control is not that it is superior to centralized control, but that it is the best the bees can do given the limited communication processes that have developed through evolution. As we will see, the mechanisms of communication in colonies of social insects are rather rudimentary, at least relative to what exists in human organizations or in multicellular organisms. Colonies of army ants, fungus-growing termites, honey bees, and other superorganisms have yet to invent anything like a mail system, telephone, or computer network. Such technologies make it possible for information to flow rapidly and efficiently between the different parts of a human organization with centralized control. One piece of evidence in support of this second hypothesis is that at the level of organization just below the superorganism, the multicellular organism, where a sophisticated intercellular communication system has evolved, there exists centralized control, with the brain taking in information about the whole organism and issuing commands to cells of the body. Whatever the underlying reason, the fact of decentralized control in honey bee colonies tells us that understanding how colonial coordination arises follows from understanding how each worker acquires information about her colony's needs.

Pathways of information

To what do worker bees respond when patrolling? The answer to this question is complex because evolution has been highly opportunistic in building pathways for information in honey bee colonies. It has shaped the workers so that they are sensitive to virtually all variables and stimuli that contain useful information: the temperature of the nest interior, the degree of crowding at a food source, the moistness of larvae, the recruitment dances of nestmates, the shape of a beeswax cell, the odor of dead bees. Furthermore, given the close alignment of the genetic interests of a colony's workers, we can expect that natural selection has molded the workers to be skilled at generating signals for information transfer. Within colonies there are various tappings, tuggings, shakings, buzzings, strokings, wagglings, crossing of antennae, and puffings and streakings of chemicals, all of which seem to be communication signals. The result is that within a honey bee colony there exists an astonishingly intricate web of information pathways, the full magnitude of which is still only dimly perceived.

Information can flow between colony members in two ways: directly, through signals and cues, as we will discuss below, or indirectly, through some component of the shared environment. An example of the latter process is the transfer of information through the process of comb building. The construction of a particular cell in a beeswax comb may involve several bees, yet these bees never need to come together and exchange information directly. The building activities can be completely and efficiently coordinated by information embodied in the structure of the partially completed cell. Thus one bee might begin a cell wall by depositing a small ridge of beeswax; a second bee might finish sculpting the wall, guided by the shape of the wax ridge left by the first bee. Another example of information flow through the shared environment is thermoregulation of the nest. A colony maintains the central broodnest at 34 to 36°C in the face of ambient temperatures that may range from −20 to 40°C. The coordinated heating and cooling of a nest occurs automatically: each bee responds to the temperature of her immediate environment by appropriately heating it (by making intense isometric contractions of her flight muscles) or cooling it (by fanning her wings to draw cooler air into the area) (Heinrich 1985). In effect, the temperature of the air and comb inside a hive provides a communication network regarding the colony's heating and cooling needs.

Several authors have expressed the concept of information flow through the shared environment in social insect colonies. These include Pierre-Paul Grassé (1959), who coined the term "stigmergy" to explain coordination in nest construction by termites, and Charles D. Michener (1974), who pointed out that "indirect social interactions," such as transfers of information through the food stored in the nest, are an important integration mechanism in colonies of social bees. Future studies of information flow in social insect colonies may reveal that more information is transmitted indirectly than directly. The use of the shared environment as a communication pathway has certain attractions, including easy asynchronous transfer of information between individuals and virtually automatic transfer of information between any two individuals sharing some portion of the nest environment. It also has the important feature whereby information can pass from a group to an individual whenever an individual responds to the environmental effects of a group. Because the process of integration of a group is largely a matter of information flow from group to individual, it may be that information flow through the shared environment has been natural selection's principal technique of integration in building superorganisms.

Signals and cues

There are two types of direct communication channels: signals and cues (Lloyd 1983). Signals are stimuli that convey information and have been shaped by natural selection to do so, whereas cues are stimuli that contain information but have not been shaped by natural selection specifically to convey information. Cues carry information only incidentally. The distinction between signals and cues deserves emphasis because studies of information flow in social insect colonies have tended to overlook cues and have focused instead on conspicuous visual, tactile, acoustical, and chemical signals (reviewed by Wilson 1971; Hölldobler 1977). The emphasis on signals reflects the fact that information transfer via signals is relatively easily detected by humans because in the mutualistic setting of a social insect colony natural selection will have shaped signals to be powerful and unam-

members is achieved through surprisingly rudimentary information transfer. Traditionally, studies of communication in the social insects have emphasized sophisticated and conspicuous communication processes that involve signals honed by natural selection, such as the dance language behavior. There is no question that these processes are important. Nevertheless, I predict that the relatively subtle communication mechanisms of cues and the shared environment will prove even more important than the more obvious signals. If so, then the impressive feats of internal coordination shown by superorganisms will often prove to be built of rather humble devices. This should not surprise us, for as Colin Pittendrigh (1958) so nicely put it, adaptive organization is "a patchwork of makeshifts pieced together, as it were, from what was available when opportunity knocked, and accepted in the hindsight, not the foresight, of natural selection."

References

Alexander, R. D. 1974. The evolution of social behavior. *Ann. Rev. Ecol. Syst.* 5:325–83.

Bonner, J. T. 1974. *On Development.* Harvard Univ. Press.

————. 1988. *The Evolution of Complexity.* Princeton Univ. Press.

Bourke, A. F. G. 1988. Worker reproduction in the higher eusocial Hymenoptera. *Quart. Rev. Biol.* 63:291–311.

Burnham, L. 1978. Survey of social insects in the fossil record. *Psyche* 85: 85–133.

Dawkins, R. 1982. *The Extended Phenotype.* Freeman.

Free, J. B. 1987. Pheromones of Social Bees. Cornell Univ. Press.

Grassé, P.-P. 1959. La reconstruction du nid et les coordinations interindividuelles chez *Bellicositermes natalensis* et *Cubitermes* sp. La théorie de la stigmergie: Essai d'interprétation du comportement des termites constructeurs. *Insectes Sociaux* 6:41–83.

Heinrich, B. 1985. The social physiology of temperature regulation in honeybees. In *Experimental Behavioral Ecology and Sociobiology*, ed. B. Hölldobler and M. Lindauer, pp. 393–406. Sinauer.

Hogendoorn, K., and H. H. W. Velthuis. 1988. Influence of multiple mating on kin recognition by worker honeybees. *Naturwissenschaften* 75:412–13.

Hölldobler, B. 1977. Communication in social Hymenoptera. In *How Animals Communicate*, ed. T. A. Sebeok, pp. 418–71. Indiana Univ. Press.

Hull, D. L. 1980. Individuality and selection. *Ann. Rev. Ecol. Syst.* 11: 311–32.

Lindauer, M. 1948. Über die Einwirkung von Duft- und Geschmacksstoffen sowie anderer Faktoren auf die Tänze der Bienen. *Zeitschrift für vergleichende Physiologie* 31: 348–412.

————. 1952. Ein Beitrag zur Frage der Arbeitsteilung im Bienenstaat. *Zeitschrift für vergleichende Physiologie* 34:299–345.

Lloyd, J. E. 1983. Bioluminescence and communication in insects. *Ann. Rev. Entomol.* 28:131–60.

Lüscher, M. 1962. Sex pheromones in the termite superorganism. *Gen. Comp. Endocrinol.* 2:615.

Margulis, L. 1981. *Symbiosis in Cell Evolution.* Freeman.

Markl, H. 1985. Manipulation, modulation, information, cognition: Some of the riddles of communication. In *Experimental Behavioral Ecology and Sociobiology*, ed. B. Hölldobler and M. Lindauer, pp. 163-94. Sinauer.

Michener, C. D. 1974. *The Social Behavior of the Bees.* Harvard Univ. Press.

Miller, J. G. 1978. *Living Systems.* McGraw-Hill.

Morse, P. M. 1958. *Queues, Inventories, and Maintenance.* Wiley.

Noonan, K. C. 1986. Recognition of queen larvae by worker honey bees (*Apis mellifera*). *Ethology* 73:295–306.

Page, R. E., Jr. 1986. Sperm utilization in social insects. *Ann. Rev. Entomol.* 31:297–320.

Page, R. E., Jr., and F. H. Erickson, Jr. 1988. Reproduction by worker honey bees (*Apis mellifera* L.). *Behav. Ecol. Sociobiol.* 23:117–26.

Page, R. E., Jr., G. E. Robinson, and M. K. Fondrk. 1989. Genetic specialists, kin recognition and nepotism in honey-bee colonies. *Nature* 338:576–79.

Pittendrigh, C. S. 1958. Adaptation, natural selection, and behavior. In *Behavior and Evolution*, ed. A. Roe and G. G. Simpson, pp. 390–416. Yale Univ. Press.

Ratnieks, F. L. W. 1988. Reproductive harmony via mutual policing by workers in eusocial Hymenoptera. *Am. Naturalist* 132:217–36.

Seeley, T. D. 1985. *Honeybee Ecology.* Princeton Univ. Press.

————. 1986. Social foraging by honey bees: How colonies allocate foragers among patches of flowers. *Behav. Ecol. Sociobiol.* 19:34–54.

————. 1989. Social foraging in honey bees: How nectar foragers assess their colony's nutritional status. *Behav. Ecol. Sociobiol.* 24:181–99.

Simon, H. A. 1962. The architecture of complexity. *Proc. Am. Philosoph. Soc.* 106:467–82.

Southwick, E. E. 1983. The honey bee cluster as a homeothermic superorganism. *Comp. Biochem. Physiol.* 75A:641–45.

Trivers, R. 1985. *Social Evolution.* Benjamin/Cummings.

Visscher, P. K. 1986. Kinship discrimination in queen rearing by honey bees (*Apis mellifera*). *Behav. Ecol. Sociobiol.* 18:453–60.

————. In press. A quantitative study of worker reproduction in honey bee colonies. *Behav. Ecol. Sociobiol.*

Visscher, P. K., and T. D. Seeley. 1982. Foraging strategy of honeybee colonies in a temperate deciduous forest. *Ecology* 63:1790–1801.

von Frisch, K. 1967. *The Dance Language and Orientation of Bees.* Harvard Univ. Press.

West-Eberhard, M. J. 1981. Intragroup selection and the evolution of insect societies. In *Natural and Social Behavior*, ed. R. D. Alexander and D. W. Tinkle, pp. 3–17. Chiron.

Wheeler, W. M. 1928. *The Social Insects: Their Origin and Evolution.* Kegan Paul, Trench, and Trubner.

Wiener, N. 1961. *Cybernetics: Or Control and Communication in the Animal and the Machine.* MIT Press.

Wilson, D. S., and E. Sober. 1989. Reviving the superorganism. *J. Theor. Biol.* 136:337–56.

Wilson, E. O. 1971. *The Insect Societies.* Harvard Univ. Press.

————. 1975. *Sociobiology.* Harvard Univ. Press.

Winston, M. L. 1987. *The Biology of the Honey Bee.* Harvard Univ. Press.

The Essence of Royalty: Honey Bee Queen Pheromone

A queen bee induces thousands of worker bees to submit to her hegemony by secreting a potent chemical blend, which is spread by physical contact throughout the colony

Mark L. Winston and Keith N. Slessor

Beekeeping, like most skills, has developed its own shorthand language, by which a word or simple phrase can evoke complex concepts that quickly are understood by those who practice the craft. One such word is "queenright," meaning that a colony's queen is present and thousands of her worker offspring are collaborating diligently on communal tasks. Hidden in this beekeeper jargon, however, lies the most fundamental mystery of social insects: the mechanisms of coordination and integration that mediate the tasks of thousands of individuals into the smoothly functioning unit of the colony. In this article, we shall describe some new findings concerning one aspect of these integrative mechanisms, the honey bee queen's pheromonal influence over worker bees. The discovery of the nature and function of this "essence of royalty" has provided profound insights into the functioning of social-insect colonies, and also has led to some significant commercial applications for crop pollination and beekeeping.

Mark L. Winston, who earned B.A. and M.Sc. degrees at Boston University and his Ph.D. at the University of Kansas, is the author of two recent books about honey bees, The Biology of the Honey Bee *(1987) and* Killer Bees: The Africanized Honey Bee in the Americas *(1992), both published by Harvard University Press. Keith N. Slessor, who received his Ph.D. at the University of British Columbia, is a native British Columbian who finds organic chemistry an excellent medium for both teaching and exploration, and regards it as a challenge to decipher the communication skills of economically important insects. Winston and Slessor are professors at Simon Fraser University, Burnaby, B.C. V5A 1S6, Canada, in the departments of Biological Sciences and Chemistry/Biochemistry, respectively.*

The study of social-insect pheromones, or sociochemistry, is not a new discipline. Many pheromones are known throughout the social insects that are used in myriad tasks such as alarm behavior, orientation, trail marking, nest recognition, mate attraction and others (Bradshaw and Howse 1984; Duffield, Wheeler and Eickwort 1984; Free 1987; Hölldobler and Wilson 1990; Howse 1984; Wilson 1971; Winston 1987). The honey bees, for example, are estimated to produce at least 36 pheromones, which together constitute a chemical language of some intricacy. However, virtually all of the known chemicals are secretions that "release," or elicit, a specific behavior. The queen sociochemicals belong to another class, called primer pheromones, that exercise a more fundamental level of control by mediating worker and colony reproduction and influencing broad aspects of foraging and other behaviors. Further, they can be both stimulatory and inhibitory, depending on the specific function. The existence of primer sociochemicals is well known, but their identification and synthesis have proved difficult.

Indeed, the only primer pheromone that has been identified comes from the queen honey bee (*Apis mellifera* L.); it is secreted by the mandibular glands, a paired set of glands on either side of the queen's head (Figure 2). Here we discuss the research that led to the pheromone's identification, new findings concerning its effects on worker bees and colony functions, transmission mechanisms within colonies, and some applications for crop pollination and beekeeping.

Figure 1. Retinue response is a stereotyped honey bee behavior by which various pheromones are transferred from the queen to the worker bees. In the photograph on the opposite page workers turn to face a queen, touching her with their antennae and licking her. By this means they pick up the pheromone that she exudes on her body. For the next half hour, the workers act as dispersers of pheromone, travelling throughout the nest and contacting other workers more frequently than normal. It is the pheromone itself—including the mixture of substances secreted by the queen's mandibular glands—that induces the retinue response. The effectiveness of the pheromone is demonstrated by experiments in which worker bees form a retinue around a glass lure coated with mandibular-gland extract, as in the photograph above. The response has formed the basis of a bioassay for synthetic pheromones. (Photograph opposite courtesy of Kenneth Lorenzen; photograph above by Keith N. Slessor.)

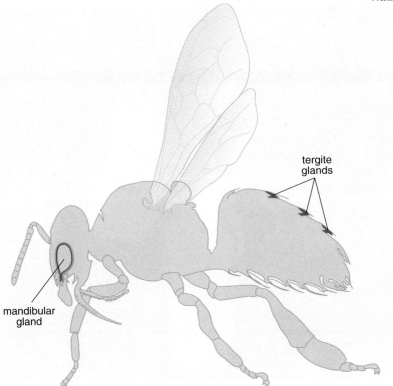

Figure 2. Two sets of glands in the queen honey bee are thought to secrete primer pheromones, the class of pheromones that regulate reproductive behavior in the colony. The best-studied primer pheromones are those produced in the mandibular glands on either side of the head. Each gland is connected to the mandible by a duct; a valve allows the bee to regulate the discharge of secretions. The queen may also synthesize primer pheromone in the tergite glands, which are large subepidermal complexes of glandular cells near the rear of some of the tergites, or dorsal plates. The functions of the secretions from the tergite glands have not been definitively established.

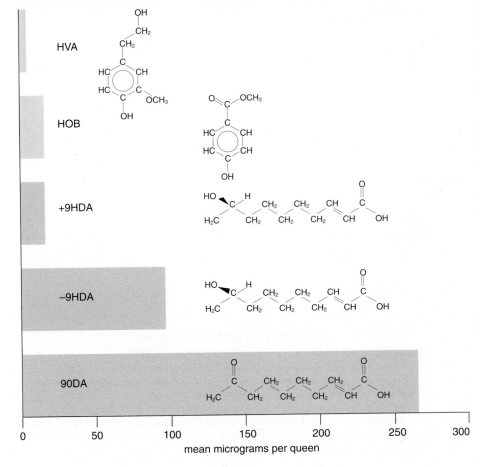

Early Research

For more than 30 years it has been recognized that queen pheromones mediate many colony activities. The pheromones were known to inhibit the rearing of new queens and the development of ovaries in worker bees; they were also observed to attract workers to swarm clusters, stimulate foraging, attract workers to a queen and allow workers to recognize their own queen.

Surprisingly, only two active substances had been identified. These two most abundant compounds in queen mandibular pheromone are the organic acids (E)-9-keto-2-decenoic acid, or 9ODA, which was identified in 1960 (Callow and Johnston 1960, Barbier and Lederer 1960) and (E)-9-hydroxy-2-decenoic acid, or 9HDA, which was identified in 1964 (Callow, Chapman and Paton 1964; Butler and Fairey 1964).

Experiments with the synthetic acids, however, showed that they did not fully duplicate the effects of mandibular extracts or of the queen's presence (reviewed by Free 1987; Winston 1987). Although analysis of crushed bee heads produced laundry lists of chemicals, none was shown on closer examination to be active, and progress stalled.

One reason it took so long to identify queen mandibular pheromone is that bioassays for primer pheromones are difficult and time-consuming. The potency of candidate releaser pheromones can be evaluated by relatively simple means. Orientation pheromones, for example, can be evaluated by counting the number of bees entering a hive after pheromone is deposited at the hive opening. Demonstrating that a synthetic blend or a natural extract is fully

Figure 3. Queen mandibular pheromone is a complex of five substances. Shown here is the average composition of one queen equivalent (Qeq) of mandibular pheromone obtained by analyzing mandibular-gland extracts from 55 queens. (The composition of the pheromone changes over the queen's life span and varies considerably even among mature, mated queens.) A queen equivalent is the average amount of pheromone in a queen's glands at a given time. The most abundant ingredient is (E)-9-keto-2-decenoic acid (9ODA). Another decenoic acid has two optical isomers, both of which must be present to achieve full activity. The isomers are designated (R,E)-(–)-9-hydroxy-2-decenoic acid (–9 HDA) and (S,E)-(+)-9-hydroxy-2-decenoic acid (+9 HDA). The remaining two ingredients are aromatic compounds: methyl *p*-hydroxybenzoate (HOB) and 4-hydroxy-3-methoxyphenylethanol (HVA).

equivalent to a primer pheromone, however, may require a battery of bioassays, some of which may take months to complete. To further complicate matters, there is often disagreement over a primer's functions.

The chemical composition of queen mandibular pheromone and its biological effects were not the only controversial topics. Even the means by which the pheromone reaches worker bees was unknown. Is the pheromone transmitted when the workers exchange food? Does it volatilize and spread through the air? Or is it spread by body contact? Here the main obstacle was the minute amount involved: A queen produces less than four ten-thousandths of a gram per day, and a worker bee is able to sense a ten-millionth of the queen's daily production.

Discovery

One of the stereotyped behaviors thought to be elicited by queen pheromones is the retinue response. Worker bees turn toward the queen, forming a dynamic retinue around her. About 10 workers at a time contact her with their antennae, forelegs or mouthparts for periods of minutes. Then the workers leave the queen, groom themselves and move through the nest, making frequent reciprocated contacts with other workers for the next half-hour (Allen 1960; Seeley 1979).

Bioassays involving retinue behavior had given particularly confusing results. Mandibular extracts seemed to be important to worker recognition of the queen and attraction to her. However, the most abundant component of the glands, 9ODA, did not by itself elicit retinue behavior. To add to the confusion, bees had been shown to exhibit the retinue response to queens whose mandibular glands had been surgically removed.

This is how matters stood in 1985 when, quite by accident, we put a glass lure coated with queen mandibular extract next to some stray workers on a laboratory bench. To our surprise, the workers formed a retinue around the lure. Clearly the extract had properties that its major known constituent did not. This was exciting because the retinue response is easily recognizable and takes place immediately. For both reasons it might serve as the basis for a practical bioassay for mandibular pheromone.

Indeed we were able to develop a bioassay based on the retinue response that was a particularly sensitive indica-

Scott Camazine (Photo Researchers, Inc.)

Figure 4. Queen rearing is one of the behaviors inhibited by queen mandibular pheromone. Queens are reared in large cells that hang from the comb. Workers, in contrast, are reared in cells that are capped at the level of the comb. The first sign of queen rearing is the construction of new cups suitable for rearing queens. Most eggs in these cups are laid there by the queen, but workers sometimes move fertilized eggs or very young larvae from worker cells into queen cups. Once the eggs hatch, the larvae are fed a specialized diet that induces them to develop into queens. As the larval queens grow, the cells are elongated downward. They are finally sealed at the end of the larval feeding period.

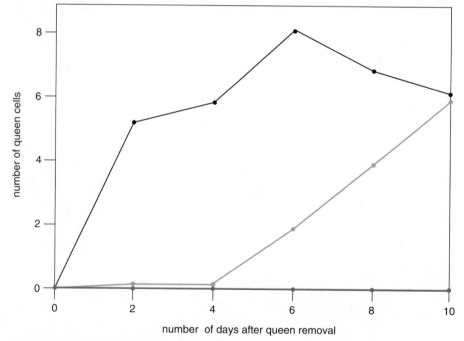

Figure 5. Queen mandibular pheromone suppresses queen rearing. This effect was demonstrated by comparing the number of queen cells in queenless colonies receiving various dosages of synthetic mandibular pheromone with the number of cells in queenright colonies. Shown here are three cases: a queenless colony receiving a glass slide coated only with the solvent used in preparing pheromone *(black)*; a queenless colony receiving one Qeq of synthetic pheromone *(gold)*; and a queenright colony *(red)*. Brood reared more than six days after the queen was removed would have developed into queens of inferior quality or into workers rather than queens.

tor of pheromonal activity. In a series of experiments we used this bioassay to evaluate complete or fractionated mandibular extracts. Interestingly, fractions that contained a single substance were relatively inactive. We began to see retinue behavior around the lure only when we combined certain fractions. Eventually, we identified the components of the active fractions and devised a synthetic blend of five components that duplicated the activity of the natural mandibular secretion (Figure 3) (Kaminski et al. 1990, Slessor et al. 1988).

The ingredients include two forms of 9HDA. This substance is chiral: It has a left-handed and a right-handed form. Earlier, we had demonstrated that one form, or enantiomer, is significantly more effective in evoking some behavioral responses than the other (Winston et al. 1982). We had also determined that one enantiomer predominates in the natural pheromone (Slessor et al. 1985). But it turned out that both enantiomers also have to be present to evoke the full retinue response.

The remaining missing ingredients were two small, aromatic compounds. One of the aromatic compounds,

methyl *p*-hydroxybenzoate, or HOB, had been identified earlier (Pain, Hugel and Barbier 1960) but had not been considered to be active as a pheromone. The other aromatic molecule, 4-hydroxy-3-methoxyphenylethanol, or HVA, was first identified in the course of our experiments.

The pheromonal activity of the aromatic molecules was quite unexpected. Most insect pheromones are aliphatic compounds, built up out of straight or branched chains of carbon atoms; they are derived from fatty acids or terpenes. The decenoic acids in mandibular gland pheromone fit this description, for example. The aromatics, which feature closed rings of carbon atoms, are a different class of chemical entirely. Indeed we probably found them only because we reversed the customary experimental procedure. Instead of identifying substances in gland extracts and then screening them for pheromonal activity, we screened for activity and then identified the compounds we had isolated.

Our research into the mandibular gland's composition and activity cleared up several points of confusion. First, the complete ensemble of molecules is necessary for full efficacy. Removing any of the components reduced the pheromone's effect by as much as 50 percent, and individual components are inactive when they are tested alone. This is why the pheromone's major component by itself had failed to elicit retinue behavior.

Second, although we managed to duplicate the effects of mandibular secretions, we should emphasize that we did not duplicate the queen's complete arsenal of pheromonal effects. The mandibular glands are not the only site of production for queen pheromones; they are also probably secreted by the tergite glands, which are located in some of the membranes between the queen's abdominal segments (Renner and Baumann 1964, Velthuis 1970). Moreover, the mandibular and other pheromones may have closely linked or overlapping effects, which may explain why demandibulated queens still evoked the retinue response.

Third, the pheromonal blend a queen produces varies with her age. A newly emerged virgin queen has virtually no pheromone in her glands, but when she begins to mate some six days later, she is secreting the decenoic acids, and her glands contain about half the amount of acid a mated queen's contain. The full blend, includ-

10 queen equivalents

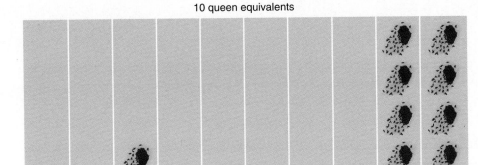

1 queen equivalent

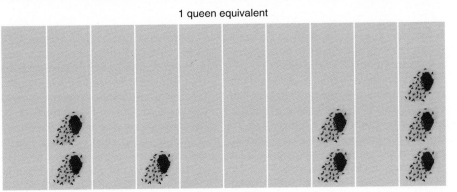

0 queen equivalents

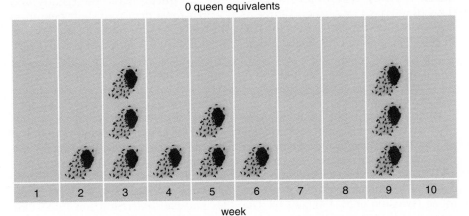

| 1 | 2 | 3 | 4 | 5 | 6 | 7 | 8 | 9 | 10 |

week

Figure 6. Queen mandibular pheromone plays a role in the timing of swarming. Experiments in which supplemental pheromone was administered on lures to queenright colonies suggest that hives swarm when congestion causes the levels of pheromone reaching workers to fall below a threshold. In the experiments summarized here, 10 Qeq of supplemental pheromone significantly delayed swarming, but one Qeq of supplemental pheromone did not. Other experiments suggest that a lower dose is effective if the pheromone is well dispersed.

Figure 7. Attracting flying workers to swarm clusters is another function of queen mandibular pheromone. The fence posts in this photograph were baited with lures containing one Qeq of synthetic pheromone. The synthetic pheromone was almost as effective as a queen in attracting workers. (Photograph courtesy of K. Naumann.)

ing the aromatic compounds, is secreted only after a queen has finished mating and begins to lay eggs (Slessor et al. 1990). This variation in the composition of the pheromone with the queen's age also may help to explain differing reports of effectiveness.

Finally, the mandibular pheromone is effective over a wide range of dosages. We defined the average amount of pheromone found in the glands of a mated queen to be one queen equivalent (Qeq). Worker bees respond to dosages as low as 10^{-7} Qeq, or a ten-millionth of her daily production (Kaminski et al. 1990, Slessor et al. 1988). This finding was particularly interesting in relation to hypotheses about pheromone transmission.

Functions

Once we were satisfied that we had completely identified queen mandibular pheromone, we began to test the effects of the pheromone on colony functions. One of the first pheromone-mediated behaviors we examined was the inhibition of queen rearing. This function can be easily demonstrated by removing a queen from her colony; the workers become agitated after half an hour and begin rearing new queens within 24 hours. Beekeepers call this

"emergency queen rearing," because failure to promptly rear a new queen results in the death of the colony.

To rear queens, workers elongate the cells around a few newly hatched larvae (Figure 4). These larvae are fed a highly enriched food called royal jelly that adult workers produce in their brood-food glands. The specialized diet directs larval development away from the worker pathway and onto the queen pathway. Brood older than six days (three days as eggs and three as larvae) lose their ability to develop into queens. Since no new eggs are laid after the queen is lost, workers have only six days to begin queen rearing before the colony loses its ability to produce a new, functional queen.

We examined the role of mandibular pheromone in emergency queen rearing by comparing queen rearing in queenright colonies, queenless colonies and queenless colonies receiving daily doses of synthetic mandibular pheromone (Winston et al. 1989, 1990). The results were dramatic. When queenless colonies received one Qeq or more of pheromone per day, they made almost no attempt to rear new queens for four days after the queen was removed. Even six days after queen loss, there was no significant difference between

the number of queen cells in queenless colonies treated with pheromone and the number in queenright control colonies (Figure 5). We concluded that the queen's mandibular pheromone is largely responsible for the inhibition of queen rearing in queenright colonies.

Colonies also rear new queens in preparation for reproductive swarming, the process of colony division in which a majority of the workers and the old queen leave the colony and search for a new nest. Left behind in the old colony are developing queens, some of which may issue with additional swarms once they emerge, and one of which will reign over the old nest once swarming is completed.

Swarming poses something of a puzzle because it does not occur unless new queens are being reared, but the old queen is present during the initial stages of queen rearing and is presumably still secreting her inhibitory chemicals (Butler 1959, 1960, 1961; Seeley and Fell 1981; Simpson 1958). One hypothesis is that the transmission of queen pheromones is slowed as colonies grow and become more congested prior to swarming. As the amounts of pheromone reaching workers diminish, the workers begin to escape the queen's control.

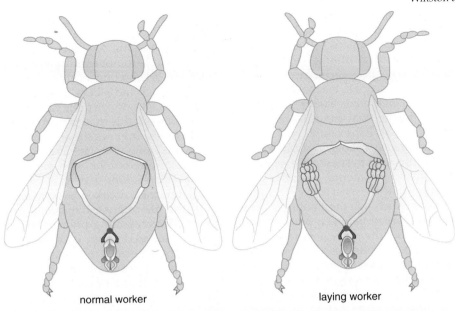

normal worker   laying worker

Figure 8. Suppression of worker ovary development, which had been thought to be a function of queen mandibular pheromone, is probably controlled by other primer pheromones, including substances produced by larval and pupal bees. Egg-laying is pre-eminently the queen's function but, under certain conditions, workers' ovaries also develop, and they can then lay eggs. In queenright colonies, a few workers have developed ovaries, but the workers lay few eggs unless the colony becomes queenless. If the queen is lost and the colony fails to rear a new queen, the workers with enlarged ovaries begin to lay eggs. Because workers lack the genital structures that allow the queen to mate and store sperm, any eggs they lay are unfertilized and usually develop into drones.

We tested the role of queen pheromone in reproductive swarming by supplementing the queen's normal secretion with additional, synthetic pheromone. Again, the pheromone had a significant effect on queen rearing. Colonies receiving 10 Qeq per day swarmed an average of 25 days later than control colonies (Figure ·6). The pheromone-treated colonies became extraordinarily crowded before they would swarm, indicating just how powerful a suppressant the queen pheromone can be.

For these experiments, the pheromone was presented on stationary lures, but queen bees move about in the nest. In another set of experiments, we tried to better mimic natural conditions by administering the pheromone as a spray. We found that a dose of one Qeq per day of supplemental pheromone was sufficient to suppress swarming. The fact that the better-dispersed spray was active at a much lower dose than the stationary lure supports the hypothesis that the slowing of pheromonal dispersal in congested colonies triggers swarming (Winston et al. 1991).

The queen mandibular pheromone plays two other roles in swarming as well: It attracts flying workers to the swarm cluster, and it stabilizes the

cluster, preventing workers from becoming restless and leaving before the swarm reaches a new nest site. We found that the synthetic mandibular pheromone is almost as effective as the queen in attracting workers to the cluster and in keeping them together, at least for the first few hours following cluster formation (Figure 7).

Next we examined the effect of mandibular pheromone on worker ovary development and egg laying. Both are known to be suppressed by queen pheromones. Indeed pheromonal suppression of worker reproduction is one of the hallmarks of advanced insect societies. Primitive social insects suppress nestmate reproduction primarily through dominance interactions. Many wasp and bumblebee queens, for example, seem to maintain dominance by biting and harassing the other females in the colony.

Honey bee workers have the potential to lay unfertilized eggs that can develop into males, or drones. However, some combination of pheromones secreted by the queen and the brood (the developing larvae or pupae) inhibits the maturation of the workers' ovaries, which usually remain small and nonfunctional (Figure 8) (Jay 1970, 1972). Nevertheless, even in queenright colonies a few workers manage to lay eggs

(Ratnieks and Visscher 1989, Visscher 1989) and many workers begin egg-laying two or three weeks after the queen and brood are removed (Winston 1987).

The ovary-suppressing substance produced by the queen was thought to be mandibular pheromone, particularly 9ODA (see reviews by Free 1987; Willis, Winston and Slessor 1990; Winston 1987). To test this hypothesis, we removed queens from groups of colonies and applied various doses of mandibular pheromone to the colonies for the next 43 days. Some colonies received daily doses as high as 10 Qeq. To our surprise, workers in queenless, pheromone-treated colonies developed ovaries at the same rate as workers in queenless, pheromone-free colonies (Willis, Winston and Slessor 1990). Apparently queen mandibular pheromone is not involved in the suppression of worker ovary development. Another primer pheromone must be responsible for this effect, possibly the unidentified queen tergite pheromone in combination with brood pheromone.

A final set of experiments demonstrated that queen mandibular pheromone is not just an inhibitor. In addition to suppressing reproductive activity, it can stimulate foraging and brood rearing. To establish this, we applied pheromone to newly founded, queenright colonies and counted the number of foragers, the size of their nectar and pollen loads and the extent of brood rearing (Figure 9). The pheromone did not increase overall foraging activity, but colonies that received one Qeq of supplemental pheromone daily had more pollen foragers, and those foragers carried heavier pollen loads. Pheromone applications increased the amount of pollen entering the nest by 80 percent, and pheromone-treated colonies reared 18 percent more brood, possibly due to the increased amounts of protein-rich pollen being brought into those colonies (Higo et al. 1992).

Transmission

The set of experiments we have just described demonstrated that queen mandibular pheromone has a wide range of functions, but we still did not know how much is produced and secreted by the queen daily or how these important compounds are transmitted to workers. The chemical components of the pheromone are not particularly volatile, and so they probably cannot spread throughout the colony by diffusion alone. The number of workers in an established colony and the rapidity with which the pheromone is lost make it unlikely that the queen is the sole disseminator of the pheromone. Some experiments had suggested that retinue workers act as messengers, transmitting the queen's pheromones to other workers when they touch antennae or mouthparts (Juška, Seeley and Velthuis 1981; Seeley 1979). Pheromone could not be found on worker bees, however, possibly because it was present in concentrations below the limit of detection of the instruments then in use.

We were able to follow pheromone secretion and transmission using the more sensitive chromatographic and spectroscopic equipment now available, as well as radiolabelled pheromone provided by Glenn Prestwich and Francis Webster of the State University of New York at Stony Brook and Syracuse (Webster and Prestwich 1988). Quantitative measurements of rates of production, transfer and loss allowed us to construct a mathematical model of pheromone dispersal. Our results essentially confirm the messenger-bee hypothesis, but they raise many interesting questions as well (Figure 10) (Naumann et al. 1991).

We first determined that a queen typically has five micrograms, or 0.001 Qeq, of pheromone on her body at any one time. To determine how much a queen secretes daily, we removed queens from colonies and measured the pheromone that built up on their cuticles. It turned out that a queen secretes between 0.2 and 2 Qeq of pheromone per day. These results fit well with our within-colony function studies, where pheromonal effects typically appeared at a dosage of about 1 Qeq.

We then quantified the processes by which pheromone is transferred or lost. For example, we allowed workers to contact the queen for different periods and then measured the amount of pheromone they had picked up. We found that each process could be approximated by a first-order rate equation. In other words, the amount of pheromone transferred during any short interval is proportional to the quantity present at the source during that interval; the constant of proportionality is called the rate constant. The first-order equation yields an exponential loss of material at the source; the "half life" is inversely proportional to the rate constant. We calculated the rate constants for all of the transmission pathways from our experimental

data and then used the value we had obtained for the queen's daily secretion and the rate equations to determine the amount of pheromone that would reach worker bees each day.

Our confidence in the model was bolstered when it predicted rates of pheromone transfer that are consistent with bee behavior. For example, the model predicts a flux of about one Qeq of pheromone through the nest per day; in our function experiments, a similar dose was typically the most active. The model also predicts that the amount of pheromone passed between workers will drop below the detectable level (10^{-7} Qeq) about 30 minutes after it is picked up from the queen, which is about how long it takes for workers to become restless after the queen is removed. Finally, the model predicts that pheromone deposited by the queen in the wax comb

are lickers, but they pick up over half the queen's pheromonal secretion. Each antennating worker picks up much less pheromone and passes on much less in subsequent contacts. Why some bees lick and others antennate is not known, nor do we understand the implications of this dual transmission system for the colony's social structure.

The wax comb also plays a role in the transfer of pheromone, although the queen deposits only 1 percent of her production there. Workers pick up a little of the pheromone the queen deposits. The remaining pheromone is probably released slowly over the next few hours. The slow release may explain why empty comb that has had brood in it is attractive to adult workers. The queen spends most of her time in the brood area, and her pheromonal footprints would be concentrated there.

Figure 9. Pollen foraging is yet another behavior influenced by queen mandibular pheromone, but in this case the pheromone stimulates rather than suppresses the behavior. A bee transports pollen in corbiculae, or pollen baskets, on the outer surfaces of the hind legs. The photograph shows a bee with heavily laden pollen baskets. Under some circumstances, queen mandibular pheromone stimulates workers to collect more pollen. (Photograph courtesy of Kenneth Lorenzen.)

will become undetectable in a few hours, which is about how long it takes the bees to begin rearing new queens.

The retinue workers do indeed remove the greatest fraction of the queen's pheromonal production. Not all bees in the retinue pick up and transfer the same amount of pheromone, however. There are two types of retinue bees, which we call licking and antennating messengers (Figure 11). The lickers touch the queen with their tongues, forelegs and mouthparts. The antennators brush the tips of their antennae lightly and quickly over her body. Only about 10 percent of the bees

One of our more surprising findings is that pheromone is lost primarily through internalization. The queen herself internalizes some 36 percent of the pheromone she produces. She swallows some of it, some is adsorbed on or bound to her cuticle, and some moves through the cuticle into the blood (hemolymph) system. What purpose internalization serves is unclear. Even if the internalized pheromone retains the same chemical identity, it is no longer available to the colony. It seems odd that such inefficient use is made of the queen's pheromone secretion. It is possible that the model somewhat overesti-

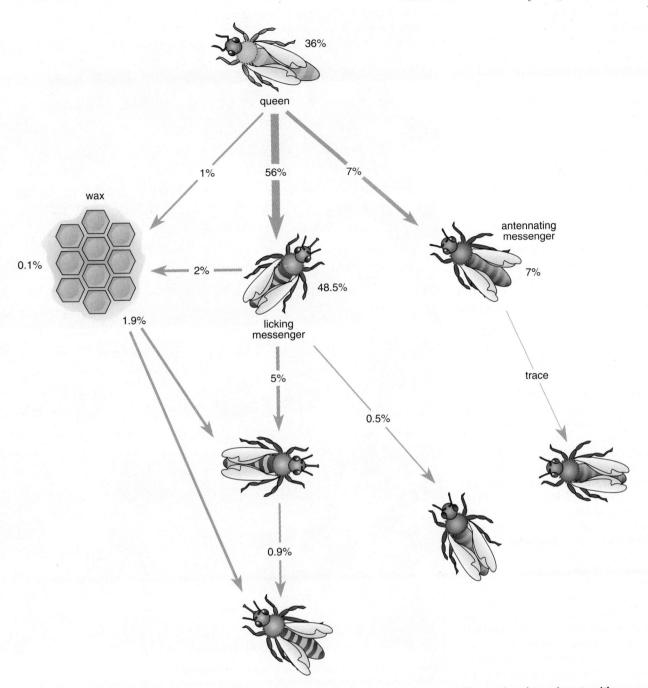

36%

queen

1%

56%

7%

wax

0.1%

antennating
messenger

2%

7%

48.5%

1.9%

licking
messenger

trace

5%

0.5%

0.9%

Figure 10. Model of pheromonal transmission within a colony was constructed from data obtained in a series of experiments with radiolabelled 9ODA as part of the synthetic pheromone blend. The pheromone is transmitted primarily by messenger bees, which gather it from the queen either by licking her body or by stroking with the antennae. Some pheromone is also deposited by the queen on the wax comb and picked up by worker bees. Only one bee in 10 is a licker, but the lickers gather much more pheromone than antennating bees and transfer more in subsequent contacts. For unknown reasons, large amounts of pheromone are internalized both by the queen herself and by the messenger bees. The numbers give the percentage of the queen's daily production that is transmitted in the indicated directions.

mates the queen's internalization of pheromone, since this was the only process that did not seem to conform to a first-order rate equation.

On the other hand, the messenger bees also internalize pheromone at a surprising rate (Figure 12). The licker bees swallow 40 percent of the pheromone they pick up. Most of the rest is quickly transferred from the worker's head to her abdomen by a combination of passive transport and active grooming (Naumann 1991). There the pheromone passes into and through the cuticle. Some of it ends up in the blood system within 30 to 60 minutes, and some remains bound to or adsorbed on the cuticle. In short, between the queen and the workers, nearly all of the pheromone is eventually internalized.

The mechanisms by which worker bees perceive the pheromone are not well understood. Sensory structures on a bee's antennae allow it to detect some odors at a concentration equal to a tenth or a hundredth of the concentration people can detect. The arrival of odorant at receptor cells inside the antennae causes associated nerve cells to fire, and the firing can initiate further behavioral and physiological changes. In most insects pheromones act by similar mechanisms.

But what about the large amounts of pheromone that are internalized? One hypothesis is that once the pheromone is translocated across the cuticle, its constituents or their breakdown products become active as hormones rather than as pheromones. We are only beginning to explore this intriguing hypothesis.

It also is unclear why the queen produces such a large amount of mandibular pheromone. The answer may be rooted in queen-worker conflict and the evolution of queen dominance in highly social insects. As we have mentioned, there are almost no dominance interactions in a normally functioning honey bee colony. Instead the queen controls the workers entirely through her sophisticated arsenal of pheromones. Perhaps over evolutionary history the workers and queen engaged in a kind of chemical arms race, where the workers began to break down the queen's compounds more rapidly and the queen responded by secreting more pheromone. Thus the swallowing and absorbing of pheromone by workers that we find so puzzling may be an attempt to catabolize the queen's pheromonal dominance and escape her control. Although this idea is speculative, it is certainly plausible. Indeed, this concept is reminiscent of most families and societies, which exhibit a complex blend of cooperation and conflict, with some objectives in common but with each individual also having its own goals.

Commercial Applications

We have been investigating commercial applications of queen pheromone for beekeeping and crop pollination. Several beekeeping applications for queen mandibular pheromone are already close to commercial realization. For example, we have shown that the pheromone allows packages of worker bees to be shipped without queens (Naumann et al. 1990). Beekeepers often establish a colony by buying a kilogram of workers and a queen in a wire package, but queens are typically produced by beekeepers who specialize in this art, and so they may come from a different source. The pheromone's ability to delay swarming also might be useful in bee management, since frequent swarming reduces honey production and can threaten colony survival. The pheromone's ability to stimulate pollen foraging and brood rearing might have commercial importance as well, since substantial increas-

Figure 11. Licking and antennating behavior are both part of the retinue response; thus actions induced by the pheromone are also responsible for its dissemination through the colony. In the upper photograph at least one worker bee can be seen licking the queen's abdomen; in the lower photograph several workers make antennal contact. (Photographs courtesy of Kenneth Lorenzen.)

es in colony growth rates can be achieved by these means. Finally, we are investigating the use of mandibular pheromone as an attractant for swarms. Such an attractant would be useful for the monitoring and control of honey bee diseases and the Africanized bee (the so-called "killer bee").

The most significant application of queen mandibular pheromone, however, may be in assisting crop pollination. Managed crop pollination is economically the most important function

of bees. Colonies of bees are moved to blooming crops in order to provide enough bees to ensure good pollination and seed set. Beekeepers receive up to $45 per colony for this service, which is essential for the production of many fruits and vegetables. Not only does pollination affect yield, it also is often closely associated with the quality of fruits, vegetables and seeds. Each year the honey bee pollinates crops worth at least $1.4 billion in Canada and $9.3 billion in the United States

(Robinson, Nowogrodzki and Morse 1989; Scott and Winston 1984). Although hundreds of thousands of bee colonies are moved to crops each year, many crops are still inadequately pollinated. Reduced yields, occasional crop failures and lowered crop quality can all be attributed to this problem (Free 1970; Jay 1986; McGregor 1976; Robinson, Nowogrodzki and Morse 1989).

Since queen pheromone is highly attractive to flying workers, we thought it possible that pollination could be improved by spraying crops with a dilute blend of pheromone. To test this hypothesis, we sprayed blocks of trees or berries with various concentrations of pheromone and monitored the number of workers attracted to the blocks and the crop yields. In almost all of the experiments, up to twice the number of bees visited treated blocks compared to untreated ones.

The effect of pheromone spraying on yield was more variable. In preliminary trials with apple trees we found no increases in yield or improvements in fruit quality. However, pear trees sprayed with pheromone produced fruit that was heavier by an average of 6 percent, which translated to an increase of 30 percent in profits, or about $1,055 per hectare once spraying costs were deducted. Cranberries treated with pheromone yielded about 15 percent more berries by weight, which increased net returns by $4,465 per hectare averaged over two years (Currie, Winston and Slessor 1992; Currie, Winston, Slessor and Mayer 1992). In some cases there was no improvement, whereas in others the yield increase was much greater than these average values. Clearly, if we are able to exercise in the farmer's field the kind of sociochemical control the queen exercises in the nest, we can dramatically expand the economic value of this already beneficial social insect.

Acknowledgments

We are grateful to all of our collaborators in the research this article describes, including: J. H. Borden, S. J. Colley, R. W. Currie, H. A. Higo, L.-A. Kaminski, G. G. S. King, K. Naumann, T. Pankiw, E. Plettner, G. D. Prestwich, F. X. Webster, L. G. Willis and M. H. Wyborn. We would also like to acknowledge the technical assistance of P. LaFlamme. Funding for the research came from the Natural Sciences and Engineering Research Council of Canada, the Science Council of British Columbia, the Wright Institute and Simon Fraser University.

Bibliography

Allen, M. D. 1960. The honeybee queen and her attendants. *Animal Behavior* 8:201–08.

Barbier, J., and E. Lederer. 1960. Structure chimique de la "substance royale" de la reine d'abeille (*Apis mellifica*). *Comptes Rendus des Séances de L'Académie de Science* (Paris) 251:1131–35.

Bradshaw, J. W. S., and P. E. Howse. 1984. Sociochemicals of ants. In: *Chemical Ecology of Insects*, W. J. Bell and R. T. Carde (eds). Sunderland, Mass.: Sinauer.

Butler, C. G. 1959. Queen substance. *Bee World* 40:269–75.

Butler, C. G. 1960. The significance of queen substance in swarming and supersedure in honey-bee (*Apis mellifera*) colonies. *Proceedings of the Royal Entomological Society* (London) (A) 35:129–32.

Butler, C. G. 1961. The scent of queen honey bees (*Apis mellifera*) that causes partial inhibition of queen rearing. *Journal of Insect Physiology* 7:258–64.

Butler, C. G., and E. M. Fairey. 1964. Pheromones of the honey bee: biological studies of the mandibular gland secretion of the queen. *Journal of Apicultural Research* 3:65–76.

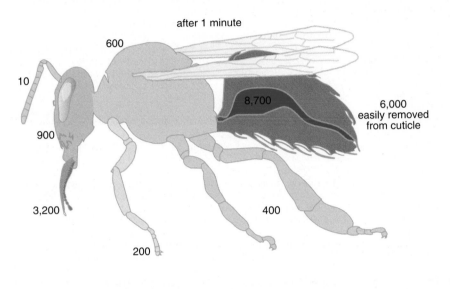

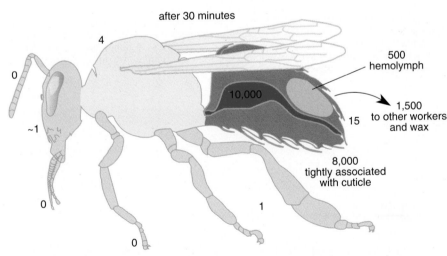

Figure 12. Maps of pheromone on a licking messenger bee illustrate one of the mysteries of pheromone transmission: Most of the pheromone is quickly internalized. The distribution of the pheromone is shown one minute (*upper diagram*) and 30 minutes (*lower diagram*) after the bee contacted the queen. To quantify the internalization of the pheromone by worker bees, 20,000 picograms of radiolabelled 9ODA (the amount a worker typically picks up in her contact with the queen) was applied topically to individual bees. Each bee was then isolated, so that the pheromone could not be transferred to other workers. After a delay, the bee was washed with methanol to dissolve any pheromone on the body surface. The crop and gut were then excised. The amounts of 9ODA in the washes, excised organs and the corpse were determined by liquid scintillation counting. The corpse scintillation count indicated how much pheromone had passed into the body or was so closely associated with the cuticle that it could not be removed by methanol. The pheromone depicted as being external remains available to the colony; it can be picked up by other workers or deposited on the wax comb. The internalized pheromone is effectively removed from circulation.

Callow, R. K., J. R. Chapman and P. N. Paton. 1964. Pheromones of the honey bee: Chemical studies of the mandibular gland secretion of the queen. *Journal of Apicultural Research* 3:77–89.

Callow, R. K., and N. C. Johnston. 1960. The chemical constitution and synthesis of queen substances of honey bees (*Apis mellifera* L.). *Bee World* 41:152–3.

Currie, R. W., M. L. Winston and K. N. Slessor. 1992. Impact of synthetic queen mandibular pheromone sprays on honey bee pollination of berry crops. *Journal of Economic Entomology* (in press).

Currie, R. W., M. L. Winston, K. N. Slessor and D. F. Mayer. 1992. Effect of synthetic queen mandibular pheromone sprays on pollination of fruit crops by honey bees. *Journal of Economic Entomology* (in press).

Duffield, R. M., J. W. Wheeler and G. C. Eickwort. 1984. Sociochemicals of bees. In: *Chemical Ecology of Insects*, W. J. Bell and R. T. Carde (eds). Sunderland, Mass: Sinauer.

Free, J. B. 1970. *Insect Pollination of Crops*. London: Academic Press.

Free, J. B. 1987. *Pheromones of Social Bees*. Ithaca: Cornell University Press.

Higo, H. A., S. J. Colley, M. L. Winston and K. N. Slessor. 1992. Effects of honey bee queen mandibular gland pheromone on foraging and brood rearing. *Canadian Entomologist* 124:409–418.

Hölldobler, B., and E. O. Wilson. 1990. *The Ants*. Cambridge, Mass.: Harvard University Press.

Howse, P. E. 1984. Sociochemicals of termites. In: *Chemical Ecology of Insects*, W. J. Bell and R. T. Carde (eds). Sunderland, Mass: Sinauer.

Jay, S. C. 1970. The effect of various combinations of immature queen and worker bees on the ovary development of worker honey bees in colonies with and without queens. *Canadian Journal of Zoology* 48:168–73.

Jay, S. C. 1972. Ovary development of worker honeybees when separated from worker brood by various methods. *Canadian Journal of Zoology* 50:661–4.

Jay, S. C. 1986. Spatial management of honey bees on crops. *Annual Review of Entomology* 31:49–66.

Juška, A., T. D. Seeley and H. H. W. Velthuis. 1981. How honeybee queen attendants become ordinary workers. *Journal of Insect Physiology* 27:515–19.

Kaminski, L.-A., K. N. Slessor, M. L. Winston, N. W. Hay and J. H. Borden. 1990. Honey bee response to queen mandibular pheromone in a laboratory bioassay. *Journal of Chemical Ecology* 16:841–49.

McGregor, S. E. 1976. *Insect Pollination of Cultivated Crop Plants*. Agricultural Handbook No. 496. Washington, D.C.: U.S. Department of Agriculture.

Naumann, K. 1991. Grooming behaviors and the translocation of queen mandibular gland pheromone on worker honey bees. *Apidologie* 22:523–531.

Naumann, K., M. L. Winston, M. H. Wyborn and K. N. Slessor. 1990. Effects of synthetic honey bee (Hymenoptera: Apidae) queen mandibular gland pheromone on workers in packages. *Journal of Economic Entomology* 83:1271–75.

Naumann, K., M. L. Winston, K. N. Slessor, G. D. Prestwich and F. X. Webster. 1991. The production and transmission of honey bee queen (*Apis mellifera* L.) mandibular gland pheromone. *Behavioral Ecology and Sociobiology* 29:321–32.

Pain, J., M.-F. Hugel and M. C. Barbier. 1960. *Comptes Rendus des Séances de L'Académie de Science* (Paris) 251:1046–8.

Ratnieks, F. L. W., and P. K. Visscher. 1989. Worker policing in the honey bee. *Nature* 342:796–97.

Renner, M., and M. Baumann. 1964. Uber komplexe von subepidermalen drusenzallen (Duftdrusen?) der bienenkonigin. *Naturwissenschaften* 51:68–9.

Robinson, W. S., R. Nowogrodzki and R. A. Morse. 1989. The value of honey bees as pollinators of U.S. crops. *American Bee Journal* 129:411–23, 477–87.

Scott, C. D., and M. L. Winston. 1984. The value of bee pollination to Canadian apiculture. *Canadian Beekeeping* 11:134.

Seeley, T. D. 1979. Queen substance dispersal by messenger workers in honey bee colonies. *Behavioral Ecology and Sociobiology* 5:391–415.

Seeley, T. D., and R. D. Fell. 1981. Queen substance production in honey bee (*Apis mellifera*) colonies preparing to swarm. *Journal of the Kansas Entomological Society* 54:192–96.

Simpson, J. 1958. The factors which cause colonies of *Apis mellifera* to swarm. *Insectes Sociaux* 5:77–95.

Slessor, K. N., G. G. S. King, D. R. Miller, M. L. Winston and T. L. Cutforth. 1985. Determination of chirality of alcohol or latent alcohol semiochemicals in individual insects. *Journal of Chemical Ecology* 11:1659–67.

Slessor, K. N., L.-A. Kaminski, G. G. S. King, J. H. Borden and M. L. Winston. 1988. Semiochemical basis of the retinue response to queen honey bees. *Nature* 332:354–56.

Slessor, K. N., L.-A. Kaminski, G. G. S. King and M. L. Winston. 1990. Semiochemicals of the honey bee queen mandibular glands. *Journal of Chemical Ecology* 16:851–60.

Velthuis, H. H. W. 1970. Queen substances from the abdomen of the honey bee queen. *Zeitschrift fuer vergleichende Physiologie* 70:210–22.

Visscher, P. K. 1989. A quantitative study of worker reproduction in honey bee colonies. *Behavioral Ecology and Sociobiology* 25:247–54.

Webster, F. X., and G. D. Prestwich. 1988. Synthesis of carrier-free tritium-labelled queen bee pheromone. *Journal of Chemical Ecology* 14:957–62.

Willis, L. G., M. L. Winston and K. N. Slessor. 1990. Queen honey bee mandibular pheromone does not affect worker ovary development. *Canadian Entomologist* 122:1093–99.

Wilson, E. O. 1971. *The Insect Societies*. Cambridge, Mass.: Harvard University Press.

Winston, M. L., K. N. Slessor, M. J. Smirle and A. A. Kandil. 1982. The influence of a queen-produced substance, 9HDA, on swarm clustering behavior in the honey bee *Apis mellifera* L. *Journal of Chemical Ecology* 8:1283–88.

Winston, M. L. 1987. *The Biology of the Honey Bee*. Cambridge, Mass.: Harvard University Press.

Winston, M. L., K. N. Slessor, L. G. Willis, K. Naumann, H. A. Higo, M. H. Wyborn and L.-A. Kaminski. 1989. The influence of queen mandibular pheromones on worker attraction to swarm clusters and inhibition of queen rearing in the honey bee (*Apis mellifera* L.). *Insectes Sociaux* 36:15–27.

Winston, M. L., H. A. Higo and K. N. Slessor. 1990. Effect of various dosages of queen mandibular gland pheromone on the inhibition of queen rearing in the honey bee (Hymenoptera: Apidae). *Annals of the Entomological Society of America* 83:234–38.

Winston, M. L., H. A. Higo, S. J. Colley, T. Pankiw and K. N. Slessor. 1991. The role of queen mandibular pheromone and colony congestion in honey bee (*Apis mellifera* L.) reproductive swarming. *Journal of Insect Behavior* 4:649–659.

INDEX